Hydraulic Fracturing Impacts and Technologies

A Multidisciplinary Perspective

Edited by

Venkatesh Uddameri
Audra Morse
Kay J. Tindle

CRC Press is an imprint of the
Taylor & Francis Group, an **informa** business

CRC Press
Taylor & Francis Group
6000 Broken Sound Parkway NW, Suite 300
Boca Raton, FL 33487-2742

Printed on acid-free paper
Version Date: 20150422

International Standard Book Number-13: 978-1-4987-2117-2 (Hardback)

Library of Congress Cataloging-in-Publication Data

Hydraulic fracturing impacts and technologies : a multidisciplinary perspective / editors, Venkatesh Uddameri, Audra Morse, and Kay J. Tindle.
pages cm
Includes bibliographical references and index.
ISBN 978-1-4987-2117-2 (alk. paper)
1. Hydraulic fracturing--Environmental aspects--United States. 2. Gas wells--Hydraulic fracturing. 3. Oil wells--Hydraulic fracturing. I. Uddameri, Venkatesh. II. Morse, Audra. III. Tindle, Kay J.

TD195.G3H945 2015
622'.3381--dc23 2014049678

Visit the Taylor & Francis Web site at
http://www.taylorandfrancis.com

and the CRC Press Web site at
http://www.crcpress.com

Contents

Foreword v
Prologue vii
Acknowledgments ix
Editors xi
Contributors xiii
Introduction to *Hydraulic Fracturing Impacts and Technologies: A Multidisciplinary Perspective* xvii

Chapter 1 Overview of Hydraulic Fracturing Operations and Technologies 1

George E. King

Chapter 2 Economic Impact of the Permian Basin's Oil and Gas Industry 21

Bradley T. Ewing, Marshall C. Watson, Terry McInturff, and Russell McInturff

Chapter 3 From Property Rights to Endangered Species: Legal Issues Surrounding Hydraulic Fracturing 65

William R. Keffer, J. Randall Miller, J. Berton Fisher, Taylor Stevenson, and Adrianne Waddell

Chapter 4 Looking into the Crystal Ball: Potential New EPA Hydraulic Fracturing Rules Impacting Unconventional Oil and Gas Production from Shale Plays 83

Ron Truelove

Chapter 5 Commentary on Health and Environmental Risks from Hydraulic Fracturing 91

David Klein, Jennifer Knaack, and Audra Morse

Chapter 6 Health and Environment Risks from Oil and Gas Development..... 101

Anne C. Epstein

Chapter 7 Addressing Concerns about Impacts from Unconventional Drilling Using Advanced Analytical Chemistry 115

Doug D. Carlton Jr., Zacariah L. Hildenbrand, Brian E. Fontenot, and Kevin A. Schug

Chapter 8 Water Availability in the Permian Basin Region of West Texas...... 133

Venkatesh Uddameri and Danny Reible

Chapter 9 Reuse and Recycling of Flowback and Produced Waters 159

John H. Williams, Danny Reible, Roxana Darvari, Tony Vercellino, and Audra Morse

Chapter 10 Impact of Hydraulic Fracturing on Transportation Infrastructure...... 175

Sanjaya Senadheera

Chapter 11 GIS-Based Assessment of Wastewater Disposal Impacts in Permian Basin, Texas 187

Elma Annette Hernandez, Sreeram Singaraju, Abdullah Karim, Jorge Ruiz de Viñaspre, and Venkatesh Uddameri

Chapter 12 Challenges and Opportunities for Increasing Guar Production in the United States to Support Unconventional Oil and Gas Production 207

Noureddine Abidi, Sumedha Liyanage, Dick Auld, Robert K. Imel, Lewis Norman, Kulbhushan Grover, Sangu Angadi, Sudhir Singla, and Calvin Trostle

Chapter 13 Characterization of the Properties of Guar Gum to Improve Hydraulic Fracturing Efficiencies 227

Noureddine Abidi and Sumedha Liyanage

Chapter 14 Communicating Fracturing Impacts and Technologies: Assessment, Public Understanding, and Theoretical Linkages 251

Shawna R. White, R. Glenn Cummins, Melanie Sarge, and Erik P. Bucy

Chapter 15 Multidisciplinary Teams as Mechanisms of Accountability: Neutralizing the Emotions and Politics of Hydraulic Fracturing Research 273

Kay J. Tindle, Daniel Marangoni, and Anna Thomas Young

Index 285

Foreword

Water: The Universe's Most Precious Resource

Water is *the* limiting resource throughout the universe, and although the earth has an abundance of it, its value and therefore its efficient use are, and will always be, paramount to humankind's survival and prosperity. As we have entered the twenty-first century, water is becoming the new *gold*, a commodity being valued like never before. The recognition of the value of water has been brought about by the demands of a rapidly expanding human population and the concomitant demands of potable water for hygiene, health, food, energy, and power. This has been exacerbated by the possibility of climate change, and as with all periods in human history, it is fair to say water is life and life is water. It is critical, in fact imperative, that we educate all humankind in the efficient use of water, in all its forms, that we develop reasonable laws governing its use, and that we continue research in water cleanup, production and discovery, and effective agricultural use. In addition, scientists and engineers must continue to develop new technologies, new methodologies, and new processes for energy production and power generation, providing for all of us a healthy, prosperous, and sustainable future.

A critical aspect of water use in the near-term and midterm is for oil and gas production utilizing hydraulic fracturing (technically correct terminology), or *fracking* as the critics of this technology like to call it. Hydraulic fracturing, and its use of potable water as a fracing medium, has been around for more than 70 years, with little public concern or notice. However, the confluence of rapid population growth and its demand for potable water, the growth of environmentalism and growing awareness of the need for sustainable development and growth (the so-called energy/food nexus), coupled with the development of horizontal drilling and the concomitant use of large amounts of potable water in hydraulic fracturing (6–8 million gallons per frac) has led to slogans such as "Leave my fracking water alone." These sentiments have been exacerbated by exploration and development companies using poor engineering techniques and, in some cases, by the use of old technology resulting in the creation of *urban myths*, such as initiating earthquakes and causing water coming from wells to *flame*, and in creating a general fear of polluting the water table and local environment. Finally, a lack of appreciation by oil and gas producers that they need to educate the public on the benefits and risks of this technology has caused confusion and further inflamed public perception. This vocal *resistance* to fracing makes little sense in light of facts. There is a great story to tell of the promise of near energy independence while providing, for the first time, a *cushion* to develop renewable energy sources along with more energy availability now for the developing world.

So, what will the future be, *fracking* or *fracing* here in the Permian Basin and elsewhere? Our *hydraulic fracturing* future will look bright if we (i.e., scientists, engineers, and policy makers) effectively educate the public on the use of this technology, while continuing to value our most precious resource, water. As energy professionals, we must increase our understanding through continuing knowledge development (research), while educating and informing the public and the policy makers on how hydraulic fracturing can be used to help realize an environmentally secure and sustainable future for all humanity. It is imperative that we take the time to educate the public on the risks and benefits of such technologies, recognizing that there are real, but very manageable, risks. We must educate and train our young engineers, scientists, and public policy makers on how to assess those risks and take advantage of new science and technologies to effectively use and reuse our precious water resources to provide the energy needs for all aspects of human life, and to do so in a way that respects the fact that water is a limited resource. Energy companies, national and state governments, and those of us at universities must make substantive investments in our efficient use of water, its discovery, and its recycle ability to build a sustainable future in which hydraulic fracturing will play a substantial role in near-term and midterm oil and gas production. Guided by ethical considerations, decisions must be made in an unbiased but educated fashion, should be data driven, and should not be guided (and likely misguided) by emotion or the possibility of rapid financial gain. However, in order to make such decisions on the use of hydraulic fracturing for oil and gas production, research will be needed on the limits of reusing flowback and produced water and on cost-effective ways to clean or effectively dispose any water used in fracturing. Also, research needs to be done on the use of alternate water sources for hydraulic fracturing, such as brackish water, and the use of other unconventional fluids for fracturing, as well as preventing casing failures.

Scientific and technological challenges and opportunities are substantial. Effective hydraulic fracturing requires a better understanding of the interactions and transport of multiple phase fluids through porous media and the associated thermodynamics of these systems under high pressure and temperature, as well as a better understanding of petrophysics associated with these plays. New environmentally friendly frac fluids need to be developed along with better casing materials, environmentally friendly drilling muds, and a better understanding of transport and rheology required for more effective cementing techniques. It will require evaluation of *big data sets* and the associated security problems with managing those data sets, and implementing them in a systems approach when laying out hydraulic fracing fields/arrays. Responsible hydraulic fracturing will require solutions to the so-called energy food nexus, which in turn requires environmentally sound and technology-friendly water, food, and energy laws, which will only come through research and education. To solve complex societal and technological problems such as these will necessitate multidisciplinary teams from industry, academia, and the government working together to develop the science, technology, and laws necessary to obtain a sustainable future for us all. A necessary first step will be to educate tomorrow's workforce to face the challenges and opportunities presented by water in general and fracturing in particular. Water is the universe's most precious resource, and it is a critical piece in the puzzle for a sustainable, productive, and healthy world.

Al Sacco
Edward E. Whitacre College of Engineering
Texas Tech University
Lubbock, Texas

Prologue

High-volume hydraulic fracturing and horizontal drilling are groundbreaking technological breakthroughs of the twenty-first century that have greatly advanced the energy security of the United States. These developments are on par with other extraordinary human technological achievements, such as space travel and digital revolution. It has not only helped the United States become the world's largest natural gas producer, but it is also transforming the economy by creating jobs and revitalizing the chemical and manufacturing sectors of the economy. Natural gas availability has reduced our reliance on less environmentally friendly fuels, and it is a major factor in reducing the emissions of carbon dioxide and other greenhouse gases. The United States had the largest reduction in carbon dioxide emissions from fossil fuel consumption of all the countries in the world between 2006 and 2011 due mainly to the availability of natural gas. Unconventional oil and gas production is a vital bridge in our transition to greener renewable energies.

Hydraulic fracturing is a complex engineering endeavor carried out under high pressure and temperature. Inadequate designs and improper operations can have catastrophic consequences. Safety is therefore of paramount importance. It is also a material-intensive operation that requires relatively large quantities of water and sand that often have to be transported over large distances. Many shale plays are located in rural areas with intensive agricultural activities and in semiarid environments where water is at a premium. Our quest for energy independence must not come at the cost of food security. Innovative water management methods that reduce the freshwater footprint and minimize waste generation are therefore of great importance. A variety of chemicals are used for, and generated due to, hydraulic fracturing. Improper handling and inappropriate disposal of these chemicals increase the risks to human health and the environment. The subsurface environment is highly heterogeneous, and some products of fracturing (e.g., methane) can be generated from other sources as well. State-of-the-art analytical methods are therefore necessary to accurately characterize the risks arising from hydraulic fracturing operations. Clearly, public policy must be grounded in best scientific principles. Policy planners and regulators must comprehend what technologies can do and where they are in need of improvements. By the same token, engineers and scientists must understand political pressures and public perceptions, in order to ensure that their solutions are pragmatic and implementable. This can only happen when traditional, discipline-specific silos are dismantled to create a multidisciplinary outlook that is focused on a comprehensive solution.

This book represents a seminal effort in fostering a multidisciplinary framework for understanding and assessing the impacts of hydraulic fracturing in a holistic manner. The editors have done an excellent job in assembling a diverse team of experts ranging from engineers, scientists, and communication researchers from both industry and academia to address various issues, challenges, and opportunities for development associated with this game-changing technology. The book will therefore appeal to a wide range of audiences who would like to get an unbiased and thorough comprehension of hydraulic fracturing technologies and their impacts on our environment and society.

Robert V. Duncan
Texas Tech University
Lubbock, Texas

Acknowledgments

The book is an outgrowth of a Fracturing Impacts and Technology Conference that was cohosted by Texas Tech University and the Air and Waste Management Association in September 2014. We would like to acknowledge the support of these institutions as well as the editorial and production team at Taylor & Francis Group in bringing this book to fruition. It is our hope that you, the reader, will come to appreciate the multifaceted nature of the hydraulic fracturing phenomenon and make meaningful contributions that will make the process safer and sustainable.

Editors

Venkatesh Uddameri, PhD, PE, is a professor and director of the Water Resources Center under the Department of Civil, Environmental and Construction Engineering at Texas Tech University, Lubbock, Texas. He has research interests in groundwater modeling, sustainable water resources management, the food–water–energy nexus, decision support systems for water resources planning, and climate change and has written over 75 technical publications. His research has been funded by NSF, NOAA, DOE, and DoD and other state and local agencies. Dr. Uddameri earned his bachelor's degree in civil engineering from Osmania University, India, in 1991 and his master's and doctoral degrees in civil engineering and environmental engineering, respectively, from the University of Maine. He is actively involved with the Water Resources Center and encourages nontraditional interdisciplinary research.

Audra Morse, PhD, PE, is the associate dean for undergraduate studies in the Whitacre College of Engineering (WCOE) and a professor in the Department of Civil, Environmental and Construction Engineering at Texas Tech University, Lubbock, Texas. She leads the Engineering Opportunities Center, which provides retention, placement, and academic support services to WCOE students. Her professional experience is focused on water and wastewater treatment, specifically water reclamation systems, membrane filtration, and the fate of personal products in treatment systems.

Kay J. Tindle, MEd, is the director of the Research Development Team in the Office of the Vice President for Research at Texas Tech University (TTU), Lubbock, Texas. In this role, she links faculty researchers with regional, state, and national partners to further develop collaborative teams to advance and achieve the strategic goals championed by the university. Among other initiatives, she works with the leader of TTU's Unconventional Production Technology and Environmental Consortium (UpTec), a multidisciplinary, collaborative research group that focuses on hydraulic fracturing research. Kay is currently pursuing a PhD in higher education research at TTU.

Contributors

Noureddine Abidi
Department of Plant and Soil Science
Texas Tech University
Lubbock, Texas

Sangu Angadi
Department of Plant and Environmental Sciences
New Mexico State University
Las Cruces, New Mexico

Dick Auld
Department of Plant and Soil Science
Texas Tech University
Lubbock, Texas

Erik P. Bucy
Department of Advertising
Texas Tech University
Lubbock, Texas

Doug D. Carlton Jr.
Department of Chemistry and Biochemistry
The University of Texas at Arlington
Arlington, Texas

R. Glenn Cummins
Department of Journalism and Electronic Media
Texas Tech University
Lubbock, Texas

Roxana Darvari
Department of Civil Engineering
The University of Texas at Austin
Austin, Texas

Jorge Ruiz de Viñaspre
Department of Civil, Environmental, and Construction Engineering
Texas Tech University
Lubbock, Texas

Anne C. Epstein
Department of Internal Medicine
Texas Tech University Health Sciences Center
Texas Tech School of Medicine
Lubbock, Texas

Bradley T. Ewing
Department of Energy, Economics, and Law
Texas Tech University
Lubbock, Texas

J. Berton Fisher
Lithochimeia, Inc.
Tulsa, Oklahoma

Brian E. Fontenot
Department of Biology
The University of Texas at Arlington
Arlington, Texas

Kulbhushan Grover
Agriculture and Home Economics
New Mexico State University
Las Cruces, New Mexico

Elma Annette Hernandez
Department of Civil and Environmental Engineering
Texas Tech University
Lubbock, Texas

Zacariah L. Hildenbrand
Inform Environmental, LLC
Dallas, Texas

Robert K. Imel
Department of Plant and Soil Science
Texas Tech University
Lubbock, Texas

Abdullah Karim
Department of Civil and Environmental Engineering
Texas Tech University
Lubbock, Texas

William R. Keffer
School of Law
Texas Tech University
Lubbock, Texas

George E. King
Apache Corporation
Houston, Texas

David Klein
Department of Environmental Toxicology
Texas Tech University
Lubbock, Texas

Jennifer Knaack
Mercer University School of Pharmacy
Atlanta, Georgia

Sumedha Liyanage
Department of Plant and Soil Science
Texas Tech University
Lubbock, Texas

Daniel Marangoni
Research and Sponsored Programs
Rogers State University
Claremore, Oklahoma

Russell McInturff
Department of Energy, Economics and Law
Texas Tech University
Lubbock, Texas

Terry McInturff
Department of Energy, Economics and Law
Texas Tech University
Lubbock, Texas

J. Randall Miller
Drummond Law, PLLC
Tulsa, Oklahoma

Audra Morse
Department of Civil, Environmental, and Construction Engineering
Texas Tech University
Lubbock, Texas

Lewis Norman
United Guar, LLC
Houston, Texas

Danny Reible
Department of Civil, Environmental, and Construction Engineering
Texas Tech University
Lubbock, Texas

Melanie Sarge
Department of Advertising
Texas Tech University
Lubbock, Texas

Kevin A. Schug
Department of Chemistry and Biochemistry
The University of Texas at Arlington
Arlington, Texas

Sanjaya Senadheera
Department of Civil, Environmental and Construction Engineering
Texas Tech University
Lubbock, Texas

Sreeram Singaraju
Department of Civil, Environmental and Construction Engineering
Texas Tech University
Lubbock, Texas

Sudhir Singla
Department of Plant and Environmental Sciences
New Mexico State University
Las Cruces, New Mexico

Taylor Stevenson
Pioneer Natural Resources
Irving, Texas

Kay J. Tindle
Research Development Team
Office of the Vice President for Research
Texas Tech University
Lubbock, Texas

Calvin Trostle
Texas A&M AgriLife Extension Service
Lubbock, Texas

Ron Truelove
ACL Combustion, Inc.
Moore, Oklahoma

Venkatesh Uddameri
Department of Civil, Environmental, and Construction Engineering
Texas Tech University
Lubbock, Texas

Tony Vercellino
Department of Civil and Environmental Engineering
Youngstown State University
Youngstown, Ohio

Adrianne Waddell
Pioneer Natural Resources
Irving, Texas

Marshall C. Watson
Department of Petroleum Engineering
Texas Tech University
Lubbock, Texas

Shawna R. White
Department of Advertising
Texas Tech University
Lubbock, Texas

John H. Williams
Xchem-Terra Services
Irving, Texas

Anna Thomas Young
National Wind Institute
Texas Tech University
Lubbock, Texas

Introduction to *Hydraulic Fracturing Impacts and Technologies: A Multidisciplinary Perspective*

While horizontal drilling and hydraulic fracturing have been around for a few decades, recent technological evolutions have made it possible for their widespread use, particularly in tight sands and shale formations. Thanks to these technologies, oil and gas production has increased significantly in the United States over the last few years. Figure I.1 depicts shale basins and shale plays with drilling and production activities. As can be seen, a majority of these basins have not been fully developed, particularly in the western United States, and unconventional oil and gas production using hydraulic fracturing has great potential in the years to come.

While the widespread use of hydraulic fracturing and directional drilling technologies is a relatively recent phenomenon, these technologies have been around for several years. Petroleum engineers have made significant strides in refining these technologies to make them more efficient and safe. They have learnt valuable lessons related to its application in various shale plays across the United States and other parts of the world. In Chapter 1, an overview of hydraulic fracturing is presented by King who has several decades of industrial experience with field-scale implementation of these technologies. He also chronicles the history of hydraulic fracturing

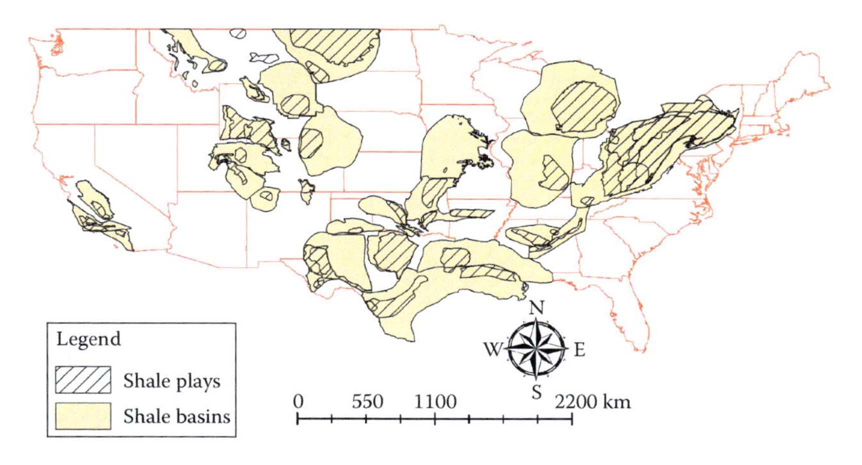

FIGURE I.1 Shale basins and active shale plays within the continental United States.

and highlights recent developments and process improvements. Understanding how hydraulic fracturing works is critical to fully characterizing its impacts.

Between the years 2005–2013, the production of natural gas in the United States increased by more than 33% and liquid fuels by 52%. Unconventional production of natural gas increased from 0.02 (0.75 trillion cubic feet or TCF) to 0.25 trillion cubic meters (8.5 TCF). Similarly, tight oil production jumped from 0.29 million to 3.48 million barrels per day.[1] The United States is now the largest producer of natural gas in the world, and our reliance on foreign oil has decreased significantly in the last few years. Unconventional oil and gas production has undoubtedly had a significant economic impact. For example, the U.S. gross domestic product (GDP) increased from $11.5 trillion in 2003 to $16.8 trillion in 2012, primarily fueled by increased oil and gas production.

Hydraulic fracturing has revitalized rural economies in North Dakota, Wyoming, and Texas. In Chapter 2, Ewing et al. discuss how unconventional oil and gas production in the Permian Basin region has created and sustained 546,000 jobs with an economic output of $137.8 billion and contributed to over $7.1 billion to the gross state products of Texas and New Mexico.

Contrary to popular belief, there are several legal statutes and regulatory requirements that affect hydraulic fracturing operations. Legal issues surrounding hydraulic fracturing cover a wide range of issues from property rights including subsurface trespass, water issues, and threats to endangered species. These issues and recent court cases are reviewed in a series of mini papers in Chapter 3. In particular, Keffer et al., in Chapter 3, use their extensive experience in oil and gas law practice to provide practical guidance to landowners considering leases for hydraulic fracturing operations. They then address the critical question of whether it is an actionable subsurface trespass when fluids injected during hydraulic fracturing operations migrate into the subsurface property of a neighbor. They further discuss issues related to water availability and argue that the water used for fracturing must be viewed in light of overall water use. Finally, they discuss the role of the Endangered Species Act in the oil and gas industry and describe how habitat fragmentation caused by hydraulic fracturing pads can further threaten sensitive ecosystems. Hydraulic fracturing activities are also subject to air quality regulations. In Chapter 4, Truelove discusses some recent white papers developed by the United States Environmental Protection Agency (USEPA) and alludes to potentially more stringent air quality regulations in the future.

Unconventional oil and gas production makes use of deep horizontal wells that many times extend for more than a mile in the subsurface. These wells have to be carefully designed, constructed, and fractured to avoid unwanted and uncontrolled movement of oil and gas to the subsurface. Toward this end, the oil and gas industry continues to improve its technology and has developed well construction and integrity guidelines.[2] Drilling and fracturing activities are regulated and states have casing, cementing, drilling, well control, and completion requirements (e.g., RRC[3]). Nonetheless, accidents and design failures cannot be completely eliminated. Spills and leaks of contaminants due to improper well design, well casing failures, and uncontrolled (or unintended) fracture development is the most cited concern associated with this activity.[4] Hydraulic fracturing operations use several hazardous chemicals that result in products (hydrocarbons) classified as known or potential carcinogens. Therefore, improper or unplanned emissions of these chemicals into the air and water pose risks to human health and ecosystems. Large-scale fuel production has once again brought forth health issues associated with hydrocarbon emissions into the environment. In Chapter 5, Klein et al. present a commentary on potential health risks associated with hydraulic fracturing, bringing to light the issue of increased health risks related to hydraulic fracturing operations.

An additional discussion on human health and environmental risks posed by oil and gas development is provided by Epstein in Chapter 6. The author reviews current literature related to toxic air emissions and presents evidence of human illnesses related to toxic air emissions from oil and gas development. She presents case studies related to occupational fatalities in the oil and gas industry as well as subsurface contamination from hydraulic fracturing activities. While the toxicity of organic compounds (i.e., benzene) that are recovered from hydraulic fracturing is well established, epidemiological studies are confounded by several other environmental factors. Our ability to effectively monitor exposures associated with hydraulic fracturing hinges on how well we can characterize trace levels of hazardous chemicals. Because certain constituents of concern such as methane can arise from either hydraulic fracturing or natural sources, better chemical characterization is critical in correctly identifying the contamination source. Chapter 7 by Carlton et al. focuses on the analytical characterization and provides a state-of-the-art review of current instrumental methods used for quantifying chemical exposures.

Unlike conventional production of oil and gas, hydraulic fracturing requires significant resources. Water, sand (proppant), and other chemicals are pumped at high pressures into formations with very low permeability to break open fractures and produce oil and gas. Water is a major component of hydraulic fracturing operations, and a typical fracturing job requires 11,000–19,000 m^3 of water. Many active hydraulic shale plays lie in semiarid and arid lands. The western United States is an area of intensive agriculture sustained by large-scale groundwater development. Hydraulic fracturing is seen as a new user of water and often in direct competition with traditional users. In Chapter 8, Uddameri and Reible discuss regional water resource issues in the Permian Basin of Texas. While the water use for hydraulic fracturing is relatively small, particularly in comparison to traditional uses, the timing of water use can be an issue. As the water quality requirements for hydraulic fracturing are not as stringent, poor-quality water such as brackish groundwater, domestic wastewater, and even flowback and produced water can be reused to reduce the freshwater footprint of hydraulic fracturing operations. In Chapter 9, Williams et al. illustrate how water is used during hydraulic fracturing operations and discuss technological solutions and challenges associates with treating flowback and produced waters.

Hydraulic fracturing is a material- and labor-intensive process. Large quantities of water, sand (proppant), and chemicals are brought into the fracturing site, resulting in the need to dispose of wastewater (produced water) and solid waste. Transportation of these materials is perhaps the most visible aspect of hydraulic fracturing. In Chapter 10, Senadheera discusses how a majority of rural roads used for transporting materials in and out of hydraulic fracturing sites are not designed to handle heavy truck loads and larger traffic volumes, which increases risks of accidents and deteriorates pavements. In Chapter 11, Hernandez et al. discuss transportation issues associated with the disposal of produced water in saltwater disposal wells in the Permian Basin region of Texas, highlighting why this disposal option is favored in Texas. Their analysis also quantifies potential air quality emissions and likely problems to the transportation infrastructure.

It may be surprising for many to know that agriculture has an important role to play in hydraulic fracturing. Guar is a drought-tolerant crop whose gum is used extensively as a gelling agent during fracturing. A significant amount of guar used by the industry is imported from India and Pakistan. In Chapter 12, Abidi et al. discuss the pros and cons of growing guar in the United States and indicate that while the crop can be grown in west Texas and Oklahoma, the lack of processing facilities is a major disincentive for farmers to grow this crop. Breeding better varieties of guar is useful to improve its material and rheological properties and thus improve the efficiency of hydraulic fracturing operations. As guar has been used in the food processing industry for a long period of time, there is an extensive body of literature related to characterizing its material properties. In Chapters 12 and 13, Abidi et al. review these techniques and discuss their application in hydraulic fracturing.

Hydraulic fracturing is no longer an isolated engineering activity and has evolved into a phenomenon that clearly affects people, their communities, and the environment. Needless to say, it has captured the attention of the public and the media. At the time of this writing, there were over 322,000 blogging sites on the Internet, and a Google search with the words hydraulic fracturing yielded 1,720,000 results. There is considerable polarization among the public with one camp focusing on the economic benefits and energy security and the other emphasizing environmental damage. Assessing public understanding of the issues surrounding hydraulic fracturing is therefore crucial to foster a healthy dialogue on this topic. In Chapter 14, White et al. note positive and negative framing that hydraulic fracturing has received in the press and assess media influence on public perceptions. They propose an integrated framework for studying the role of media influence on shaping the fracturing impacts and technology debate.

The need for unbiased and scientifically credible research cannot be emphasized enough given the politically charged nature of the topic. Industry-funded research is often viewed as tainted, and problems highlighted by environmental groups are dismissed as being alarmist calls not grounded in science. Therefore public institutions have the responsibility to serve as honest brokers to develop accurate and unbiased information needed to inform public policy. In Chapter 15, Tindle et al. highlight the importance of multidisciplinary teams in tackling these contentious issues. To this end, public institutions must foster a research environment that upholds the highest standards of integrity and transparency.

The main aim of our effort is to bring together a team of experts and present a holistic picture of hydraulic fracturing and its impacts on the environment and the society. Unconventional oil and gas production using hydraulic fracturing will not only proliferate in the United States but also is being actively pursued in many other parts of the world such as Germany, United Kingdom, China, and India.[5–8] The ideas and lessons contained in this monograph will be useful for a wide range of audiences and help them develop a multifaceted understanding of the topic.

REFERENCES

1. Gruenspecht, H. K. Annual energy outlook (early release): With projections to 2040. Presentation on behalf of U.S. Energy Information Administration for Center on Global Energy Policy, Columbia University, New York, p. 18, 2013.
2. API. *Hydraulic Fracturing Operations—Well Construction and Integrity Guidelines*. Washington, DC: American Petroleum Institute, 2009.
3. RRC. Oil and gas well records. http://www.rrc.state.tx.us/oil-gas/research-and-statistics/obtaining-commission-records/oil-and-gas-well-records/, accessed February 12, 2015.
4. Cooley, H. and K. Donnelly. *Hydraulic Fracturing and Water Resources: Separating the Frack from the Fiction*. Oakland, CA: Pacific Institute, 2012.
5. Olsson, O., D. Weichgrebe, and K.-H. Rosenwinkel. Hydraulic fracturing wastewater in Germany: Composition, treatment, concerns. *Environmental Earth Sciences* 70(8), 3895–3906, 2013.
6. Westaway, R. Quantification of potential macroseismic effects of the induced seismicity that might result from hydraulic fracturing for shale gas exploitation in the UK. *Quarterly Journal of Engineering Geology and Hydrogeology*, 47(4), 330–350, 2014.
7. Wu, L. P., T. C. Judd, S. L. Xi, L. Yin, and L. Hai. An integrated disciplinary approach towards hydraulic fracturing optimization of tight oil wells in the Ordos Basin China. Paper presented at the SPE Unconventional Resources Conference and Exhibition-Asia Pacific, Brisbane, Queensland, Australia, 2013.
8. TOI. Should India go in for shale gas from hydraulic fracturing. *The Times of India*. http://timesofindia.indiatimes.com/edit-page/Should-India-go-in-for-shale-gas-from-hydraulic-fracturing/articleshow/20820706.cms, accessed February 12, 2015.

1 Overview of Hydraulic Fracturing Operations and Technologies

George E. King

CONTENTS

1.1 Overview ... 1
1.2 Introduction ... 2
1.3 Nothing New under the Ground ... 2
1.4 Fracturing History and Scope ... 3
1.5 Impact of Fracturing ... 4
1.6 Science of Fracturing Fluids ... 7
1.7 Proppant ... 8
1.8 Volumes ... 8
1.9 Chemicals ... 9
1.10 Rate ... 9
1.11 General Completion and Frac Design ... 10
1.12 Fracture Complexity ... 11
1.13 Pumping a Frac ... 12
1.14 Fracture Pressure Response ... 12
1.15 Simultaneous and Sequential Fracturing ... 12
1.16 Refracturing ... 14
1.17 Flowback and Fracture Load Recovery ... 15
1.18 Tracers ... 15
1.19 Risks: Evaluation and Comparisons ... 16
1.20 Summary and Concluding Remarks ... 18
References ... 18

1.1 OVERVIEW

The chapter presents an overview of the hydraulic fracturing operations and technologies and chronicles the history and development of unconventional oil and gas production. The chapter also discusses the risks associated with the use of these technologies.

1.2 INTRODUCTION

The use of horizontal wells and hydraulic fracturing is so effective that it is a *disruptive* technology, essentially redefining the access to hydrocarbon energy supplies. The individual technologies of fracturing and horizontal wells are older, established technologies, but the combination of these technologies, beginning in 1974, and with widespread acceptance in conventional formations by the mid-1990s, has made possible a new era of oil and gas production.[1,2] This chapter covers several parts of hydraulic fracturing and serves as an introduction to the technology and how it is constantly evolving.

1.3 NOTHING NEW UNDER THE GROUND

Hydraulic fracturing and horizontal wells are not new tools for the oil and gas industry. The first fracturing experiment took place in 1947, and the process was commercialized by 1950.[3] The first horizontal well was drilled with difficulty in the 1930s, but horizontal wells became common by the late 1970s.[2] Millions of fracturing stimulations have been pumped, and roughly 100,000 horizontal wells have been drilled worldwide during the past 40 years.[2]

The recovery of shale gas is not new, either—especially from the Devonian shales in western Pennsylvania. The first shale gas well was drilled by William Hart in Fredonia, New York, in 1821 and found gas at a depth of 28 ft. Then in 1859, Edwin Drake, exploring the area's natural seeps of oil and gas, drilled the first U.S. oil well, and hit flowing oil at a depth of 69 ft. It is certainly not surprising that freshwater wells in this area of the northeast United States often contain methane gas, with publicly documented reports of gas in water wells reaching back over a hundred years.[4] Native Americans predated all this activity, gathering oil and tar from natural seeps more than 1500 years ago. These natural seeps, which led oil drillers to areas all over the world, are very common with thousands of natural seeps in North America and several hundred more in the seas around North America.[4,5] Oil and gas are natural products resulting from thermal degradation and occur in most areas of the world.

Because of shale gas, the United States has transformed itself from an importer of natural gas to energy independence in natural gas. During the next 20–25 years, shale gas will account for as much as 45% of all natural gas produced in the United States, and fracturing of tight gas, coal gas, and other low permeability formations will likely account for 70%–80% of all gas produced in the United States. Liquid-rich shales have had a similar effect on oil supplies, lifting the United States to the top oil-producing nation in early 2014.

The oldest modern shale gas wells are only about 13 years old. Yet many of these wells already have produced more gas than initially estimated. Shale-specific fracturing technology has increased production rates dramatically and is reversing the rapid decline witnessed in many early shale gas wells. The bottom line is that shale gas has increased gas reserves far beyond initial hopes. Today's advanced shale gas technologies help engineers place wells in the most productive areas. These technologies are enhancing the economics of shale, even with a surplus of gas now available.

1.4 FRACTURING HISTORY AND SCOPE

Fracturing uses hydraulic pressure, usually applied with water-based fluids, to crack the hydrocarbon-containing rock and produce a higher permeability flow path than that found naturally in the formation. The crack is a fraction of an inch wide (Figure 1.1) and fracture height is commonly only a few hundred feet tall at most.[6,7] Sand is carried into the crack by the fracturing fluid for the purpose of holding fractures open after hydraulic stresses have been released. Fracturing is a basic technology in concept but complicated in application. This is not surprising given that the application is made on a *structure* deposited in a succession of small layers with components washed into place from changing rock sources over millions of years and modified by earth forces and events over several more million years.

Fracturing was applied mostly in vertical wells from 1950 to the early 1970s before gradually being accepted as a stimulation in horizontal wells. As the industry has become more proficient at drilling longer horizontal wells, the lateral length has increased steadily, with laterals reaching out several miles and dozens of small fracture jobs being placed along the length of the well that is in contact with the pay zones. Longer lateral lengths usually deliver lower development costs if frac stages are increased to effectively break up the formation exposed by the extended wellbore reach, although the increase in rate is not exactly proportional to increased lateral length. As wells become longer, some energy is lost in friction, especially at higher fracturing pump rates.

Wells are fractured either to increase formation stability in soft sand formations or to accelerate or enable production from moderate to very low permeability reservoirs. Multifractured wells were successfully placed in 1952, just 2 years after

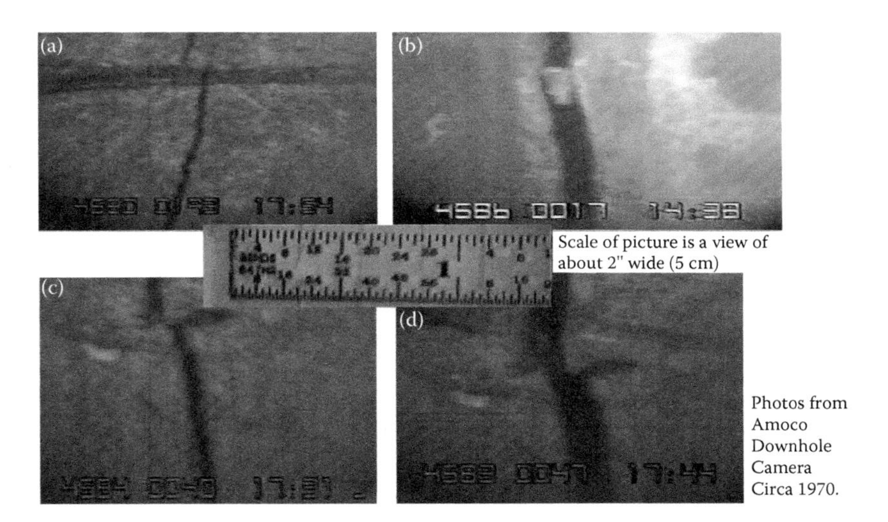

FIGURE 1.1 Downhole camera pictures of hydraulic fractures in open hole completions in vertical wells.

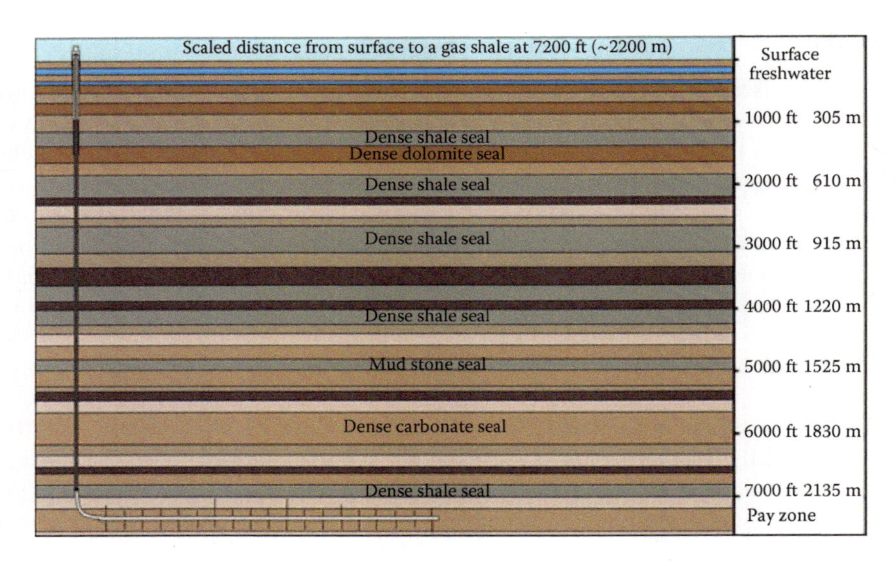

FIGURE 1.2 Multifractured horizontal well, drawn to a depth scale with reference to common usable water depths.

hydraulic fracturing was offered commercially. The targets of these wells were both offshore and onshore wells where stacked pay zones offered multiple targets.

Hydraulic fracturing technology has made a number of unconventional formations into producers with a very large impact on the supply of energy on a worldwide scale. Tight gas, marginally consolidated deep water formations, highly naturally fractured formations, coal bed methane recovery, North Slope wells, soft chalks, and now gas and hydrocarbon liquid-rich shales have been developed into stable supply sources by multifractured horizontal wells.

The importance of the horizontal multifractured well in the development of unconventional resources cannot be overstated. This type of completion, often a mile or more beneath the deepest freshwater sands, uses numerous small fracture treatments to create access to very low permeability formations (Figure 1.2).[5]

Documentation of the separation between fracture tops and the deepest freshwater was provided in microseismic records of over 2000 fracture treatments in the Barnett shale (Figure 1.3).[7]

There were over a million fracture treatments in wells from 1950 to 1995. In the era of horizontal multifractured wells, the number of fracture treatments, often smaller individually, has grown by another million as shale development has progressed. As successful adaptations, innovations, and inventions were found, the gas recoveries from shale have sharply increased (Figure 1.4).

1.5 IMPACT OF FRACTURING

Development of oil and gas wells in the United States started in New York and Pennsylvania over 150 years ago, with early forms of drilling and virtually no pollution

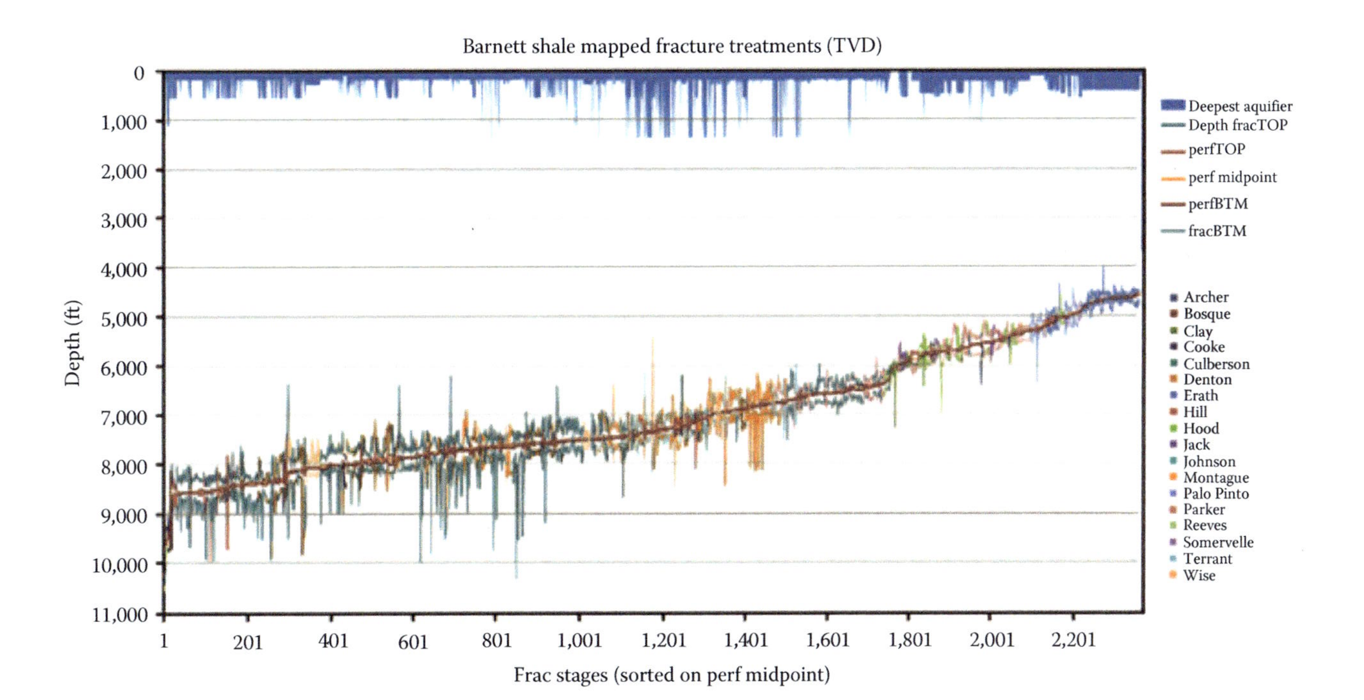

FIGURE 1.3 A multithousand well record of microseismic location of the tops of fractures in the Barnett shale with respect to the deepest freshwater in the area. (From Fisher, K. and Warpinski, N., *The American Oil and Gas Reporter*, May 2010, http://www.aogr.com/magazine/frac-facts, accessed February 12, 2015.)

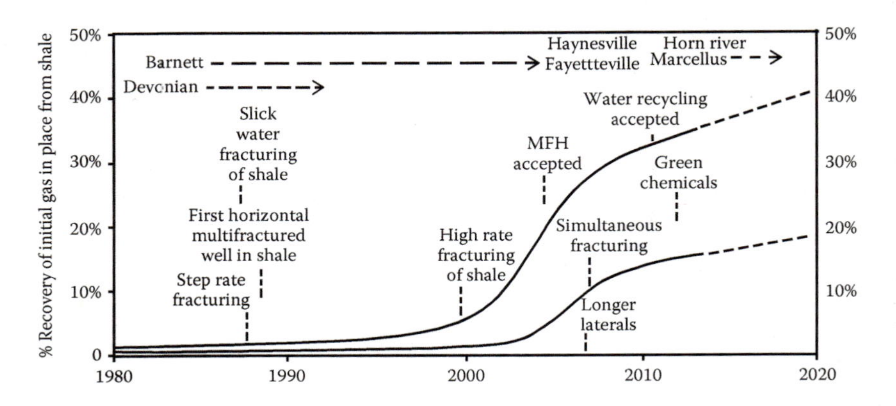

FIGURE 1.4 Percent recovery of initial gas in place versus year of the development and sequence of technology introduction.

control. By the 1890s, Pennsylvania was the oil capital of North America, and within a few years, oil enabled the mobilization of America and substantially cleared the air in many large cities by displacing coal as a primary home heating source. Oil and gas sources currently provide over 60% of total energy requirements in the United States.

Of the many forms of energy that satisfy U.S. demand, only the fossil energies have proven to supply sufficient energy to provide light, heat, cooling, and transport. Oil and gas provide over 60% of the total U.S. energy supply when power and transport are considered. The impact of fracturing (Figure 1.5) on this overwhelming majority of supply is simple: about 75% of U.S.-produced gas comes from wells that must be fractured and roughly 80% of U.S.-produced oil is

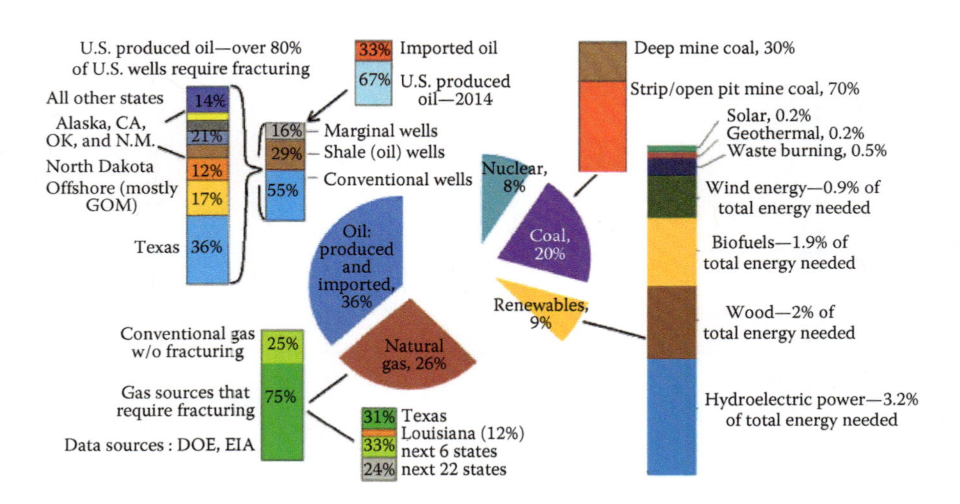

FIGURE 1.5 U.S. energy supply and effect of hydraulic fracturing.

from wells that use fracturing to stabilize soft sandstones or enable flow from very low permeability formations.

1.6 SCIENCE OF FRACTURING FLUIDS

Although slickwater fracture treatments have dominated shale gas fracturing applications, they should not be assumed to be the frac fluid of choice, particularly where higher flow capacity needs are proven. The choice of slickwater, gelled fracs, gas-assisted, or hybrid fracs should be made based on individual shale characteristics and requirements for stable production. The choices depend on increasing shale contact area, meeting proppant placement needs, and achieving production results.

Slickwater fracs are very simple fracturing treatments with minimum polymer and a lower sand-carrying capacity than either linear or cross-linked gels. Failure to create conductivity over the full fracture height can sharply limit productivity from some formations. In some cases with high modulus, the upper part of an unpropped fracture in low stress environments (like the Barnett) may still have sufficient total flow capacity to flow gas into the lower, propped area of the fracture but may not have the flow capacity to flow much gas horizontally along the unpropped fracture.

Slickwater fracs can break down the fissures, microcracks, natural fractures, and bedding boundaries in shale, opening up very large-shale face contact areas. In shales with natural fractures, development of natural fractures into flow paths (complex fracturing) is usually preferred over planar fracs. Shale characteristics and economics often favor a slickwater frac, which perform well in very low permeability formations and can induce complex fracturing, opening up very large access to the formation (Figure 1.6).

Formations without natural fractures, where planar fractures are common, may use gelled fracturing fluids in a more conventional fracturing design.

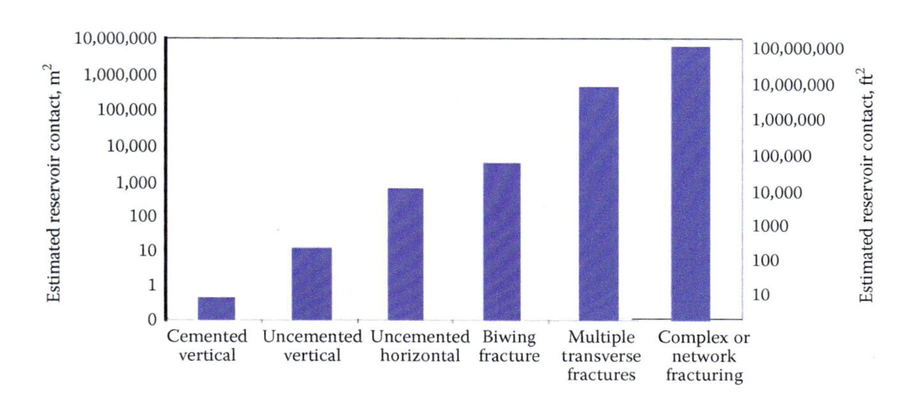

FIGURE 1.6 Completion designs and stimulation type impact on reservoir contact. (Modified from Vincent, 2011.)

1.7 PROPPANT

Proppant type, amount, and placement control the flow capacity of the main fractures and influence some of the smaller fractures. The proppants used in the majority of shale gas fracs are sands of various qualities. Sand has been found to be adequate for many shale and low permeability reservoirs, particularly in the smaller size ranges like 100 mesh (–70 to +140 U.S. mesh). Some of the larger sizes, notably 40/70, have been used where conductivity is important, but field testing and effects on both cost and production are needed. As deeper shales are investigated, sand proppant may not give adequate conductivity at the higher stresses and may give way to higher strength proppants and more high-tech fracturing designs for placement.

One-hundred mesh sand has also been used in fracturing to stop downward fracture growth, particularly when the size of the pad (no sand) is curtailed and the rate is reduced. It has also been used in mixtures with other sands to bridge-off larger fractures. The mechanism behind reducing downward frac growth may have many explanations, but is generally thought to form a wedge with 100 mesh sand slugs in the fracture, preventing excessive downward or, sometimes, outward growth.[8] Application of slugs of proppant for control of leak-off is an older technique brought to both oil-rich shale and gas shale fracturing. A typical slug is 60–180 kg/m^3 (0.5–1.5 lb/gal) over the programmed sand loading level and continues for 12–18 m^3 (100–150 bbls). The effect of the slugs may not be seen immediately but are often noticed a few minutes after the slugs hit the perforations.

The loading or ramping of sand in a shale gas frac job is very shale play dependent. Frac width and the amount of open natural fractures are primary controls on bridging potential in the near-wellbore part of the fracture. Width depends on rate, fluid viscosity, formation brittleness, local stresses, and presence of effective frac barriers. Typically, an initial loading of 24–40 kg/m^3 (0.2–0.25 lb/gal) is a starting point with common increase steps of 40 kg/m^3 (0.25 lb/gal) possible after the pressure stabilizes. Upper limits on the proppant concentration in the slickwater depend on the proppant size (both carrying capacity and frac width limited) and frac water velocity, and are often in the 300 kg/m^3 (2.5 lb/gal) range for 100 mesh sand and 240 kg/m^3 (2 lb/gal) for 40/70 mesh. Actual loadings will vary with application.

Alternative proppants (nonsand) have included small mesh bauxite for creation of erosion on the formation and bottom barrier formation, medium-strength man-made proppants, and light-weight proppants.

1.8 VOLUMES

The total water volume used in all the fracture stages in a well may be several million gallons depending on the requirements of the shale, with the volume of each individual fracture stage typically being between about 50,000 and 400,000 gal. Although freshwater was initially used for much of the fracturing, recycled produced water, nonpotable brackish water, treated effluent from cities, and other wastewater streams are being increasingly used for fracturing.

Size of the frac fluid stages depends on the application (shale thickness, amount of natural fracs, presence of frac barriers, etc.).

1.9 CHEMICALS

There are generally one to five purchased chemicals used in a slickwater frac job, with descriptions shown in Table 1.1.

Chemicals have steadily become safer over the past several years with the introduction of greener technologies, with many companies selecting chemicals from the EPA's Designed For Environment (DFE) list or the North Sea's Gold Band listed chemicals.

Trace chemicals used in product preparation, as carriers and impurities, can be found in some fracturing fluids. Even the freshwater supplies used in fracturing often contain a group of common minerals and metal ions, plus several *tag-along* trace chemicals, such as by-products of manufacturing or other trace pollutants found in drinking water, that have nothing to do with the petroleum industry (see: EPA Drinking Water Contaminants, EPA Water on Tap). The EPA lists common sources of drinking water pollution such as: bacteria and nitrates (human and animal wastes—septic tanks and large farms), heavy metals (mining construction, old fruit orchards), fertilizers and pesticides (anywhere crops or lawns are maintained), industrial products and wastes (local factories, industrial plants, gas stations, dry cleaners, leaking underground storage tanks, landfills, and waste dumps), household wastes (cleaning solvents, used motor oil, paint, paint thinner), lead and copper (household plumbing materials), and water treatment chemicals (wastewater treatment plants).[9] Chemicals from oil and gas operations rarely are mentioned.

1.10 RATE

In the most general terms, the frac injection rate produces the pressure to drive the type of fractures needed in a completion, but to drive fractures from multiple initiation points requires some thought about how to both initiate and drive a fracture. If only a single fracture was created, the design would be much simpler. However, multiple initiation points needed for increased fracture complexity require carefully

TABLE 1.1
Common Additives Used in Slick Water Fracturing in Shales

Most Common Slickwater Frac Additives	Composition	CAS Number	Percentage of Shale Fracs That Use This Additive	Alternate Use
Friction reducer	Polyacrylamide	9003-05-8	Near 100% of fracs use this additive	Adsorbent in baby diapers, flocculent in drinking water preparation
Biocide	Glutaraldehyde	111-30-8	80% or less	Medical disinfectant
Scale inhibitor	Phosphonate and polymers	6419-19-8 and others	~10%–25% of all fracs use this additive	Detergents
Surfactant	Various	Various	~10%–25% of all fracs use this additive	Dish soaps and cleaners

spaced cluster perforating, and sufficient rate to initiate and drive fractures. In some cases, 10–20 bpm (barrels per minute) per perforating cluster is used as a minimum, but as the fracture grows and leak-off increases, any injection rate will reach a point where the injected fluid is completely lost in leak-off and the fracture no longer grows. Whereas, the upper fracturing rate limit may not have been found yet, fracturing surface injection rates in the order of 15–25 bpm or less do not trigger many microseismic signals.[4]

As more and more fissures are opened, rate will need to be raised to sustain some level of fracture development. At some point of the fracture treatment, the rate available through the perforation cluster will be so dispersed in the fracture network that the velocity is reduced and a screen-out may occur at multiple points along the flow path. If this happens in the primary fracture system near the wellbore, the pressure rise will be quick. If multiple small screen-outs occur at a few of the growing tips, pressure rise may be slow and relatively constant, with lateral growth in areas of natural fracture density along the primary frac. This effect may be seen as semisteady net surface pressure rise, or a sharp pressure rise, depending on how many blockages are formed, where they are, and if enough unopened and optimally oriented natural fractures remain in the area of the frac to act as a pressure relief valve.

1.11 GENERAL COMPLETION AND FRAC DESIGN

Completion/frac designs require establishing a completion that allows the frac to make the best possible connection between the best parts of the reservoir and the wellbore. The specifics of the frac include the following:

1. How to initiate a primary frac from each stage.
   a. Locating the clusters in optimum sites, not just mathematical points
   b. Spacing of the clusters to achieve minimum stress interference from adjacent fracs, but still allow fracture complexity development
   c. Optimizing the number of perforating clusters to take advantage of design rate (matching clusters to rate)
   d. Establishing a fracture in each cluster during rate ramp up, even when the rate is still below the best hydraulic diversion rate
2. How to drive the frac outward, creating a frac of optimum length when using pad locations and well patterns, but retaining ability to fill in the areas between the staged fracs with a network of effective and open natural fractures. Some variables are
   a. Optimum rate ramp-up and when and how to stage a job for proppant ramp-up
   b. Optimum spacing of offset wells
   c. Thickness of net and gross pay (and brittle and ductile shale—how each behaves)
   d. Frac boundaries and how to avoid activities that breach these boundaries
   e. Presence, location, density, and opening pressure of primary, secondary, and even tertiary natural fractures
   f. Geological hazards and/or boundary interferences

There are many good fracture stimulators developed primarily for sandstones with reasonably constant leak-off from a planar frac. However, as shale frac leak-off enters fissures and cracks, the stresses created by enlarging these fissures place an entirely different and constantly changing stress state on the shale. This inflation of the reservoir, especially to a volume created by injection of large volumes of fluid, appears to transmit stresses in a different manner than is seen in porous and permeable formations. The bulk of information derived from fracturing in the shales indicates that the fractures often follow the natural fracture systems, branching when stresses, friction, internal frac diversion, or other factors create higher than normal fluid flow friction and offer an alternate path down a crack at a different angle to the stresses. The fracture may switch from the primary natural fracture system to the secondary natural fracture system and back again very quickly and very frequently. In viewing real-time microseismic signals from the on-site treating van, several active fracture growth patterns have been recognized as occurring simultaneously or sequentially in a single fracture stage.

Some parts of very anisotropic rocks fracture at a much lower pressure than other parts; thus, the frac will have a preferential path through the shale fabric. The impact of anisotropic rock fabric and the flow paths that they create is one of the major unknowns for any shale. Even fracs in shale with extensive natural fracture systems have shown preferential fracture development that is traceable from well to well. In addition to the natural fracture systems, regional fractures and faults also play an important role in both frac development and production capacities.[10]

Some operators found that ramping up frac rate too quickly in the Barnett shale (in areas without the lower frac barrier) could actually drive the fracture out of zone in the first minute of pumping, whereas a slow ramp-up of rate could help keep the frac in zone. This was documented with a set of experiments that showed that a low breakdown rate followed by 5 or 10 bpm increase steps during ramp-up kept many fracs in zone and increased complexity in the frac. Several fracs in the Devonian shales in the 1980s documented that low rate fracturing tend to open natural fracture pathways, while higher rates were more likely to create hydraulic fractures.[11]

1.12 FRACTURE COMPLEXITY

Complexity or network fracturing is a combination of establishing main fractures with high conductivity and adding immense contact areas created by opening and stabilizing natural fractures in the rock. How orderly the network is may be doubtful since local and regional stresses and postdepositional modifications will strongly influence which fractures open and in what order. Only a few formations, mostly shales and coals, have shown significant fracture complexity development on a regular basis. Warpinski states that the entire premise for the success of slickwater fracs in the Barnett is the ability of the low viscosity fluid of a water frac to activate, dilate, and shear offset natural fractures.[12] Complexity appears related to the use of slickwater fracs and is influenced by rate and injection pressure.

Olsen lists four conditions that make frac complexity in the Barnett a desired and possible outcome: (1) orthogonal regional tensile fractures, (2) low horizontal stress and stress anisotropy, (3) low Poisson's ratio, and (4) extremely low matrix permeability.[13]

The goal of increasing fracture complexity is increasing recovery of the gas in place in the shale. Optimizing the recovery, especially in ultralow permeability shales, requires generating both extensive reservoir contact to access the natural fractures and keeping the fracture pathway open after fracturing stresses have been released. If the fractures are unpropped, the matrix permeabilities of 100–200 nD (0.0001–0.0002 md) and the permeability of the healed natural fractures (variable perms) are simply too low to feed much gas into the main fractures. As has been found in lower permeability reservoirs, large areas of the reservoir are often largely undrained with wells on conventional spacing. In most of these reservoirs, infield drilling has tapped largely virgin pressures and previously stranded reserves. In shales with the possibility of fracture complexity, opening and increasing linkage of natural fractures can achieve the same result with a smaller surface footprint of fewer wells. By combining a hybrid approach of rate variation and sand slugs, it may be possible to custom *build* a complex network that selectively drains specific parts of the rock. This would be a benefit in any initial stimulation as well as remedial (refrac) stimulations.

1.13 PUMPING A FRAC

Surface injection pressure may change during a job for many reasons: opening up a new area of the rock, initiating a new fracture from the wellbore, encountering a plane of weakness (laminations or fault), or fracturing out of zone. Surface pressure records are often the only data available on a frac, so experience with pressure trends is important. Although measured bottom hole pressures (BHPs) are preferred over calculated BHP values, the reality of reliably obtaining BHPs in a multistage frac makes gauging BHP an elusive target.

Fracture intersection with natural fractures can create very complex occurrences in conventional rocks, but in unconventional reservoirs, natural fracture interaction may be a primary stimulation. Knowing how a frac will behave when it encounters a group of natural fractures that are either open or easily opened and what actions to take makes at least the initial frac work in shales a hands-on activity. Natural fractures are not just leak-off sites but serve as alternate frac pathways and can build quick stress points in the formation. Because they are linked in behavior, frac-induced stresses and natural fracture openings must be considered jointly.

1.14 FRACTURE PRESSURE RESPONSE

There is a considerable amount of information available from monitoring sources during and after fracturing, but the lowest cost information is available immediately in the form of the pump chart (Figure 1.7). The explanation of each of the generalized actions in the fracturing recording is explained in only minor detail.

1.15 SIMULTANEOUS AND SEQUENTIAL FRACTURING

In a few reservoirs, fracturing of multiple parallel wells has proven useful by using stresses created by fracturing one stage to divert another frac stage direction and even increase complexity development in subsequent fracturing stages. The effect

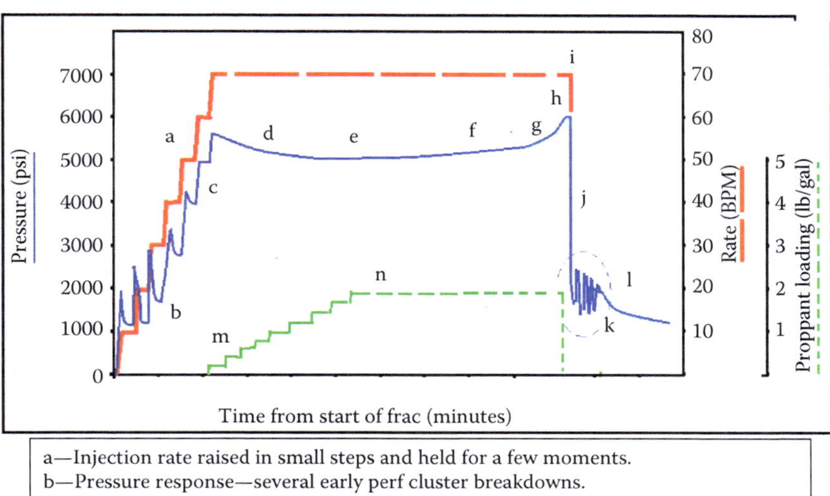

a—Injection rate raised in small steps and held for a few moments.
b—Pressure response—several early perf cluster breakdowns.
c—Pressure response controlled by hydraulic diversion.
d—Early frac extension, leakoff, and slight pressure decline.
e—Surface pressure flat but net pressure gradually increasing.
f—Frac reorientation possible—net pressure increasing more rapidly.
g—Slope change—press increase nearly double previous press trend.
h—Unit slope pressure rise as proppant feed is stopped.
i—Pumps are shut down after post flush—this job not overflushed.
j—Immediate pressure drop—friction pressure during injection.
k—Expanded grid—pressure *hammer*—fract-to-wellbore connection ok.
l—Instantaneous shut in pressure (ISPI)—hold to get frac closure pressure.
m—Proppant started at minimum rate that proppant can be transported in wellbore.
n—Proppant held constant (single proppant size shown in this frac).

FIGURE 1.7 Generalized pump chart from a gas shale frac using data from Barnett shale fracture treatments.

was reported by Warpinski et al. as altered stress fracturing where a frac direction was modified by a previous frac in the area.[14] Simultaneous fracturing has been successfully used in a number of shale developments but does not appear to have much effect in more ductile reservoirs. Two to five parallel wells have been fractured at once by one frac unit (blender and pumps) for each well. Perhaps the most famous Barnett shale simultaneous fracs are the EOG Resources Fowler fracs in the Barnett shale in northeast Johnson County. Five wells were fractured with five frac equipment sets. The total initial production (IP) rate for these five wells at 114 m (375 ft) well-to-well spacing was nearly 962 × 10^3 m^3/day (34 mmscf/day). Estimated ultimate recovery (EUR) was projected as 894 × 10^6 m^3 (31.6 BCF) and EOG's EUR of the gas in place was 54%, more than double the early GIP recovery expectation.

Candidate requirements for simultaneous or sequential fracture operations are not well defined. Most companies that have used the processes in the more brittle shales (Barnett and Woodford) have indicated good production responses; however, the distances between most of the paired wells have been in the order of 300 m (1000 ft) or less with extreme cases of 450 m (1500 ft) separation. The maximum distance will probably depend on the time between the fracs, the specific formation, the initial and after-frac stresses, and the fracture-induced stresses that may be linked to volume of fluid, pump rates, and diverting methods.

1.16 REFRACTURING

When dealing with formations that have unstable flow paths, it will be necessary to re-establish the flow paths as these natural fractures, fissures, and microcracks are not sufficiently propped to maintain stable flow channels. As fracturing technologies improve, the need for refracs may decline. Proppant placement improvement is the key.

Refracs of shale wells have often been successful for several reasons:[15–17]

- Enlarged fracture geometry and reservoir contact
- Improved pay coverage—increased frac height in vertical wells
- More thorough coverage in laterals (more fracs)
- Increased frac conductivity
- Restoration of fracture conductivity (various)
- Propping previously unpropped fractures
- Improved production profile
- More suitable fracture fluids
- Re-energizing or re-inflating natural fissures
- Reorientation of fracs-field stress alternations—new rock contacted
- Repair of wells overflushed after frac jobs
- Repair of wells negatively impacted by frac hits

Failures do happen in refracs, mostly because poor geology (low potential for production) is not always distinguished from poor completions. A few of the reasons for failures are also worth examining:[15,16]

- Low-pressured, depleted wells
- Low press or fault isolated wells—limited reserves
- Wells in which diagnostics indicate effective fractures and drainage to boundaries
- Wells with undesirable existing perforations or uncertain mechanical integrity of tubulars or cement
- Wells that do not allow access to better parts of wellbore
- Where off-set wells have recovered more than their *share* of reserves

Refracs of even good wells increased the recovery and re-established near IP rate. Increasing stimulated reservoir volume should increase both the IP and the EUR. When new areas of the shale are exposed in a refrac, there should also be a gain in reserves. Increases in stimulated reservoir volume could be accomplished by opening many of the microcracks and laminations within the undisturbed matrix blocks that were not stimulated by previous fracturing attempts. Re-opening natural and hydraulic fractures that had closed due to overburden and confining stress created by depletion would re-establish matrix area contact.

1.17 FLOWBACK AND FRACTURE LOAD RECOVERY

The amount of fracture fluid recovery in gas shales varies with the shale character, the frac design, and the type of fluid as the main drivers. Fractures that are more conventional with long reach and minimum complexity often flow back quickly, and the percentage of frac fluid recovered is high. In shale fracs where extensive complexity is developed or the shale is mildly reactive, the amount of fluid recovered may be on the order of 10%–50% of the total pumped and the time for fluid recovery may stretch over several weeks. This relationship depends on system energy and closure stresses. Controlling backpressure to use available formation gas energy to most efficiently remove the load water from the frac may have significant benefits.

Smaller natural fractures that are the source of fracture complexity may be the main cause of delays in water recovery. Relative permeability effects in the narrow fractures, related wetting phenomena, and the tortuous path from the far reaches of the frac fluid penetration are the main causes of these delays.

1.18 TRACERS

Field measurements that enable understanding hydraulic frac behavior and water recovery include microseismic monitoring while fracturing; radioactive tracers to mark frac entry points; tracers that allow tracing of fluid return efficiency for different stages; quantitative records of volume recoveries; salinity; solids; gas rates and pressures; and finally, tools that can qualitatively differentiate the type of fluid entering the well on backflow, the relative rates of return, and the entry points.

Chemical and gamma ray isotope tracers are useful in both gas shale stimulation and tracing fluid flowback. Chemical tracers, both oil soluble and water soluble, are used throughout the treatment and return as produced oil and water. Using DNA-like fingerprinting of fluids may be the next technology for tracking fluid and proppant movements.

By combining the timing of the volumes of marked load recovery, a picture begins to emerge of stage-by-stage load fluid backflow and water input from other formation-fed sources of water. Tracing the timing, volume, and contamination of the frac fluid return in a stage-by-stage manner can assist in differentiating the performance of the frac and the ability of a section of the formation to return fluids. In addition, monitoring the dilution of the tracer in a mass balance model can improve understanding of fracture fluid breakout, confirm effect of frac barriers, and directly measure flowback efficiency of each frac stage. Variability in the rate of return of fluids may be caused by gel damage, relative permeability effects, the complexity of the fracture system, frac behaviors that strand fluids (fracture closure), and other factors.

Gamma ray tracers (tagged proppant) are used to mark frac points, diagnose shallow fracture and cement isolation problems, and can be useful in determining vertical wells' frac height.

1.19 RISKS: EVALUATION AND COMPARISONS

There are no human endeavors that are free of risk; however, risks can be mitigated or lessened by performing risk analyses. One frequently used risk assessment with many variations is the ALARP (or as low as reasonably practicable) model originated in the North Sea area for the evaluation of industrial operations.[18] ALARP recognizes the evolving nature of any technical activity, be it medicine, mining manufacturing, or oil and gas production (Figure 1.8).

The unique character of the ALARP model is that there is an implied requirement in the *practicable* term for continuous improvement of the understanding, techniques, and application of the technology it is assessing, hence the learning loop in Figure 1.8. The more a person becomes educated about a specific activity, the more they will understand about actual risk and the safer that operation can become.

Risk in fracturing was described from literature sources and reported by King on the basis of research into the occurrence and environmental impact of twenty possible negative outcomes of a fracturing operation.[6] Nine possible negatives described transportation and storage effects and 11 possible negatives described perceived negative outcomes of the act of fracturing. The results were illustrated in a grid comparing reported occurrence with environmental impact. The results (Figure 1.9) indicated ways to improve the process, mostly in transportation and storage.

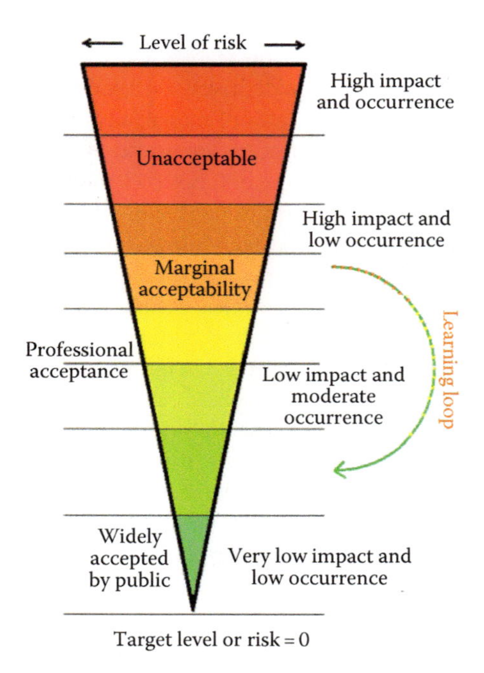

FIGURE 1.8 The ALARP risk triangle.

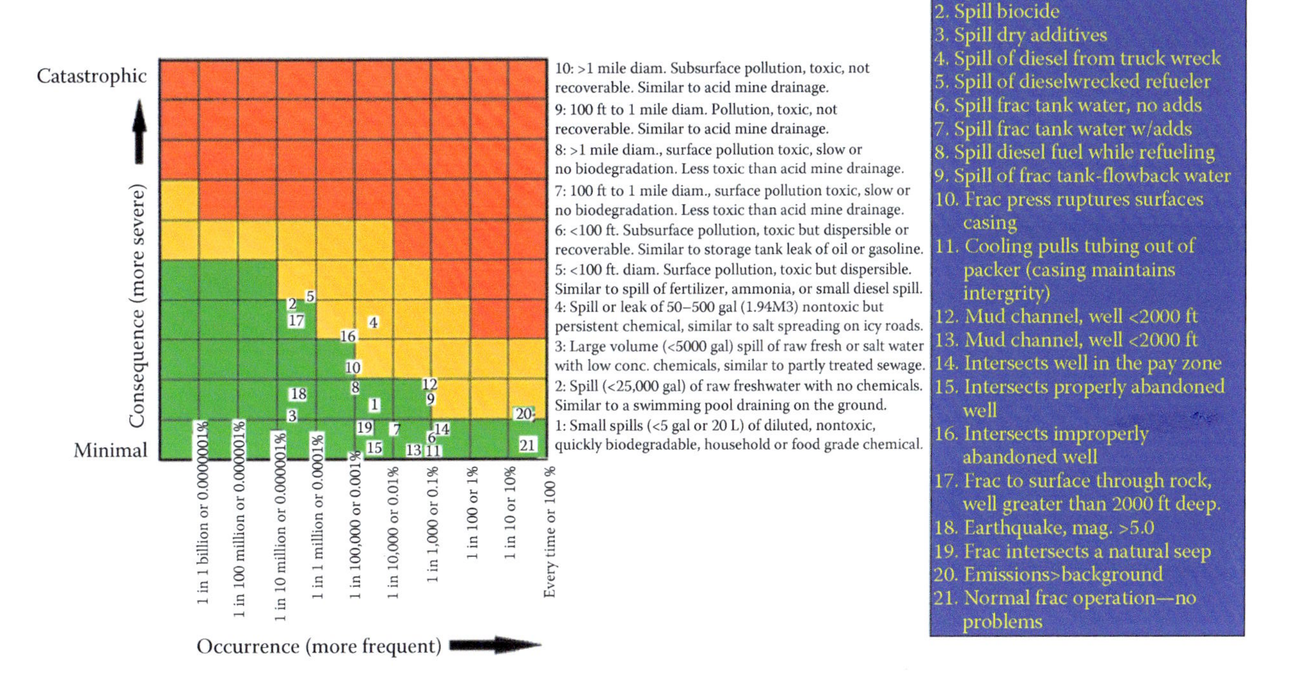

FIGURE 1.9 Risk as consequence versus occurrence for actions in hydraulic fracturing.

The risk matrix must be dynamic, and using technology to address challenges is a critical element of the application.

1.20 SUMMARY AND CONCLUDING REMARKS

The overall goal of this chapter was to provide a broad overview of hydraulic fracturing and discuss the evolution of this technology over time. Understanding how the process works is critical to evaluating the risks associated with it. Improvements in storage and transportation processes can greatly mitigate actual and perceived risks associated with hydraulic fracturing. The use of hydraulic fracturing in oil and gas wells can be a safe, highly effective way to efficiently produce reserves, but continuous technical improvement is required to optimize the stimulation to specific formations.

REFERENCES

1. Strubhar, M.K., Fitch, J.L., Glenn, E.E. Jr.: Multiple, vertical fractures from an inclined wellbore—A field experiment, Paper 5115, presented at *SPE-AIME 49th Annual Fall Meeting*, Houston, TX, October 6–9, 1974.
2. King, G.E.: 60 Years of multi-fractured vertical, deviated and horizontal wells: What have we learned? SPE 170952, *SPE Annual Technical Meeting and Exhibition*, Amsterdam, the Netherlands, November 26–28, 2014.
3. Montgomery, C.T., Smith, M.B.: Hydraulic fracturing: History of an enduring technology, *Journal of Petroleum Technology*, 63, December 2010, 26–32.
4. King, G.E.: Thirty years of shale gas fracturing: What have we learned? SPE 133456, *SPE Annual Technical Conference and Exhibition*, Florence, Italy, September 19–22, 2010.
5. King, G.E., King, D.E.: Environmental risk arising from well-construction failure—Differences between barrier and well failure, and estimates of failure frequency across common well types, locations, and well age, *SPE Production and Operations*, 28(4) (SPE 166142), November 2013, 323–344.
6. King, G.E.: Hydraulic fracturing 101: What every representative, environmentalist, regulator, reporter, investor, university professor, neighbor and engineer should know about estimating frac risk and improving frac performance in unconventional gas and oil wells, SPE 152596, *SPE Hydraulic Fracturing Technology Conference*, The Woodlands, TX, February 6–8, 2012.
7. Fisher, K., Warpinski, N.: *The American Oil and Gas Reporter*, May 2010, http://www.aogr.com/magazine/frac-facts, accessed February 12, 2015.
8. King, G.E., Haile, L., Shuss, J., Dobkins, T.A.: Increasing fracture path complexity and controlling downward fracture growth in the Barnett shale, Paper 119896, presented at the *2008 SPE Shale Gas Production Conference*, Fort Worth, TX, November16–18, 2008.
9. EPA: Water on tap: What you need to know, Office of Water, Washington DC, December 2009.
10. Warpinski, N.R.: Integrating microseismic monitoring with well completions, reservoir behavior, and rock mechanics, Paper SPE 125239, presented at the *2009 SPE Tight Gas Completions Conference*, San Antonio, TX, June 15–17, 2009.

11. Overbey, W.K., Yost II, A.B., Wilkins, D.A.: Inducing multiple hydraulic fractures from a horizontal wellbore, Paper SPE 18249, *1988 SPE Annual Technical Conference and Exhibition*, Houston, TX, October 2–5, 1988.
12. Warpinski, N.R.: Microseismic monitoring: Inside and out, SPE Distinguished Author Series, *Journal of Petroleum Technology*, 61, November 2009, 80–85.
13. Olsen, T.N., Bratton, T.R., Thiercelin, M.J.: Quantifying proppant transport for complex fractures in unconventional formations, Paper SPE 119300, presented at the *2009 SPE Hydraulic Fracturing Technology Conference*, The Woodlands, TX, January 19–21, 2009.
14. Warpinski, N.R. et al.: Altered stress fracturing, Paper SPE 17553, *Journal of Petroleum Technology*, 41, September 1989, pp. 990–997.
15. Vincent, M.C.: Refracs—Why do they work, and why do they fail in 100 published field studies? *SPE Annual Technical Conference and Exhibition*, Florence, Italy, September 19–22, 2010.
16. Vincent, M.C.: Restimulation of unconventional reservoirs: When are refracs beneficial, CSUG/SPE 136757, *Canadian Unconventional Resource & International Petroleum Conference*, Calgary, Alberta, Canada, October 19–21, 2010.
17. Vincent, M.: Restimulation of unconventional reservoirs: When are refracs beneficial? *Society of Petroleum Engineers*, May 2011, doi:10.2118/136757-PA.
18. Taylor, M., Israni, C.: Understanding the ALARP Concept: Its origin and application. *Society of Petroleum Engineers*, March 2014, doi:10.2118/168486-MS.

2 Economic Impact of the Permian Basin's Oil and Gas Industry

Bradley T. Ewing, Marshall C. Watson, Terry McInturff, and Russell McInturff

CONTENTS

2.1 Overview 22
2.2 Introduction 22
2.3 Objectives and Methodological Approach 23
2.4 A Brief History of the Oil and Gas Industry in the Permian Basin 23
2.5 Economic Landscape of the Permian Basin: Current Conditions 25
2.6 Plays and Reservoir Description within the Permian Basin 26
2.7 Assessment of Wells Drilled to Date and Production by Basin 29
2.8 Drilling Activity 33
2.9 Drilling/Completion Technology and Production Trends 37
2.10 Economic Impacts from the Production of Oil and Gas in the Permian Basin: 2013 42
2.11 Industry Taxation 54
2.11.1 Texas 55
2.11.1.1 Severance Tax 56
2.11.1.2 Sales and Use Tax 56
2.11.1.3 Oil and Gas Well Servicing Tax 56
2.11.1.4 Franchise Taxes 56
2.11.2 New Mexico 60
2.12 Summary and Concluding Remarks 61
Acknowledgments 61
Appendix A: Counties Included in the Permian Basin IMPLAN Models 61
Appendix B: Society of Petroleum Evaluation Engineers (SPEE) Parameters for Evaluating Resource Plays 63
References 63

2.1 OVERVIEW

The oil and gas industry has revitalized the economy of the Permian Basin region. This industry has created and sustained more than a half million jobs in the region, generated almost $137.8 billion in output, and contributed more than $71 billion in total gross state product in Texas and Mexico.

2.2 INTRODUCTION

This study provides estimates of the economic impact of the Permian Basin's oil and gas industry. The study examines these impacts at the county level as well as in the context of the overall Permian region. The Permian Basin region covers parts of both New Mexico and Texas. The region is rich in oil and deposits and includes the legendary Delaware and Midland basins and a number of well-known plays such as Wolfberry/Spraberry, Bone Springs, San Andres, Clearfork, Cline Shale, and Wolfcamp. The industry, as a whole, inherently provides economic benefits to the region and beyond. The oil and gas industry is characterized by a highly integrated supply chain that results in substantial employment, high-paying jobs, investment, and economic growth. Moreover, the Permian's oil and gas industry contributes significantly to state and local governments through taxation.

The focus of this study is on the value creation and economic sustainability that lies in upstream through midstream oil and gas industry activities. Ultimately, economic impacts derive from the exploration, drilling, and production of oil and gas, which require a multitude of support activities for oil and gas operations. These core activities, in turn, lead to a number of noncore but very critical midstream supply chain activities such as pipeline, transportation, refining, and equipment manufacturing. The secondary effects of the oil and gas industry include the numerous expansions and continuing operations of suppliers to the industry as well as wholesale, retail, real estate and housing, and financial services that benefit from the increased dollars generated.

This chapter provides a brief overview of the history and importance of the oil and gas industry in the development and growth of the Permian Basin region. This chapter also includes a review of the economic landscape and current conditions that define the state of the economy. Additionally, a unique feature of this study is the inclusion and analysis of the petroleum engineering and geophysical factors that characterize this region. Permian Basin well productivity has improved dramatically since 2011 due to improved technology in horizontal drilling and multistage hydraulic fracturing. Drilling efficiencies are being realized in all U.S. resource plays, and the Permian Basin is the least mature, thus vast efficiency improvements are not only being realized but expected in the Permian Basin. It is these factors that have led to the present state of production that make possible the economic benefits of the industry, which will play a vital role in the future developments of the Permian Basin's economy. The economic benefits arising from the existing and continuing operations of the oil and gas industry are quantified in terms of employment, labor income (including proprietor's income), value added, and output. Additionally, this study provides a comprehensive analysis of taxation associated with the industry.

A specific list of the counties comprising the Permian Basin and which are examined in this study is contained in Appendix A.

2.3 OBJECTIVES AND METHODOLOGICAL APPROACH

The major objective of this study presented in this chapter is to quantify the economic impacts of the Permian Basin's oil and gas industry. Specifically, these impacts are in the form of jobs created and sustained, economic output as measured by the value of all industrial production in an area or region, value added to state gross domestic product, and various forms of tax-related revenues generated from the production and sale of oil and gas. The methodological approach of this research combines elements from the fields of both energy economics and petroleum engineering.

The research utilizes the IMPLAN economic impact modeling software. To fully capture the underlying factors of economic activity in the Permian Basin, historical economic and petroleum engineering data are collected and analyzed. The study includes analyses of oil and gas production, assessment of wells, drilling activity, drilling/completion technology, and production trends, combined with economic analyses of oil and gas, to estimate the economic impacts of the industry. Results are presented for the Permian Basin as a whole, as well as for the New Mexico and Texas portions of the Permian Basin, and the counties comprising the Permian Basin.

The study is unique in that it blends expertise in energy economics and petroleum engineering to provide an engineering-based economic impact model that takes into account the geological, engineering, and economic nature of the industry.

2.4 A BRIEF HISTORY OF THE OIL AND GAS INDUSTRY IN THE PERMIAN BASIN

The Permian Basin is located in West Texas and southeastern New Mexico and is one of the world's leading oil- and gas-producing areas. However, defining the exact boundaries of the Permian Basin is a difficult task. Due to a better understanding of geological formations and new engineering technologies, what we know as the Permian Basin has actually expanded over the years and today covers an area nearly 402 km (250 mi) wide and 482 km (300 mi) long. The region includes thousands of fields over several producing formations (e.g., Yates, Spraberry, Wolfcamp, Yeso, and Bone Spring) and spans nearly 50 counties, most of which are in Texas. The total cumulative production for just the Texas portion of the Permian Basin exceeds 29 billion barrels of oil and approximately 2.12 trillion cubic meters (75 trillion cubic feet) of natural gas (*Source*: Railroad Commission of Texas, December 17, 2013). It's been said that, "as the Permian Basin's oil and gas industry goes, so goes the region's economy." That statement is not far off from reality, either. The historical production growth has led to the development of important centers of population and commerce, particularly the Texas cities of Midland and Odessa, but also Eddy and Lea counties in New Mexico. For example, Midland was the fastest growing metro area from July 2011 to July 2012, while Odessa was the fifth fastest growing metro area in the

United States during the same time period. Likewise, Lea County experienced a population growth of 16.6% from 2000 to 2010.

The first major well in the Permian Basin (located in Mitchell county) was completed in the early 1920s, the Santa Rita No. 1, at a total depth of 762 m (2498 ft). This well produced for several decades before being capped in 1990, giving an indication of the type of production that would come from the Permian Basin. This discovery led to additional drilling fields including World field in Crockett County, McCamey field, and the Yates field in Pecos County. The Yates field is still producing today and is one of the top 50 fields for proved oil reserves in the United States. These discoveries were made as a result of random drilling or surface/subsurface mapping. In 1928, the Hobbs field discovery came from magnetometer and torsion balance survey. The seismograph was being used around this same time as an exploratory tool and maps were providing outlines of the various basins.

Prior to 1928, most oilfields were at depths less than 1372 m (4500 ft) due to deep tests not being economically feasible. However, large flows of oil and gas were discovered in the Big Lake Oilfield in Reagan County at a depth of 2598 m (8525 ft) in 1928. While this discovery increased the prospects of the Permian Basin, it was the need for oil during World War II that provided the incentive to drill more and at greater depths. As such, several major fields were discovered during this time including Wasson, Slaughter, and Seminole, which are all still producing today. In fact, these fields are still ranked by the Energy Information Administration among the top 20 in the United States for remaining proved reserves.

In the 1940s, with the help of more advanced scientific techniques, several additional structures were uncovered. One such example was the Horseshoe Atoll, which yielded several significant fields. However, the Spraberry has been the major play and continues to be a profitable play with its predictable reserves. The Spraberry lies above the Wolfcamp in the Midland Basin and today is about 241 km (150 mi) long and up to 56 km (35 mi) wide.

The Permian Basin and the various operators have experienced production swings over the past several decades from fluctuations in oil prices to the development of new technologies. For example, in the 1970s, total production from the Permian Basin was around two million barrels a day. In the past, oilfield operators would only drill and complete the Spraberry/Dean. Recently, operators started deepening these wells into the Wolfcamp and comingle to gain extra production. In the early 2000s, multistage hydraulic fracturing techniques were refined, which led producers to go deeper and to commingle even more productive zones with the Spraberry. The combination play of the Spraberry, Wolfcamp, and other formations has been nicknamed the *Wolfberry*. On the Delaware Basin side of the Permian, another combination play of the Bone Spring sands with the Wolfcamp led to the nickname *Wolfbone*. These are some of the examples illustrating the potential of multistage hydraulic fracturing and horizontal drilling that has led to the increase in recent activity in the Permian.

Today, total crude oil production in the Permian Basin is around 1,400,000 barrels a day (Energy Information Agency (EIA), Drilling Productivity Report). Over the past couple of years, there has been a resurgence in drilling activity in the Permian

Basin. For example, the number of drilling permits issued has more than doubled since 2005, while the rig count has more than tripled since 2005 from 129 rigs to 415 as of 2012 and to over 500 at the time this report was written. A large portion of the aforementioned rigs in the Permian Basin are now drilling horizontally. Twelve percent of the drilling permits issued in 2005 were for horizontal wells compared to 41% in 2013. The total crude oil production has gone from 253 million barrels in 2005 to 312 million barrels for the calendar year 2012 and was even greater at year-end 2013.

Growth in drilling and production has spearheaded the economic growth of the Permian Basin for decades. The Texas Permian Basin's crude oil production is 57% of Texas's annual crude oil production and about 14% of the total annual U.S. crude oil production. Historically, the Texas portion of the Permian Basin represents about 90% of the oil production and about 80% of the gas production. However, over the past decade, the New Mexico portion has increased as the Texas share of oil production now compromises only about 70% of the total Permian Basin, and the gas production compromises about 65%. Most recently, the Cline Shale has gained attention because it extends farther north and east across the Eastern Shelf totaling 112 km (70 mi) wide and 225 km (140 mi) long. While the recoverable reserves are uncertain, producers are still drilling many Cline horizontal producers to assess the commerciality of the Cline Shale.

2.5 ECONOMIC LANDSCAPE OF THE PERMIAN BASIN: CURRENT CONDITIONS

The region known as the Permian Basin has experienced phenomenal economic changes in recent years. The area was not entirely immune from the worst national recession in decades (2008–2010), a national housing crisis, and a financial crisis. However, the region has retained its reputation for withstanding adverse economic conditions, and as the national recovery is well underway, the energy sector leads the way toward a bright future. In fact, output in oil and gas is approaching the record levels of several decades ago. Improvements in technology—hydraulic fracturing, horizontal drilling—coupled with geological discoveries make the economy of 2013 one of the best years ever for the Permian Basin.

It is the combination of the application of technology with economic factors (operations, energy prices, material and input costs, supply chain linkages, to name a few) that forms the basis for the economic impacts of the oil and gas industry. The larger metropolitan statistical areas in the region have lower unemployment rates than state and national averages. The preliminary end-of-year 2013 unemployment rate for New Mexico is 6.3% and for Texas is 6.0%. Unemployment rates for the Midland MSA and Odessa MSA are 3.1% and 3.6%, respectively, while the Lubbock MSA is 4.7%. At the county level, the U.S. Bureau of Labor Statistics (BLS) reports the unemployment rate (August 2013) for Eddy County (NM) is 3.9% and for Lea County (NM) is 3.8%. The employment picture is similar in many of the Permian Basin's lesser populated counties, also, with Andrews County at 3.6% and Sterling County at 2.7%, to name a few.

Crude oil and natural gas prices have always been volatile; however, the landscape has changed a bit since the last recession. First, both crude oil and natural gas prices reached a peak shortly after the start of the recession prior to dramatic drops in price. However, while natural gas prices have remained relatively low, crude oil prices have generally trended upward. This change in pricing patterns has led the oil and gas industry to substitute drilling and production activities from gas to oil.

2.6 PLAYS AND RESERVOIR DESCRIPTION WITHIN THE PERMIAN BASIN

A map showing all the current plays in the Permian Basin is shown in Figure 2.1. To effectively assess the plays/reservoirs in this study, the counties in the Permian Basin (Appendix A) are divided into three regions: the Midland Basin, the Delaware Basin portion of Texas, and the Delaware Basin portion of Southeast New Mexico.

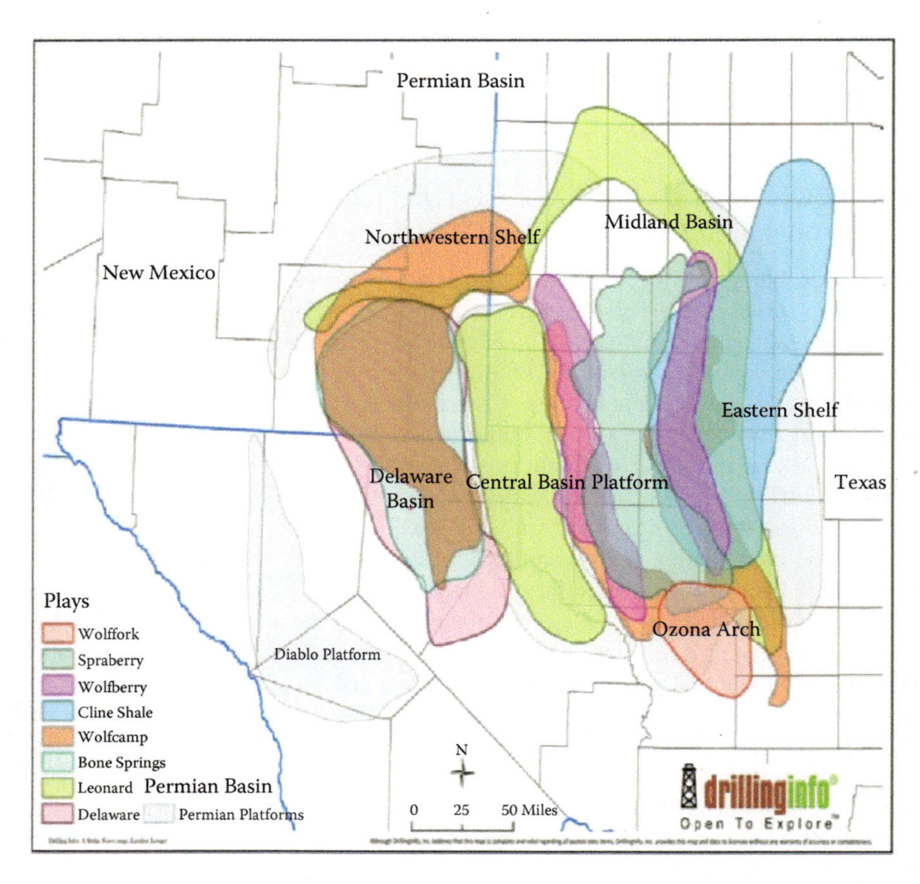

FIGURE 2.1 Map of plays/reservoir distribution in the Permian Basin. (From DrillingInfo and DI Desktop™, http://www.drillinginfo.com/.)

Oil and gas productions in these three regions are shown in Figures 2.2 through 2.6. Detailed assessments of the most active plays/reservoirs in each Texas Railroad commission district and Southeast New Mexico were conducted. Based on these surveys, the Bone Spring and Yeso Plays, located in the Delaware basin are the most active plays. The most active wells targeting these plays are predominantly located in southeast New Mexico. In addition, the Spraberry/Wolfberry Trend is the next largest play.

All of these plays are either considered or may be potentially a resource play as defined by the Society of Petroleum Evaluation Engineers (SPEE) in Watson et al. (2010) and shown in Appendix B.

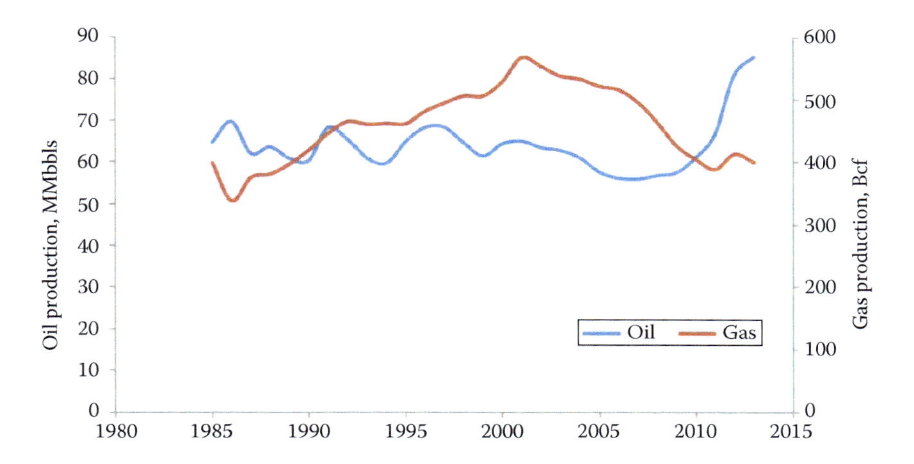

FIGURE 2.2 Delaware Basin South East New Mexico historical production.

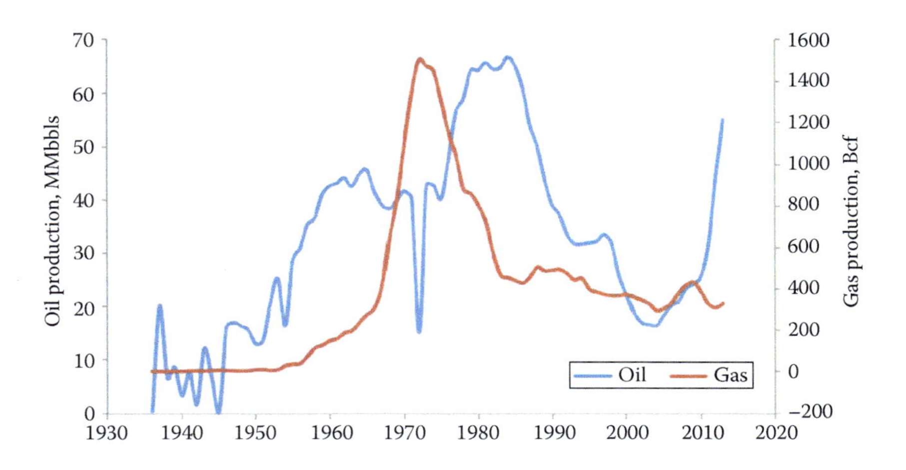

FIGURE 2.3 Delaware Basin Texas historical production.

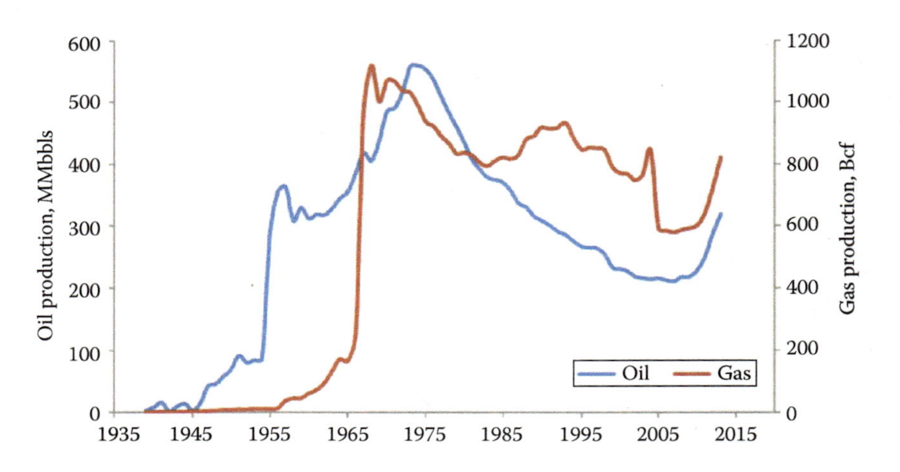

FIGURE 2.4 Midland Basin historical production.

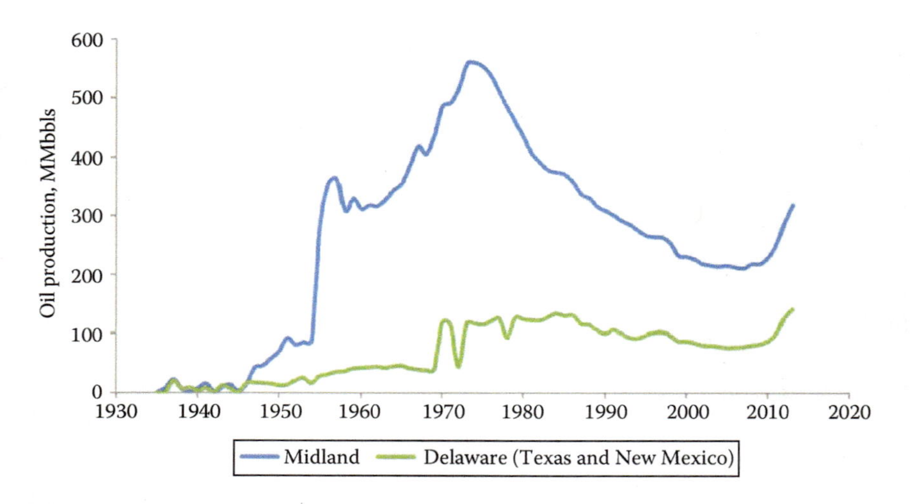

FIGURE 2.5 Entire Permian Basin historical annual oil productions.

Resource plays have also been called *unconventional plays* or *unconventional reservoirs*. These resource plays are important to this study as they result in long-term sustained activity. However, they are very price sensitive, as will be addressed later in this report. The Spraberry activity, located in the Midland Basin, is attributed to downhole reservoir commingling and hydraulic fracturing techniques recently being employed by operators in the Permian Basin. The Spraberry wells are predominantly vertical. Activities of the Bone Spring (Delaware Basin) and Wolfcamp play (Midland Basin) can be attributed to horizontal drilling and hydraulic fracturing technology. Within these districts, the Spraberry Play is the predominant play, followed by the Wolfbone/Wolfcamp.

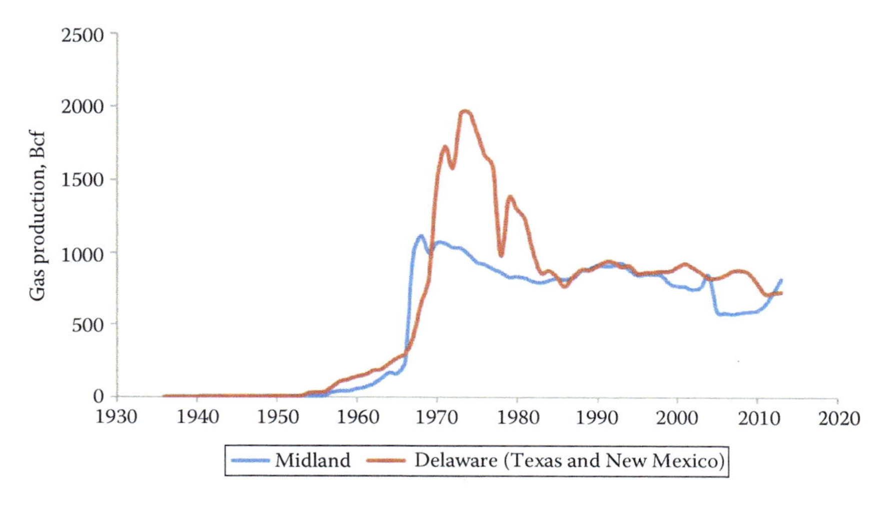

FIGURE 2.6 Entire Permian Basin historical annual gas productions.

2.7 ASSESSMENT OF WELLS DRILLED TO DATE AND PRODUCTION BY BASIN

The Permian Basin is spread across the states of Texas and New Mexico, and a comparison is made herein between this basin and other basins located within the two aforementioned states in Figures 2.7 through 2.10. The Permian Basin has the highest number of wells drilled, with over 392,000 wells (37%), followed by Texas Coast Basin, which contains the Eagle Ford Play, with 257,938 wells (24%) as shown in Figure 2.7. The Texas Gulf Coast and Ft. Worth Basins, the latter of which contains the Barnett Shale Play, have the highest daily average gas production of 8.5 and 6.3 Bcf, respectively, followed by the Permian with over 5.1 Bcf/day (17.3%) as seen in Figure 2.8. However, as shown in Figure 2.9, the Permian Basin has the highest

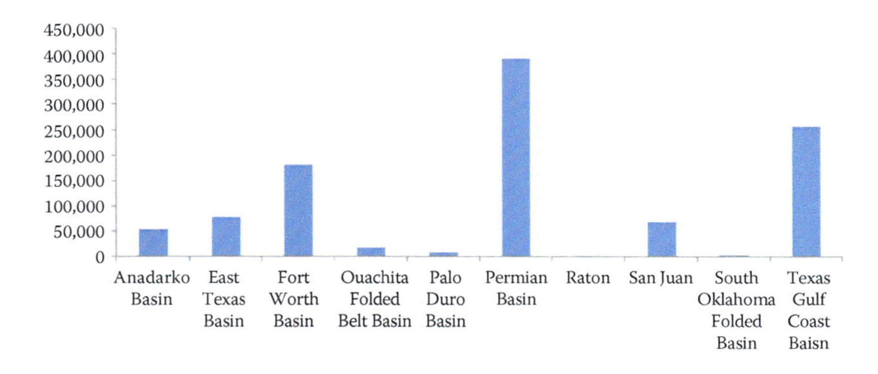

FIGURE 2.7 Number of wells drilled in the Permian Basin versus other basins in Texas and New Mexico from inception through December 2013. (From DrillingInfo and DI Desktop™, http://www.drillinginfo.com/.)

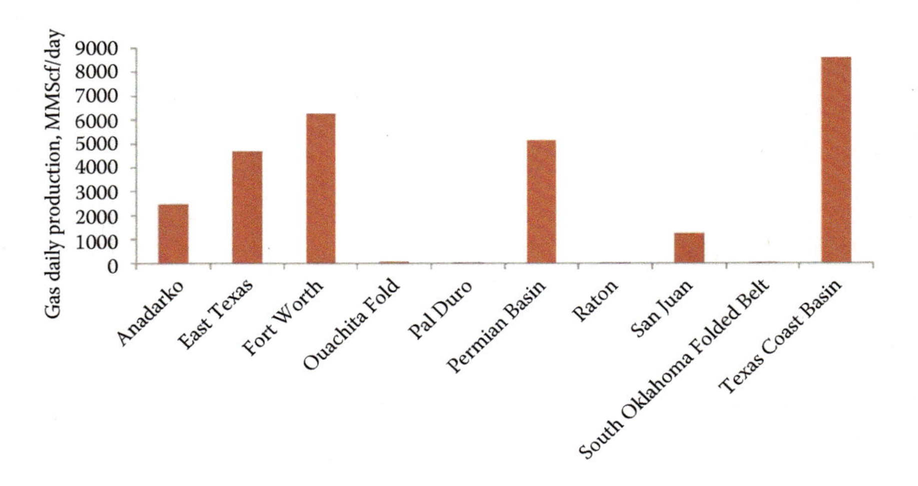

FIGURE 2.8 Daily gas production from all basins in Texas and New Mexico. (From DrillingInfo and DI Desktop™, http://www.drillinginfo.com/.)

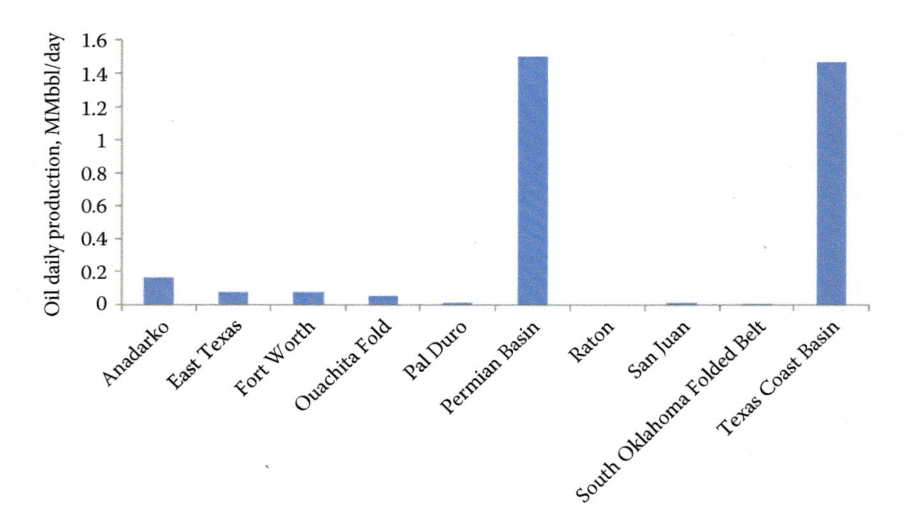

FIGURE 2.9 Daily oil production from all basins in Texas and New Mexico. (From DrillingInfo and DI Desktop™, http://www.drillinginfo.com/.)

liquid daily average production with over 1.49 MMbbl (44.6%), followed by Texas Gulf Coast Basin with 1.46 MMbpd. Note that the Eagle Ford Shale Play is located in the Texas Gulf Coast Basin, and the Barnett Shale Play is located in the Ft. Worth Basin.

Since early 2012, as shown in Figure 2.10, average breakeven cost has declined due to improved drilling/completions efficiency and liquid production in the Permian, Bakken, and northwest portion of the Eagle Ford. Gas-prone plays, like the

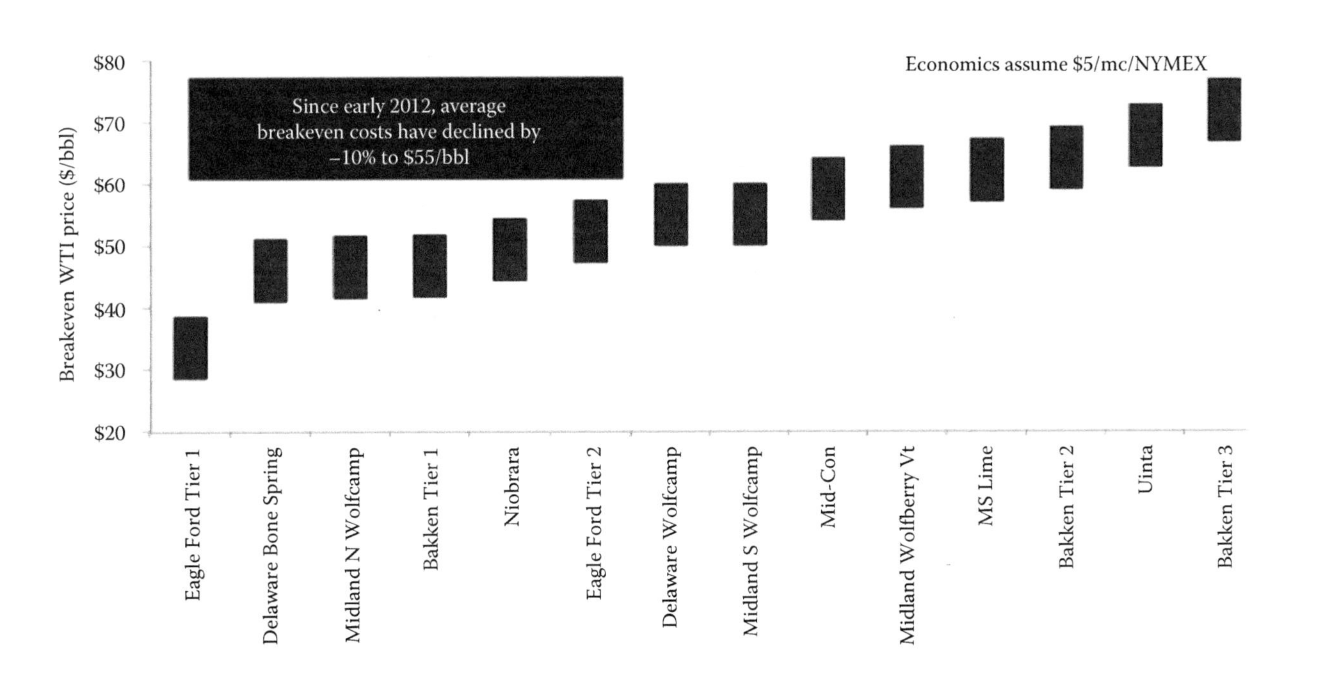

FIGURE 2.10 Breakeven WTI price of the Permian Basin versus other plays since early 2012. (From Tudor, Pickering, Holt & Co Research, http://www.tphco.com/reports.)

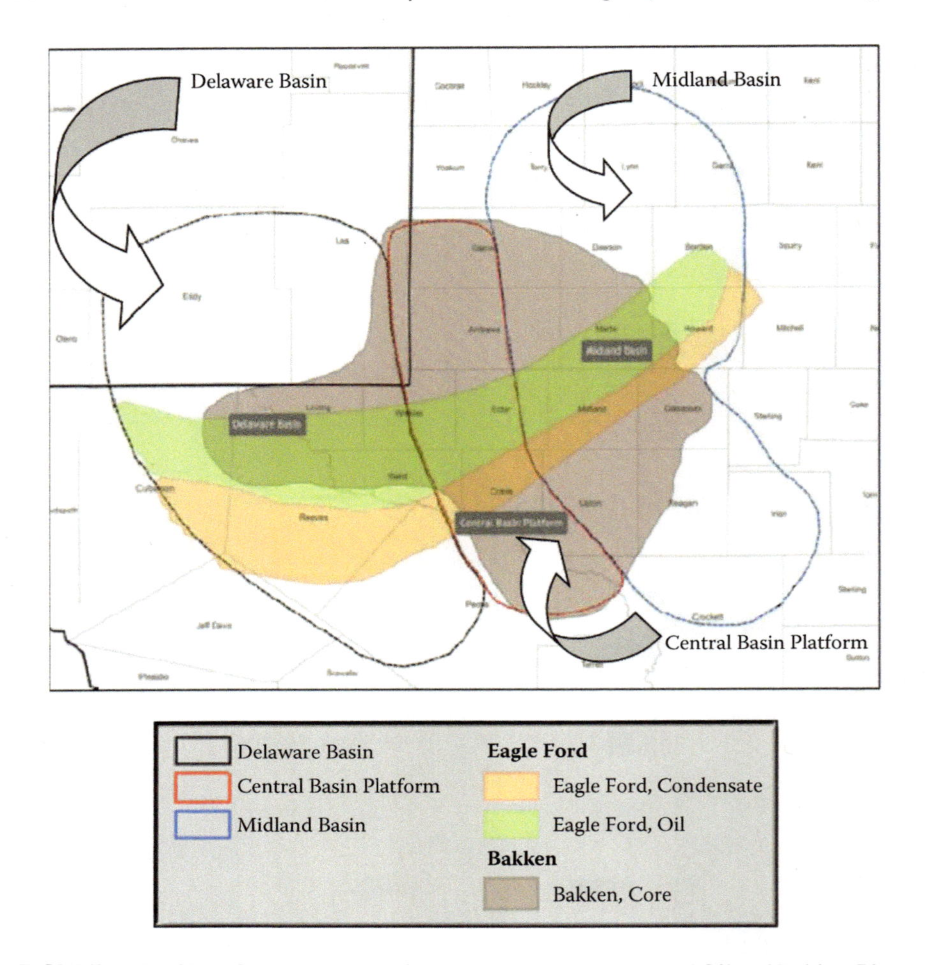

FIGURE 2.11 Size of the Permian Basin compared to the Eagle Ford Oil and Bakken Plays. (From Tudor, Pickering, Holt & Co Research, http://www.tphco.com/reports.)

Barnett Shale Play, will probably experience no growth in drilling activity and may even decline in the near term due to the drop in gas prices relative to sustained high oil prices. The northwest portion of the Eagle Ford is liquid prone, as is the Bakken and Permian Basins, thus will continue to see an expanding drilling program. As seen in Figure 2.10, the Eagle Ford, Bakken, and Permian Basin Plays are economical in the sub-$50/bbl oil price environment. As seen in Figure 2.11, the Permian Basin Plays located in the Midland and Delaware Basins are larger than the Bakken and Eagle Ford Plays. In addition, the Permian Basin has more multilayer targets in a given area than either of the two other plays.

Over 182,000 wells are currently reported active in the Permian Basin, with approximately 130,000 (71%) oil wells and over 26,000 (15%) gas wells and the remaining 14% for injection purposes (Figures 2.12 and 2.13). Water and CO_2

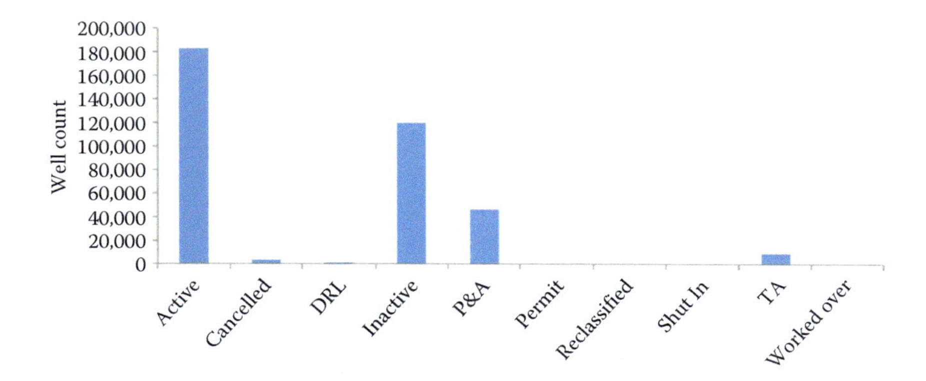

FIGURE 2.12 Well status distribution of the Permian Basin since inception. DRL, drilled; P&A, plugged and abandoned; TA, temporarily abandoned. (From DrillingInfo and DI Desktop™, http://www.drillinginfo.com/.)

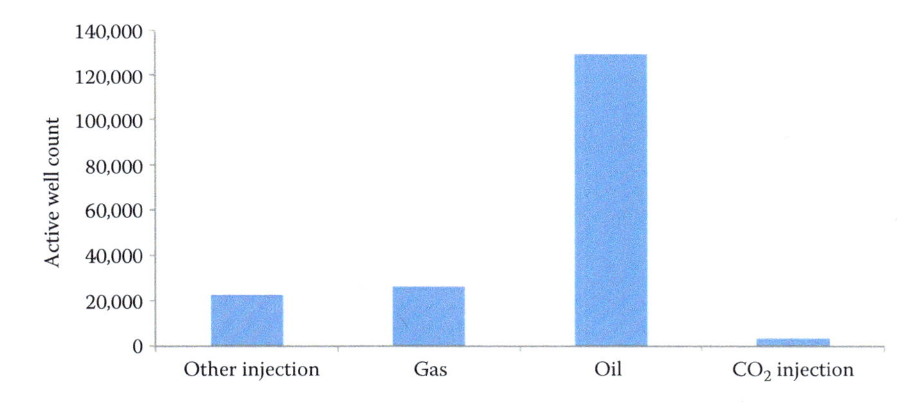

FIGURE 2.13 Active well-type distribution of the Permian Basin. (From DrillingInfo and DI Desktop™, http://www.drillinginfo.com/.)

injection, for the purpose of secondary and tertiary oil recovery, have substantially influenced oil production and operations since 1960. CO_2 flooding commenced in the late 1970s and has recently seen more activity as a result of increased oil prices. Almost 26,000 wells (14% of active wells) are being utilized for injection purposes. Of the 156,000 active, producing oil and gas wells in the Permian Basin, over 34,000 (22%) are directional and horizontal wells, and the remaining are vertical wells, as shown in Figure 2.14.

2.8 DRILLING ACTIVITY

The United States has the highest number of rotary rigs with 1757 or 53% of the world's total rigs in operation as of December 2013 (Figure 2.15). The Permian Basin has the highest number of rigs than any other basins/region of the world, with 469 as

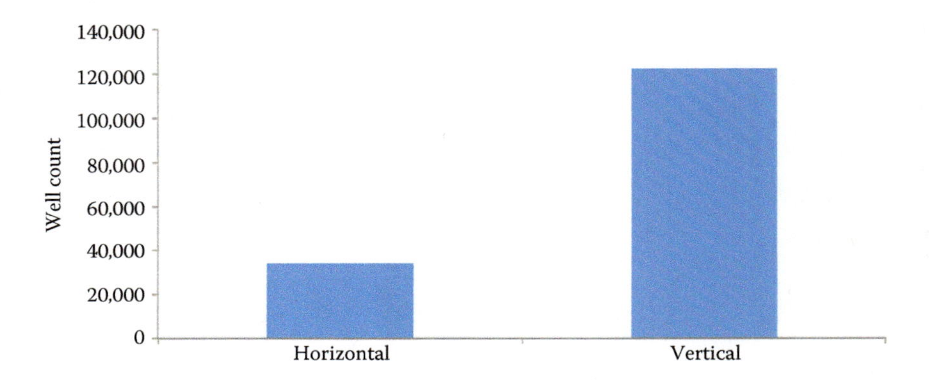

FIGURE 2.14 Active drill-type distribution of the Permian Basin since inception (Assumed that *Unknown* Wells are most likely Verticals). (From DrillingInfo and DI Desktop™, http://www.drillinginfo.com/.)

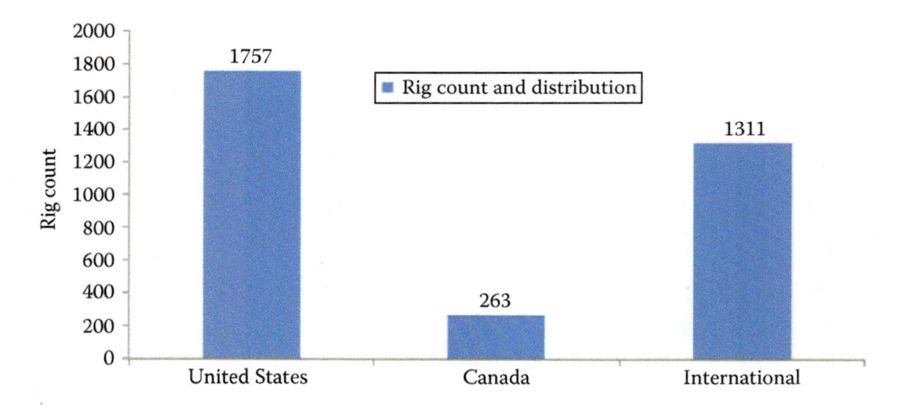

FIGURE 2.15 Global land rig count, week ending December 2013. (From Baker Hughes, http://gis.bakerhughesdirect.com/Reports/RigCountsReport.aspx.)

of December 27, 2013. This represents 56% of the active rigs currently running in Texas, 27% of total rigs running in the United States, and 14% of total world rig count.

An increasing percentage of the wells are being drilled horizontally as a result of the advancement of hydraulic fracturing and horizontal drilling technology. In the last 2 years, most of the rigs in the United States are drilling directional/horizontally versus vertically. Over the last year, the number of horizontal rotary rigs increased by 84 rigs and the number of rotary rigs drilling vertically decreased by 90 as shown in Figure 2.16.

As shown in Figure 2.17, there are 835 rigs (48% of the United States) in Texas and 81 rigs (5% of the United States) in New Mexico. As can be seen in Figure 2.18, the Permian Basin has the highest drilling activity as compared to other

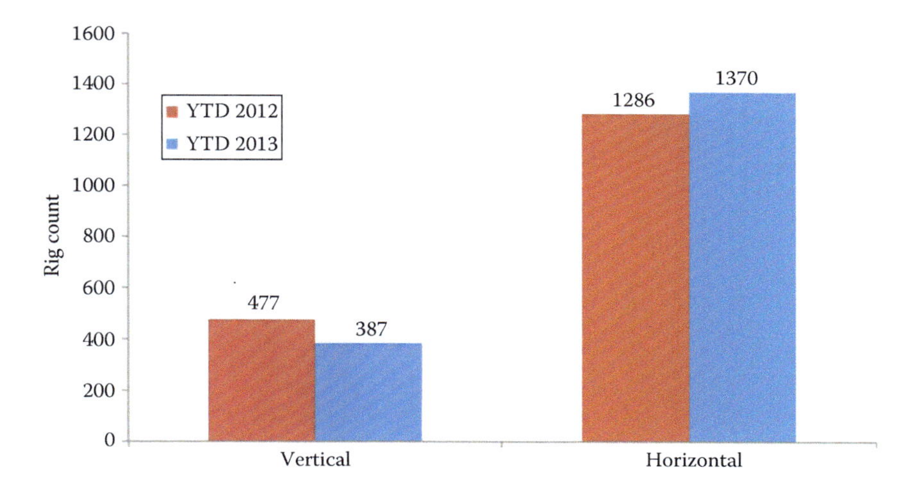

FIGURE 2.16 U.S. rotary rig by drill type and change from last year. (From Baker Hughes, http://gis.bakerhughesdirect.com/Reports/RigCountsReport.aspx.)

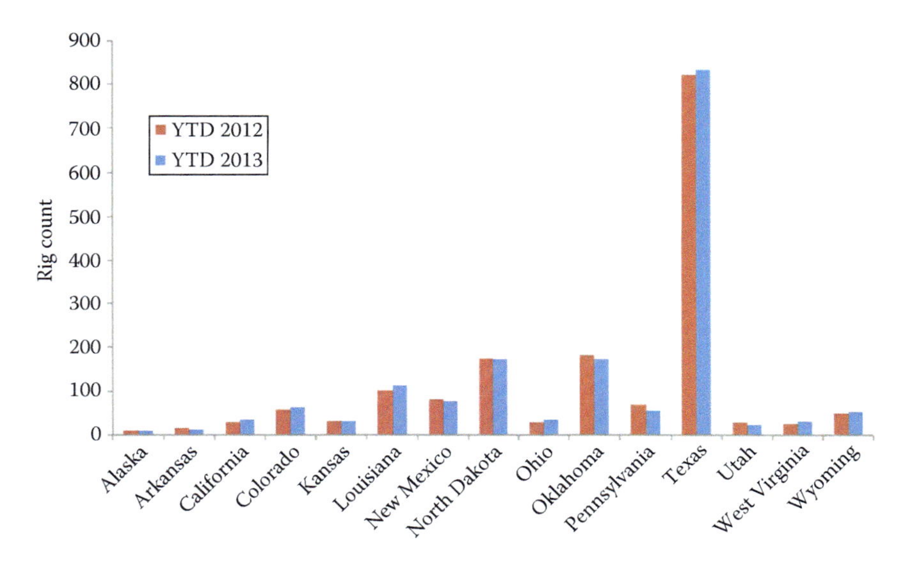

FIGURE 2.17 U.S. states rig count and change from last year. (From Baker Hughes, http://gis.bakerhughesdirect.com/Reports/RigCountsReport.aspx.)

plays/basins in the United States. In addition, the Permian Basin saw an increase of six rigs over the last year. The total number of rigs in Texas increased by 13, while New Mexico's decreased by 3 rigs from last year (2012). Figure 2.19 shows a consistent historical increase in the number of rotary drilling rigs in Texas and the Texas portion of the Permian Basin since 1987. Also, as seen in Figure 2.19, there was a

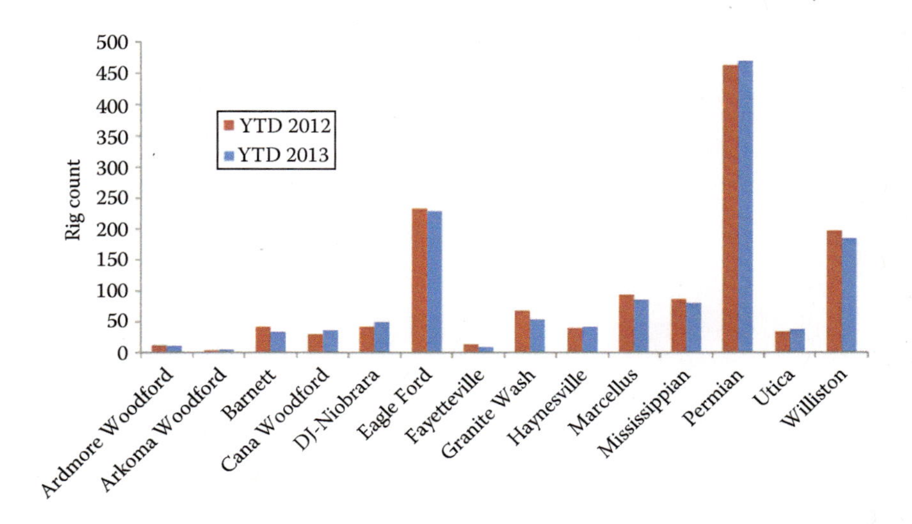

FIGURE 2.18 Basin/plays rig count and change from last year. (From Baker Hughes, http://gis.bakerhughesdirect.com/Reports/RigCountsReport.aspx.)

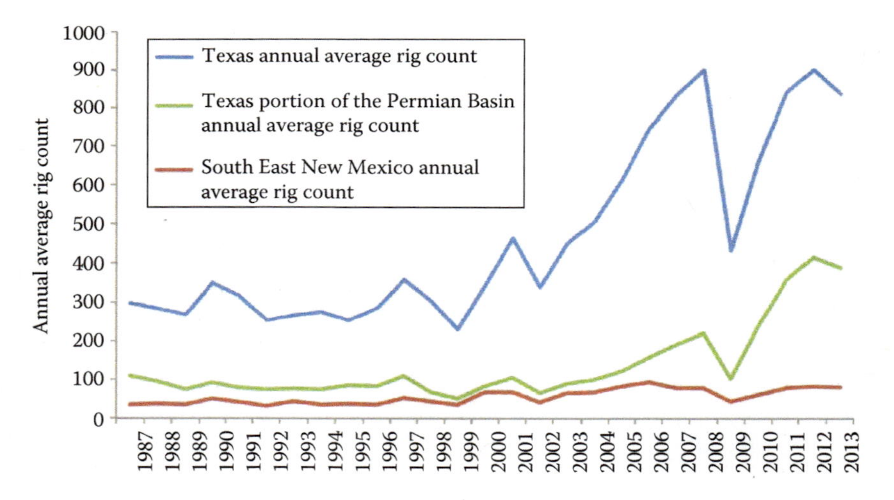

FIGURE 2.19 Historical annual average rig count of Texas and Texas portion of the Permian Basin. (From Baker Hughes, http://gis.bakerhughesdirect.com/Reports/RigCountsReport.aspx.)

dramatic drop in Rig count due to the 2008 recession; however, rig counts quickly recovered and were back to pre-2008 levels by 2012. The reason for this is after the 2008 recession, natural gas prices never recovered, while oil prices quickly recovered back to pre-2008 prices and then continued to increase. Since oil and gas companies have refocused capital budgets to oil (liquid)-prone basins, areas that benefited the most were the Permian Basin, Bakken Play (Williston Basin), and the

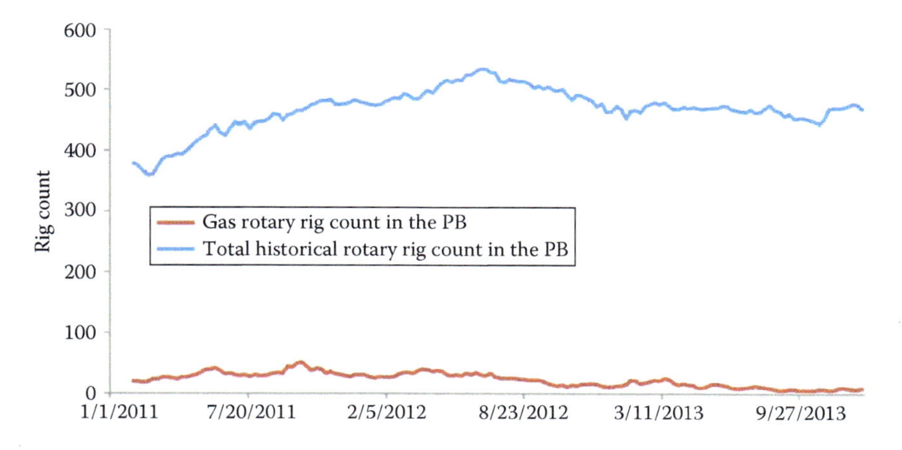

FIGURE 2.20 Detailed annual average rig count of the Permian Basin from 2011. (From Baker Hughes, http://gis.bakerhughesdirect.com/Reports/RigCountsReport.aspx.)

Eagle Ford (Texas Coast Basin). While Texas shale gas plays, such as the Barnett and the Haynesville, saw a sustained drop in rig count, the Permian Basin and the Eagle Ford more than made up for that drop.

Figure 2.20 shows a decline in drilling rigs running in the Permian Basin since early 2012. Figure 2.20 is weekly data, whereas Figure 2.19 shows annual averages of Rig counts. The reason for this decline is that operators are switching from vertical to horizontal drilling, which requires more personnel per well. In addition, horizontal wells are more efficient in draining reservoirs (less wells per unit area). As reported in the 2013 SPE Forum on Petroleum Engineering Education, it takes three times the personnel for unconventional reservoir/horizontal well development as opposed to that required for conventional/vertical wells (Figure 2.21).

2.9 DRILLING/COMPLETION TECHNOLOGY AND PRODUCTION TRENDS

Permian Basin completion efficiencies have dramatically improved with the application of new technologies. As seen in Figures 2.22 and 2.23, efficiencies, which are defined as initial oil rate per well drilled, have improved since 2010.

In addition to productivity, technology is enabling operators to be more efficient. As shown in Figures 2.24 and 2.25, time to drill horizontal wells has been drastically reduced resulting in substantial cost savings. More importantly, the Permian Basin is the least mature relative to horizontal drilling experience. Thus, it would be reasonable to assume that the Permian Basin has great potential for further increases in drilling efficiency. With the recent increase in horizontal drilling permits, as shown in Figures 2.26 and 2.27, and increased well productivity, we expect continued increases in efficiency and activity levels in the Permian Basin.

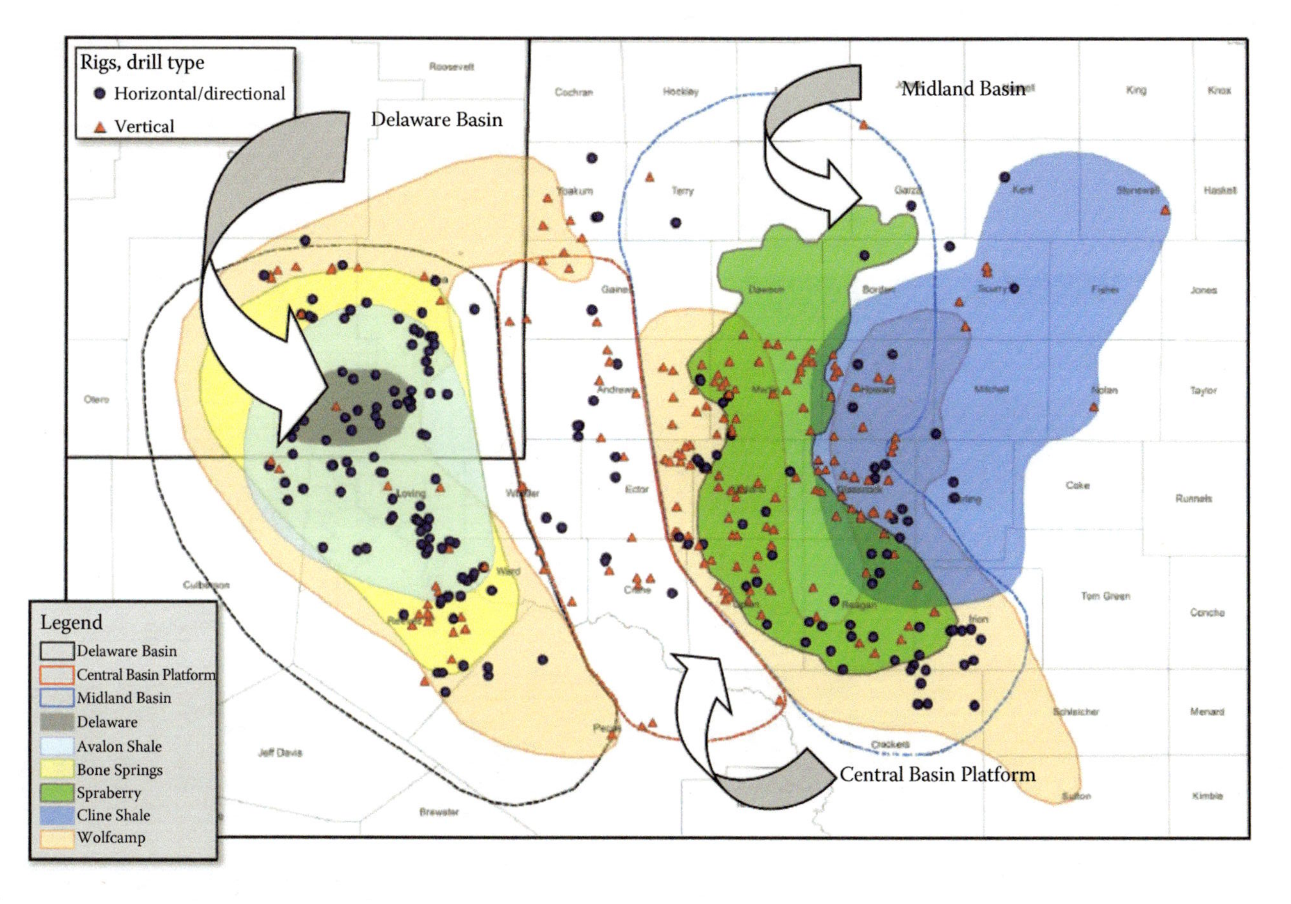

FIGURE 2.21 Rig drill type on different plays. (From Tudor, Pickering, Holt & Co Research, http://www.tphco.com/reports.)

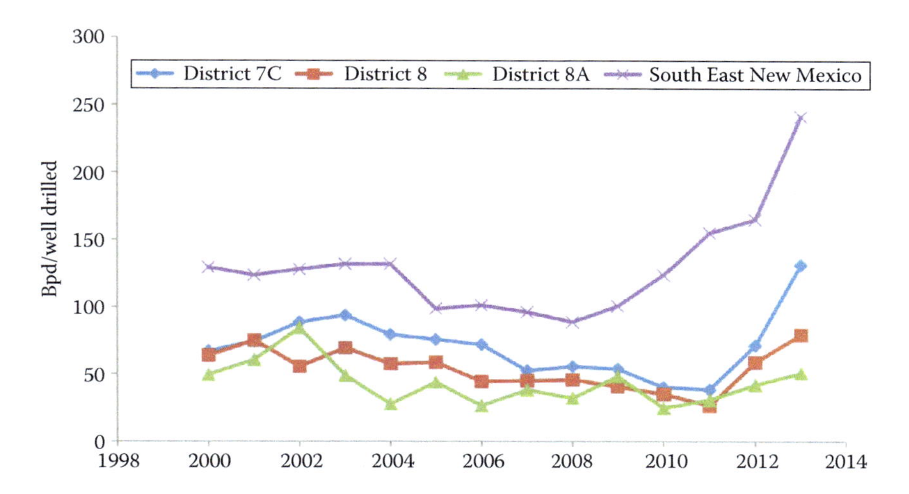

FIGURE 2.22 Drilling/completions efficiency comparison for different districts in Texas and South East New Mexico. (From DrillingInfo and DI Desktop™, http://www.drillinginfo.com/.)

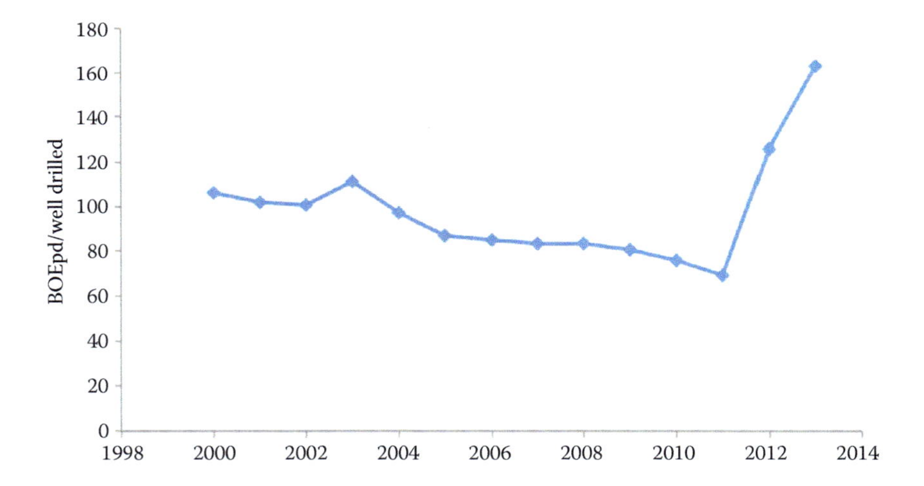

FIGURE 2.23 Drilling/completions efficiency comparison for the entire Permian Basin. (From DrillingInfo and DI Desktop™, http://www.drillinginfo.com/.)

Permian Basin technological trends include the following (Engler et al., 2011):

- Horizontal drilling
- Changes in hydraulic fracturing technology
- Slick-water fracing
- Changes in fluid type and amount
- Increased use of 3D seismic surveys
- Downhole commingling/multi-zone completions

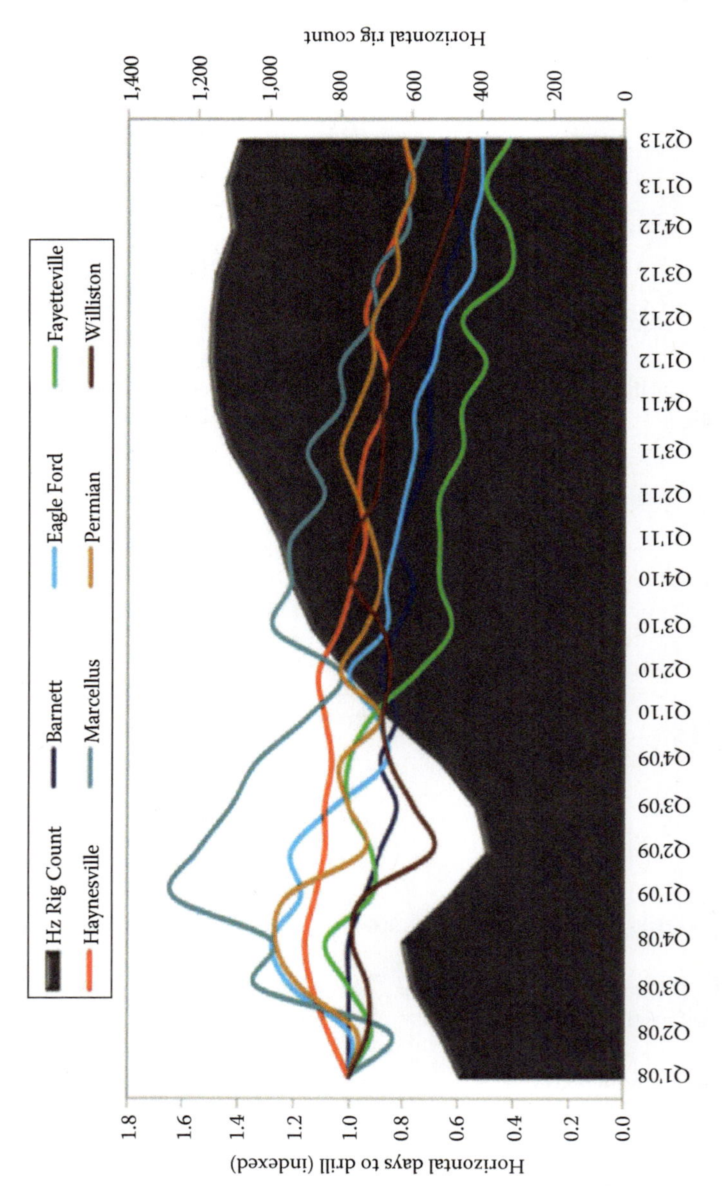

FIGURE 2.24 Horizontal drilling efficiency comparison for different basins. (From Tudor, Pickering, Holt & Co Research, http://www.tphco.com/reports.)

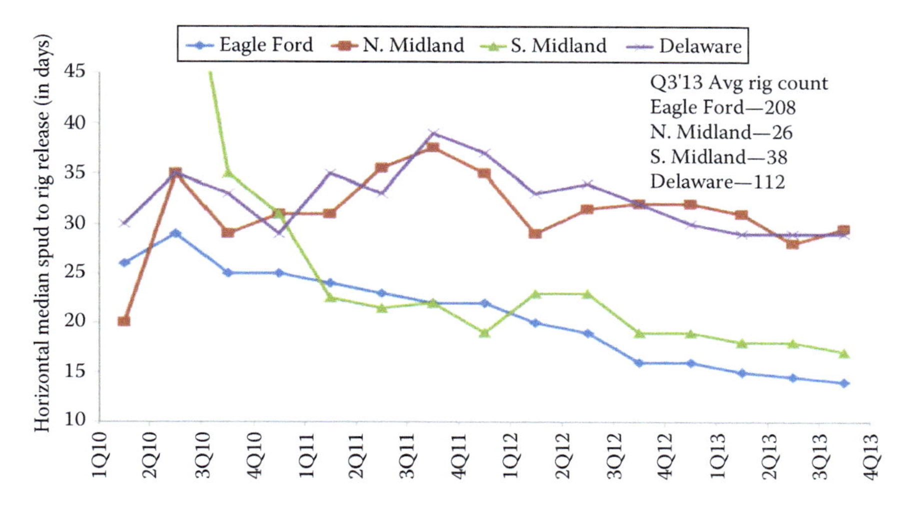

FIGURE 2.25 Horizontal median spud to rig release comparison between Eagle Ford and plays in the Permian Basin. (From Tudor, Pickering, Holt & Co Research, http://www.tphco.com/reports.)

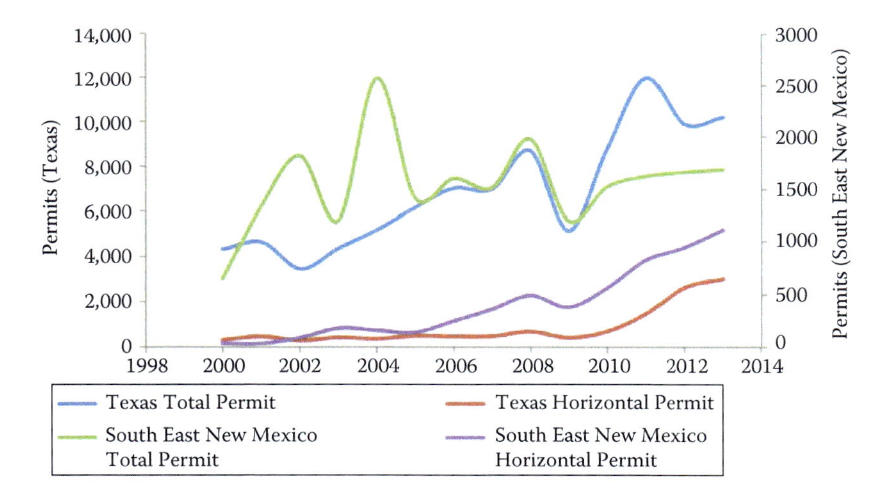

FIGURE 2.26 Drilling permit versus time plot for Texas and South East New Mexico.

- Multiple well pads
- Changes in rules that allow for down-spacing of particular fields to tap undrained areas in existing pools

The number of drilling permits acquired is an important indicator that drilling activities will likely be carried out in the region of interest. This influences the future production of oil and gas in that county or region. A strong increase in drilling activity will often result in higher future production levels.

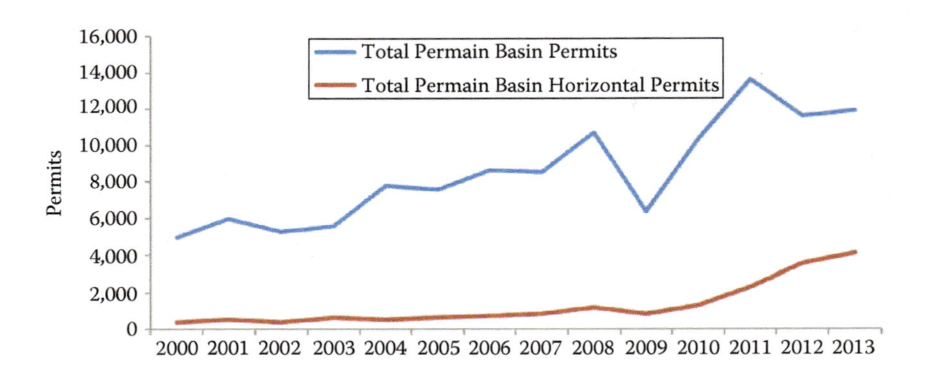

FIGURE 2.27 Drilling permit versus time plot for the entire Permian Basin.

Moreover, technology shifts and newly developed plays have even a greater impact on production and level of activity.

2.10 ECONOMIC IMPACTS FROM THE PRODUCTION OF OIL AND GAS IN THE PERMIAN BASIN: 2013

The myriad of activities associated with the drilling for and extraction of oil and gas in the Permian Basin entails a number of upstream- and some midstream-level businesses. These activities create significant economic benefits by creating and sustaining jobs, income, value added, and output. In addition, the industry provides important tax revenues that benefit the citizens of the region and state.

A set of economic models, referred to as *input–output* (I–O) models by economists, were constructed to measure the economic impact that the oil and gas industry has on the Permian Basin economy. The basis of an economic impact model is the spending patterns of individuals and businesses in the region. In particular, expenditures by firms engaged in the oil and gas industry on equipment and supplies occur within the region and elsewhere, while oil and gas employees tend to spend the majority of their income more locally. Economists generally categorize the economic impacts from these expenditures into two types of effects: direct and secondary. Direct effects represent those expenditures within the region of the industry being studied. Direct effects lead to secondary effects in the form of business-to-business transactions in the region (e.g., to restore inventory) referred to as indirect effects and also to new income in the form of wages and salaries, rent and interest payments, payments to proprietors and stockholders for investment, etc., also known as induced effects. For purposes of this study, economic output (i.e., gross revenues) refers to the value of all industrial production (i.e., mining, services, retail trade, manufacturing, etc.) in a region; following the convention used by the US BLS and Bureau of Economic Analysis (BEA), jobs are defined as the average annual number of jobs in a sector, industry, or region, while labor income consists of all forms of employment income, including employee compensation (wages and benefits) and proprietor income, and value added indicates the addition to gross state product.

The regional economic model identifies the *linkages* within the economy that exist between businesses (or enterprises) and other businesses, and businesses (or enterprises) and final consumers. From the regional economic model, a set of industrial sector *economic multipliers* unique to the regional economy are calculated. These multipliers are used to provide a comprehensive assessment of the economic impact of the oil and gas industry. Specifically, the economic impact analysis provides information as to the number of jobs created and sustained by the ongoing operation of the industry, the income added to the local economy from the industry's operations, which includes household income or earnings, and the total output (in dollars) that industry contributes to the economy.

For purposes of this report, the economic models were constructed for the Permian Basin region, the New Mexico portion of the Permian Basin, the Texas portion of the Permian Basin, and for each individual county in the Permian Basin region, for a total of 51 models constructed. As previously noted, the list of counties considered to be in the Permian Basin for this study is contained in Appendix A.

To estimate the economic impact of the Permian Basin's oil and gas industry in 2013, we follow the convention of first estimating the production of oil and gas. Production data generally come from the Railroad Commission of Texas and the New Mexico Oil Conservation Division. Price information is from the US EIA and used to quantify revenues associated with the extraction of oil and gas. Information on costs associated with other core oil and gas activities was obtained from several sources including recent annual reports of major operators in the Permian Basin and the EIA trends on operating and drilling costs. The analysis was conducted using the IMPLAN Version 3.0 software package. Values of 2011 were updated to reflect the information obtained and estimated for 2013. The conversion to 2013 values was accomplished through a set of extrapolations (economic time series analysis) and validated using the production-, drilling-, and engineering-related data. For purposes of this study, and following convention of economic impact research, the core sectors were defined as oil and gas extraction, drilling for oil and gas, and support activities for oil and gas operations. Noncore-related sectors include economic activity related to transportation (e.g., by pipeline, truck, and rail), specific oil and gas manufactured parts and machinery, and refining. While these activities are imperative for the proper and efficient operation of the industry, they are generally labeled noncore in economic impact studies for the reason that they would not exist if it were not for the drilling and production of oil and gas.

Table 2.1 summarizes the 2013 impacts for the entire Permian Basin region. There were nearly a quarter million jobs in the oil and gas industry in 2013 and over $95 billion of production value (i.e., commodity price multiplied by volume of production). These direct impacts generated and/or sustained more than a half million jobs in the Permian Basin. Moreover, the industry generated almost $137.8 billion in output and contributed more than $71 billion in total gross state product (across Texas and New Mexico), that is, value added. The existence of the oil and gas industry in the Permian Basin generates substantial economic activity and economic benefits to both New Mexico and Texas.

The oil and gas activity in the Permian Basin impacts various sectors differently. Table 2.2 illustrates the impacts for the top 10 sectors in the Permian Basin, ranked

TABLE 2.1
Permian Basin

Impact Type	Employment	Labor Income	Total Value Added	Output
Direct effect	244,074	19,160,541,128	47,442,572,577	95,080,000,000
Indirect effect	155,766	7,537,767,130	13,459,362,924	25,394,643,742
Induced effect	146,376	5,064,342,129	10,215,221,057	17,305,934,221
Total effect	546,216	31,762,650,388	71,117,156,558	137,780,577,963

Note: Labor income, total value added, and output are measured in current dollars ($).

TABLE 2.2
Impacts for the Top 10 Sectors in the Permian Basin

Sector	Total Employment
Support activities for oil and gas operations	130,307
Extraction of oil and natural gas	87,121
Drilling oil and gas wells	36,176
Maintenance and repair construction of nonresidential structures	31,394
Food services and drinking establishments	26,137
Architectural, engineering, and related services	17,043
Wholesale trade businesses	11,608
Securities, commodity contracts, investments, and related activities	8,728
Civic, social, professional, and similar organizations	8,688
Monetary authorities and depository credit intermediation activities	8,411

TABLE 2.3
Permian Basin—Texas

Impact Type	Employment	Labor Income	Total Value Added	Output
Direct effect	190,714	15,706,086,363	40,086,447,398	77,880,000,000
Indirect effect	130,728	6,238,975,069	11,405,088,258	21,103,160,550
Induced effect	123,311	4,297,026,885	8,724,073,294	14,646,352,568
Total effect	444,753	26,242,088,318	60,215,608,950	113,629,513,118

Note: Labor income, total value added, and output are measured in current dollars ($).

by employment. The values clearly illustrate the demand for workers in sectors related to construction, engineering and design, food, wholesale trade, and financially related activities. These sectors benefit greatly from the presence of the oil and gas industry across the Permian Basin.

Table 2.3 summarizes the 2013 impacts for the Texas portion of the Permian Basin region. There were over 190,000 jobs in the oil and gas industry in 2013 and

TABLE 2.4
Permian Basin—New Mexico

Impact Type	Employment	Labor Income	Total Value Added	Output
Direct effect	53,496	3,462,805,294	7,351,417,620	17,200,000,000
Indirect effect	21,441	1,063,371,089	1,619,372,471	3,311,307,057
Induced effect	19,834	642,904,720	1,254,387,233	2,185,253,445
Total effect	94,771	5,169,081,103	10,225,177,324	22,696,560,502

Note: Labor income, total value added, and output are measured in current dollars ($).

$77.9 billion of production value. These direct impacts generated and/or sustained over 444,000 of the jobs in the Texas portion of the Permian Basin. Moreover, the industry generated $113.6 billion in output and contributed more than $60.2 billion in total gross state product (across Texas), that is, value added. The existence of the oil and gas industry in the Texas Portion of the Permian Basin generates substantial economic activity and economic benefits to this area.

Table 2.4 summarizes the 2013 impacts for the New Mexico portion of the Permian Basin region. There were over 53,000 jobs in the oil and gas industry in 2013 and $17.2 billion of production value. These direct impacts generated and/or sustained over 94,000 of the jobs in the New Mexico portion of the Permian Basin. Moreover, the industry generated $22.7 billion in output and contributed more than $10.2 billion in total gross state product (across New Mexico), that is, value added. The existence of the oil and gas industry in the New Mexico portion of the Permian Basin generates substantial economic activity and economic benefits to this area.

Tables 2.5 and 2.6 summarize the economic impacts of the Permian Basin's oil and gas industry by county. Texas counties are listed first, followed by New Mexico counties. Labor income, total value added, and output are measured in current dollars ($).

The preceding discussion and tables highlighted the economic benefits created and sustained by the ongoing operations and continued growth of the Permian Basin's oil and gas industry. While the same methodology is used for the county-level analyses as was used for the larger regional analyses, it should be noted that the sum of the county impacts may sum to the total impacts neither for the larger Permian region as a whole, nor for the Texas and New Mexico portions of the Permian Basin. This is generally due to the following reasons. First, numbers and values are rounded. Second, leakages occur in all economies in that not all monies spent are entirely contained within the study area. Generally speaking, the larger the study area, the more able the model is to capture the spending and consequently reduce leakages. A third reason is that while data are generally available for each county (e.g., employment), it is quite possible that, for example, an oil and gas worker in one county may work in that county for just a portion of the year and then work in another county (or other counties) for the remainder of the year. Thus, it is possible that this worker may be counted twice, once in each county worked, or possibly not counted at all. In larger models, this type of estimation or measurement *error* tends to wash out,

TABLE 2.5
Texas County Impacts

Impact Type	Employment	Labor Income	Total Value Added	Output
Andrews				
Direct effect	6,870	507,552,965	1,025,804,304	2,039,700,000
Indirect effect	1,858	98,294,600	165,435,994	291,913,704
Induced effect	1,578	52,695,954	125,814,301	198,192,179
Total effect	10,306	658,543,520	1,317,054,599	2,529,805,884
Cochran				
Direct effect	171	8,567,977	48,804,058	86,050,000
Indirect effect	31	1,038,820	2,926,715	5,524,247
Induced effect	18	614,832	1,707,044	2,666,069
Total effect	219	10,221,630	53,437,817	94,240,316
Coke				
Direct effect	193	5,848,334	47,092,995	87,090,000
Indirect effect	125	2,450,888	4,880,094	13,097,185
Induced effect	21	417,601	1,438,521	2,689,884
Total effect	339	8,716,823	53,411,610	102,877,069
Crane				
Direct effect	2,332	217,407,431	399,495,982	726,770,000
Indirect effect	690	22,349,152	40,540,546	92,876,380
Induced effect	613	17,789,207	46,655,433	75,287,073
Total effect	3,634	257,545,790	486,691,961	894,933,452
Crosby				
Direct effect	93	3,727,889	27,438,481	48,640,000
Indirect effect	23	644,822	3,134,338	4,785,379
Induced effect	11	422,145	1,011,265	1,640,190
Total effect	127	4,794,855	31,584,084	55,065,568
Culberson				
Direct effect	40	32,116	2,374,923	11,580,000
Indirect effect	28	581,809	1,384,264	2,704,197
Induced effect	2	37,375	118,307	206,018
Total effect	70	651,300	3,877,493	14,490,216
Dawson				
Direct effect	1,166	71,533,278	245,417,108	479,410,000
Indirect effect	438	14,770,026	30,441,034	58,364,026
Induced effect	316	8,005,428	22,445,258	35,904,562
Total effect	1,919	94,308,732	298,303,400	573,678,588
Dickens				
Direct effect	70	1,587,170	19,321,938	35,180,000
Indirect effect	10	358,375	1,252,019	2,313,979
Induced effect	5	85,266	315,139	575,705
Total effect	84	2,030,811	20,889,097	38,069,684

(*Continued*)

TABLE 2.5 (*Continued*)
Texas County Impacts

Impact Type	Employment	Labor Income	Total Value Added	Output
Ector				
Direct effect	34,255	3,119,807,793	6,629,516,105	13,199,000,000
Indirect effect	17,317	896,677,543	1,527,839,414	2,506,761,149
Induced effect	17,057	600,079,518	1,237,851,908	1,934,553,175
Total effect	68,629	4,616,564,855	9,395,207,427	17,640,314,324
Edwards				
Direct effect	40	444,604	2,693,246	11,580,000
Indirect effect	30	530,966	803,903	2,284,100
Induced effect	2	65,492	172,648	276,209
Total effect	72	1,041,062	3,669,797	14,140,309
Floyd				
Direct effect	1	213,190	452,771	667,000
Indirect effect	0	13,078	48,759	71,508
Induced effect	1	14,693	41,357	69,150
Total effect	2	240,960	542,887	807,658
Gaines				
Direct effect	2,550	276,144,227	670,180,225	1,266,400,000
Indirect effect	1,226	39,864,186	68,167,779	138,717,744
Induced effect	831	24,038,072	63,589,052	101,581,383
Total effect	4,607	340,046,485	801,937,056	1,506,699,127
Glasscock				
Direct effect	227	1,512,472	14,382,947	65,170,000
Indirect effect	31	1,088,576	1,645,313	4,363,445
Induced effect	3	77,267	250,161	439,870
Total effect	261	2,678,315	16,278,421	69,973,315
Hale				
Direct effect	86	10,486,842	20,538,209	35,500,000
Indirect effect	50	1,564,327	2,524,812	5,686,452
Induced effect	57	1,634,874	3,479,365	5,853,241
Total effect	193	13,686,043	26,542,387	47,039,692
Hockley				
Direct effect	8,986	641,989,590	2,030,888,259	4,310,000,000
Indirect effect	3,174	152,418,706	279,143,973	527,098,420
Induced effect	2,645	78,500,956	179,719,247	301,932,605
Total effect	14,805	872,909,252	2,489,751,480	5,139,031,024
Howard				
Direct effect	5,245	264,565,536	999,039,058	2,235,700,000
Indirect effect	1,813	69,309,737	140,920,980	283,839,246
Induced effect	1,360	43,827,128	102,560,466	171,272,198
Total effect	8,418	377,702,401	1,242,520,503	2,690,811,444

(*Continued*)

TABLE 2.5 (*Continued*)
Texas County Impacts

Impact Type	Employment	Labor Income	Total Value Added	Output
Irion				
Direct effect	1,097	73,461,267	188,679,272	426,630,000
Indirect effect	283	6,610,326	13,813,166	38,524,793
Induced effect	100	2,524,223	8,592,258	15,215,896
Total effect	1,480	82,595,816	211,084,696	480,370,690
Jeff Davis				
Direct effect	70	331,955	4,391,139	20,340,000
Indirect effect	21	650,944	1,396,276	2,945,722
Induced effect	3	69,252	215,680	360,878
Total effect	93	1,052,152	6,003,096	23,646,600
Kent				
Direct effect	74	5,722,421	7,196,439	15,300,000
Indirect effect	20	200,119	700,883	2,483,292
Induced effect	11	216,128	907,659	1,566,310
Total effect	105	6,138,668	8,804,981	19,349,603
Kimble				
Direct effect	86	1,048,277	5,634,157	24,000,000
Indirect effect	73	1,851,696	2,960,486	6,882,279
Induced effect	11	269,420	691,214	1,134,236
Total effect	170	3,169,393	9,285,857	32,016,515
Lamb				
Direct effect	203	5,035,912	56,754,397	103,000,000
Indirect effect	105	3,178,154	6,239,561	12,622,231
Induced effect	25	651,079	1,908,823	3,109,292
Total effect	332	8,865,145	64,902,782	118,731,523
Loving				
Direct effect	60	25,307	3,529,955	17,300,000
Indirect effect	10	269,902	2,932,187	4,203,221
Induced effect	0	2,360	13,474	23,239
Total effect	71	297,569	6,475,616	21,526,459
Lubbock				
Direct effect	5,705	113,783,050	1,187,776,431	2,314,300,000
Indirect effect	5,976	251,523,435	450,127,859	810,537,296
Induced effect	2,637	94,712,316	181,383,295	304,162,564
Total effect	14,319	460,018,801	1,819,287,585	3,428,999,859
Lynn				
Direct effect	67	151,969	4,036,675	19,300,000
Indirect effect	23	377,820	937,609	2,221,926
Induced effect	1	23,622	91,332	159,879
Total effect	91	553,410	5,065,617	21,681,806

(*Continued*)

TABLE 2.5 (*Continued*)
Texas County Impacts

Impact Type	Employment	Labor Income	Total Value Added	Output
Martin				
Direct effect	353	29,130,728	108,030,144	183,080,000
Indirect effect	126	6,196,732	8,984,536	15,989,353
Induced effect	103	2,721,974	7,169,160	12,053,629
Total effect	582	38,049,435	124,183,839	211,122,982
Midland				
Direct effect	67,867	7,203,273,090	17,687,942,264	31,899,000,000
Indirect effect	45,699	2,415,746,033	4,071,252,600	6,784,509,114
Induced effect	36,324	1,354,959,307	2,586,321,716	4,093,269,566
Total effect	149,889	10,973,978,430	24,345,516,580	42,776,778,680
Mitchell				
Direct effect	753	39,563,668	155,794,074	300,300,000
Indirect effect	200	5,860,324	10,376,508	22,773,747
Induced effect	112	2,809,405	7,479,279	12,351,489
Total effect	1,065	48,233,398	173,649,861	335,425,236
Motley				
Direct effect	23	305,142	1,427,383	6,030,000
Indirect effect	17	355,347	557,499	1,481,947
Induced effect	2	33,736	115,095	197,923
Total effect	43	694,226	2,099,978	7,709,871
Nolan				
Direct effect	1,887	77,143,888	163,608,549	411,880,000
Indirect effect	825	25,812,532	48,003,074	92,585,109
Induced effect	461	12,243,693	28,228,095	46,485,331
Total effect	3,172	115,200,114	239,839,717	550,950,440
Pecos				
Direct effect	5,734	491,466,938	703,446,744	1,382,100,000
Indirect effect	1,104	42,371,647	81,248,020	153,856,546
Induced effect	1,737	46,466,997	115,028,414	189,568,905
Total effect	8,575	580,305,582	899,723,178	1,725,525,451
Reagan				
Direct effect	2,773	133,236,549	329,298,656	717,600,000
Indirect effect	417	16,195,062	30,911,645	62,211,615
Induced effect	155	4,052,662	13,731,605	23,134,880
Total effect	3,346	153,484,273	373,941,906	802,946,496
Reeves				
Direct effect	1,480	98,943,162	139,532,380	307,600,000
Indirect effect	276	8,205,098	20,298,206	37,970,700
Induced effect	343	8,220,097	24,109,851	39,787,042
Total effect	2,099	115,368,358	183,940,437	385,357,742

(*Continued*)

TABLE 2.5 (*Continued*)
Texas County Impacts

Impact Type	Employment	Labor Income	Total Value Added	Output
Scurry				
Direct effect	6,575	517,122,846	1,007,369,620	1,998,600,000
Indirect effect	2,087	107,858,267	178,280,725	307,572,411
Induced effect	1,907	58,841,632	138,990,975	218,853,904
Total effect	10,569	683,822,745	1,324,641,320	2,525,026,315
Sterling				
Direct effect	392	40,550,699	125,628,573	207,500,000
Indirect effect	152	4,339,943	9,946,356	18,843,863
Induced effect	81	1,608,802	6,843,095	10,787,930
Total effect	625	46,499,444	142,418,024	237,131,793
Terry				
Direct effect	1,894	190,531,542	379,791,469	809,630,000
Indirect effect	418	16,197,002	28,234,955	57,482,068
Induced effect	732	20,973,245	46,356,923	76,624,213
Total effect	3,044	227,701,789	454,383,346	943,736,281
Tom Green				
Direct effect	11,170	299,733,806	2,198,247,533	5,188,000,000
Indirect effect	7,792	277,399,611	552,964,821	1,030,567,084
Induced effect	3,368	114,574,191	226,600,965	374,660,103
Total effect	22,329	691,707,608	2,977,813,319	6,593,227,187
Upton				
Direct effect	2,384	171,208,914	497,554,358	931,400,000
Indirect effect	534	23,251,966	45,096,300	91,611,446
Induced effect	323	8,696,957	27,554,786	46,891,837
Total effect	3,241	203,157,836	570,205,444	1,069,903,283
Val Verde				
Direct effect	1,091	17,977,963	76,343,745	309,700,000
Indirect effect	745	19,374,187	32,164,609	71,873,678
Induced effect	160	4,485,087	10,012,555	16,145,297
Total effect	1,996	41,837,237	118,520,909	397,718,976
Ward				
Direct effect	5,565	393,839,173	884,451,598	1,876,100,000
Indirect effect	2,211	100,766,270	155,715,145	304,099,742
Induced effect	1,129	33,679,633	92,226,024	147,686,017
Total effect	8,906	528,285,076	1,132,392,767	2,327,885,759
Winkler				
Direct effect	3,501	226,009,852	570,920,638	1,096,700,000
Indirect effect	799	35,629,279	79,930,054	150,739,125
Induced effect	621	15,456,004	53,388,130	87,645,655
Total effect	4,920	277,095,135	704,238,821	1,335,084,781

(*Continued*)

TABLE 2.5 (*Continued*)
Texas County Impacts

Impact Type	Employment	Labor Income	Total Value Added	Output
Yoakum				
Direct effect	4,652	386,553,126	957,973,049	1,729,700,000
Indirect effect	1,255	66,216,977	120,816,073	218,617,847
Induced effect	839	25,054,797	68,876,835	109,494,413
Total effect	6,746	477,824,900	1,147,665,956	2,057,812,260

Note: Labor income, total value added, and output are measured in current dollars ($).

TABLE 2.6
New Mexico County Impacts

Impact Type	Employment	Labor Income	Total Value Added	Output
Chaves				
Direct effect	4,445	137,967,873	654,007,048	1,553,650,000
Indirect effect	2,099	86,768,679	126,548,029	240,544,264
Induced effect	1,235	40,106,245	74,442,130	123,013,121
Total effect	7,780	264,842,798	854,997,207	1,917,207,385
Eddy				
Direct effect	17,988	1,287,876,681	3,043,729,349	6,861,000,000
Indirect effect	7,108	390,585,270	602,240,095	1,189,621,495
Induced effect	6,620	230,431,857	437,017,835	748,505,885
Total effect	31,716	1,908,893,809	4,082,987,279	8,799,127,380
Lea				
Direct effect	30,638	2,027,855,784	3,624,245,897	8,659,000,000
Indirect effect	9,586	517,845,084	721,659,314	1,400,588,009
Induced effect	9,574	328,137,714	614,484,518	1,056,741,944
Total effect	49,798	2,873,838,582	4,960,389,728	11,116,329,953
Otero				
Direct effect	13	332,483	1,329,297	3,500,000
Indirect effect	6	230,813	368,785	810,997
Induced effect	2	53,120	116,573	198,085
Total effect	22	616,415	1,814,655	4,509,082
Roosevelt				
Direct effect	321	2,838,301	18,813,627	100,640,000
Indirect effect	181	5,124,142	7,371,993	16,691,990
Induced effect	29	730,681	1,742,236	3,014,798
Total effect	531	8,693,124	27,927,856	120,346,788

Note: Labor income, total value added, and output are measured in current dollars ($).

for example, with one county being down a worker and another county being up a worker, with the aggregate result having no average error. Moreover, standard economic impact analysis follows the convention used by the US BLS and BEA in which jobs are defined as the average annual number of jobs in a sector, industry, county, or region over a period of time such as a month or year. For example, a 40 h/week job lasting for one full year is equivalent to two part-time jobs lasting one full year. Accordingly, for smaller counties in particular, interpretation of impacts should be viewed and interpreted with this caveat in mind.

The Permian Basin's oil and gas industry also generates other economic benefits that are not measured in terms of current jobs, income, value added, and output. Historically, the economy has been impacted by the cyclical nature of the industry. However, recent advances and changes in both plays and technology may be altering that scenario. Figure 2.28 shows the rig count over 2011–2013. Interestingly, while the rig count has held relatively steady, production has actually been quite high.

Figure 2.29 shows the oil share of total rig counts, which has trended up over this same time period. The nonlinear trend was estimated econometrically, and the data were decomposed into both the longer run trend (shown in red) and the shorter run cycle (shown in green) using the well-known Hodrick–Prescott filter. Of interest to economists is whether or not the long run characteristic is sustainable. A series of statistical tests (e.g., unit root tests) were conducted, and the results were consistent with this trend being sustainable. Moreover, examination of the cycle reveals another economic benefit, namely, the typical industry cycles are becoming less volatile. Thus, the ability of the oil and gas industry to realize changes in plays and technologies has far-reaching economic benefits to the

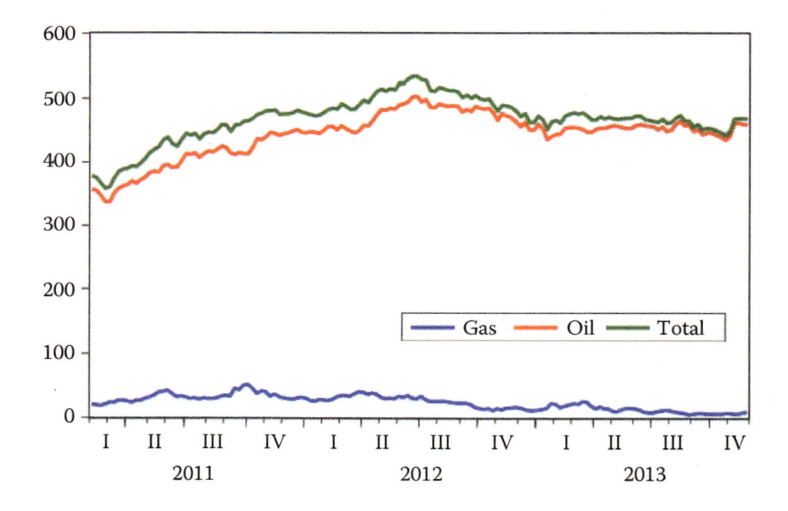

FIGURE 2.28 Permian basin rig count. (From Baker Hughes, http://gis.bakerhughesdirect.com/Reports/RigCountsReport.aspx.)

FIGURE 2.29 Oil share of total rig counts. *Note*: Left axis measures the percent deviation (in decimal) from trend. Right axis measures oil share as a percent (in decimal) of total.

Permian Basin region. Not only are jobs, income, value added, and output created and generated, but also it appears that the economy is becoming more stable and growth more sustainable as a result of the way in which the oil and gas industry now functions.

A similar analysis was conducted examining total production. Figure 2.30 shows production levels increasing since 2011 following the most recent recession.

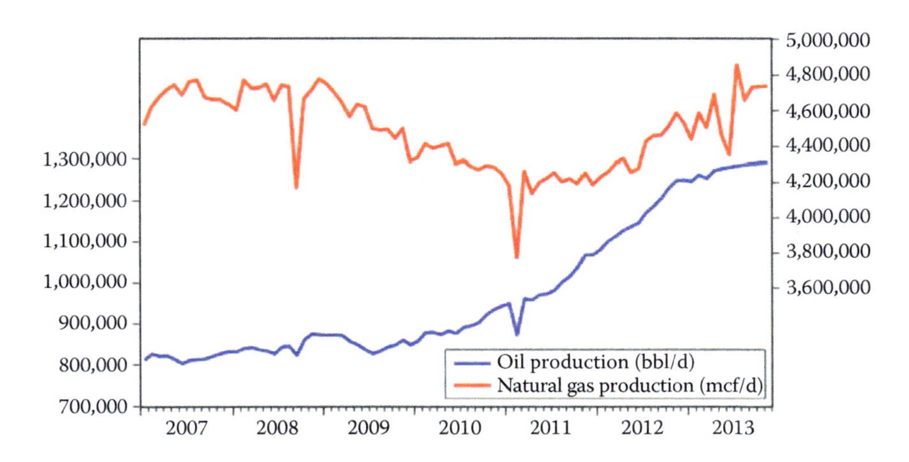

FIGURE 2.30 Total production—Permian Basin. *Note*: Left axis measures oil production. Right axis measures gas production.

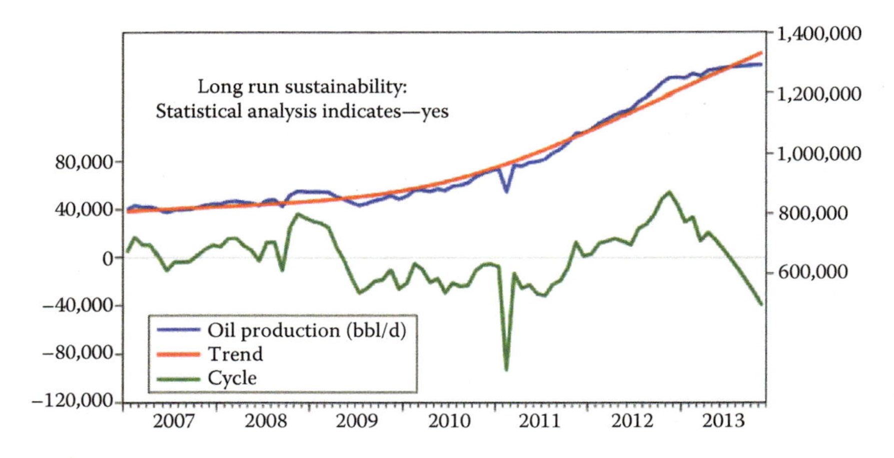

FIGURE 2.31 Oil production—Permian Basin: cycles and trends. *Note*: Left axis measures the actual deviation from trend. Right axis measures oil production.

Focusing on oil production, the time series was decomposed into its trend and cycle. Results imply the economic benefits of sustainable and stable production and are consistent with broader economic benefits associated with this industry (Figure 2.31).

Further analysis of the trend and cycles in employment levels yielded similar results. That is, recent oil and gas activities generate employment in the Permian Basin that is experiencing some reduction in cycle (i.e., a more stable labor market) but with a slightly less pronounced trend.

2.11 INDUSTRY TAXATION

The oil and gas industry is one of the most heavily taxed industries in this country. Although the scope of this section is limited to the upstream portion of the industry from discovery to production to sale of hydrocarbons, there are significant types and amounts of taxation. This section discusses the impacts of taxation of the industry and is not intended or offered as legal, accounting, or other professional advice, but only as general information affecting the oil and gas industry in those counties included in the Permian Basin. Should the information presented here appear to impact any person, or organization of entity, competent professional advice should be obtained. None of the taxes covering transmission, transportation, refining, wholesaling, and retailing are included in this study.

Taxes assessed on this portion of the industry include numerous types at the federal, state, and local levels. Additionally, while royalties are not a tax, their impact is likely to be substantial on an aggregate basis. According to Kaiser (2010), "Royalty rates are (generally) negotiable on private land but are not publicly disclosed, and will vary with time and location, sometimes dramatically." In fact, according to the

New Mexico Oil and Gas Association, total oil and gas revenue to the state was $3.8 billion, the majority of which comes from Southeast New Mexico, of which $1.7 billion goes to the State of New Mexico General Fund. However, these figures represent statewide totals from all sources (royalty, severance taxes, and property taxes) paid to all entities and are not broken down by county, and thus not Permian Basin specific.

According to the Texas Comptroller of Public accounts, the state of Texas received revenue of $1.5 billion from gas production and $2.1 billion from oil production. Previous studies have suggested that on a statewide basis, royalties to state funds may exceed $1.3 billion (Texas Oil and Gas Association). Note that these figures do not include property taxes. As with New Mexico, these amounts are not reported by county or source.

The impact on state budgets cannot be determined directly, either overall or with respect to the Permian Basin, because there is no mechanism to (1) determine the royalty payments for the states from the counties of the Permian Basin or (2) determine the other oil- and gas-related revenues to the state by county without an extensive audit of the comptroller's records. However, using the foregoing amount of oil and gas revenue ($1.7 billion) to the General Fund and the New Mexico state budget of $5.4 billion (Council of State Governments), revenue received (statewide) from the oil and gas industry is estimated to amount to 31% of the 2012 budget.

Texas differs from New Mexico in that the state budget is created for a 2-year period. To determine the impact of the industry on the Texas state budget, several simplifying assumptions must be made and are intended for illustrative purposes only. According to the Texas Comptroller of Public Accounts, the state budget for the 2012–2013 biennium was $173.5 billion. The nonfederal portion (which would include oil and gas revenues) of that total amount was $107.055 billion. Allocating one-half of this total to each of the 2 years gives an amount for 2012 of $53.5275 billion. Thus, the taxes on oil and gas production represent 8.4% of the budget.

The estimation of the amount of taxes collected by the federal government is hindered due to the lack of publicly available data for determining the federal income taxes assessed to the thousands of companies, partnerships, individuals, and royalty owners involved in the upstream sector in the Permian Basin. However, it does seem reasonable to suppose that it is significant compared to other industries and activities. Royalties to Texas and New Mexico universities and Bureau of Land Management constitute other significant benefits from the oil and gas industry; however, a county-level determination would require an extensive audit of the General Land Office records. Currently, there is not a mechanism in place for determining the leasehold royalties and production amounts paid to the states and University Lands in Permian Basin counties. However, it is recognized that these facets of the industry do contribute additional benefits and impacts to the economy.

2.11.1 Texas

State Taxes include severance taxes, sales taxes, well servicing taxes, and franchise taxes.

2.11.1.1 Severance Tax

The standard rates for Texas severance tax are

Oil: 4.6% of market value of oil produced
Natural gas: 7.5% of market value of gas produced
Condensate: 4.6% of market value

2.11.1.2 Sales and Use Tax

State Sales and Use Taxes are imposed on all retail sales, leases, or rentals of most goods and taxable services. Texas cities, counties, transit authorities, and special-purpose districts have the option of imposing an additional local sales tax for a combined total of state and local taxes of 8.25%/4% (0.0825). The ranges of these tax rates are

- State—6.25%/4% (0.0625)
- Cities—0.25%/4% (0.0025)—2% (0.02), depending on local rate
- Counties—0.5%/2% (0.005)—1.5% (0.015), depending on local rate
- Transit districts—0.25%/4% (0.0025)—1% (0.01), depending on local rate
- Special-purpose districts—0.125%/8% (0.00125)—2% (0.02), depending on local rate

2.11.1.3 Oil and Gas Well Servicing Tax

A tax of 2.42% (0.0242) of taxable services is imposed on those in the business of providing certain well services and who

- Own, control, or furnish the tools, instruments, and equipment used in providing well service
- Use any chemical, electrical, or mechanical process in providing service at any oil or gas well during the drilling, completion, or reworking or reconditioning of an oil or gas well

Services that are taxable include

- Cementing the casing seat
- Perforating the formation
- Fracturing the formation
- Acidizing the formation
- Surveying or testing the formation

In addition to these taxes, the state receives income from the industry in the form of royalties paid to the Permanent School and the Permanent University Fund as well as environmental and permitting fees.

2.11.1.4 Franchise Taxes

As with federal income tax collections, there is no public database available to use in determining the state franchise taxes assessed to and paid by the thousands of companies and royalty owners involved in the upstream sector in the Permian Basin. The basic elements of the franchise tax are as follows:

1. The current franchise tax applies to partnerships (general, limited, and limited liability), corporations, LLCs, business trusts, professional associations, business associations, joint ventures, incorporated political committees, and other legal entities.
2. The franchise tax does *not* apply to the following:
   a. Sole proprietorships (except the tax does apply to single-member LLCs filing as a sole proprietor for federal income tax purposes).
   b. General partnerships directly and solely owned by natural persons (except the tax does apply to all limited liability partnerships).
   c. Entities exempt under Subchapter B of Chapter 171 Texas Tax Code.
   d. Passive entities (as defined under TTC 171.0003). Note that some passive entities have an annual reporting requirement to affirm their passive status (see FAQ#8 under Passive Entities Rule 3.582).
   e. Unincorporated political committees organized under the Election Code or the provisions of the Federal Election Campaign Act of 1971. This provision is effective for reports originally due on or after January 1, 2012.
3. There is a $1,030,000 no-tax-due threshold, meaning entities to which the franchise tax applies do not owe a tax if their gross income is less than the threshold amount. However, a *NO Tax Due* report must be filed. Likewise, those entities to which the franchise tax *does not* apply may still be required to make periodic filings.
4. The franchise tax rates are as follows:
   - 1% (0.01) for most entities
   - 0.5% (0.005) for qualifying wholesalers and retailers
   - 0.575% (0.00575) for those entities with $10 million or less in total revenue (annualized per 12-month period on which the report is based) electing the E–Z computation

Retail Trade means the activities described in Division G, and *Wholesale Trade* means the activities described in Division F of the 1987 Standard Industrial Classification Manual published by the federal Office of Management and Budget, respectively. For more detailed information on tax rates, see the Texas State Comptroller's website (www.window.state.tx.us). For more detailed information on all aspects of the Texas Franchise Tax, see the Texas State Comptroller's website (www.window.state.tx.us/).

Local Taxes include those levied by cities, counties, school districts, and special districts such as hospital districts and utility districts. The taxable value of producing oil and gas properties is determined each year by the appraisal district in each county. The governing body of each taxing entity approves a tax rate annually. The tax rate of each taxing entity is then applied to the taxable value to obtain the tax due to each taxing entity. In Texas, the tax amount equals the tax rate per $100 of taxable value. The tax collector for each taxing entity then mails the tax bills to all owners of

TABLE 2.7
Oil and Gas Property Taxes—FY 2012

County—Population	Total Tax Base (Mill. $)	Oil and Gas Tax Base (Mill. $)	Oil and Gas Tax Levy (Mill. $)	O&G% of Total Tax Base
Andrews—14,786	$4,400	$3,509	$15.2	79.7%
Borden—641	645	574	1.7	89.1
Cochran—3,127	784	669	3.2	85.3
Coke—38,437	352	155	0.8	44.0
Crane—4,375	2,245	1,963	5.8	87.5
Crosby—6,059	393	116	0.6	29.6
Culberson—2,398	263	36	0.3	13.8
Dawson—13,833	1,089	583	3.2	53.5
Dickens—2,444	309	153	0.8	49.5
Ector—137,130	10,859	4,105	14.6	37.8
Edwards—2,002	353	70	0.3	19.7
Floyd—6,446	301	0[8]	0[8]	0.1
Gaines—17,256	5,951	4,656	16.4	78.2
Garza—6,461	845	635	2.4	75.1
Glasscock—1,226	1,403	1,134	3.4	80.9
Hale—36,273	1,974	260	1.3	13.2
Hockley—22,935	3,624	2,427	8.7	67.0
Howard—35,012	2,561	1,226	5.4	47.8
Irion—1,599	733	541	2.2	73.8
Jeff Davis—2,342	227	0	0	0
Kent—808	711	639	3.6	90.0
Kimble—4,607	395	0	0	0
Lamb—13,977	847	83	0.7	9.8
Loving—82	700	614	3.4	87.8
Lubbock—278,831	15,490	316	1.0	2.0
Lynn—5,915	318	33	0.3	10.4
Martin—4,799	2,913	2,572	6.3	88.3
Midland—136,872	13,213	3,214	6.4	24.5
Mitchell—9,403	1,135	688	2.8	60.6
Motley—1,210	88	4	0[8]	4.1
Nolan—15,216	1,603	190	0.8	11.9
Pecos—15,507	3,602	2,456	16.9	68.2
Reagan—3,367	2,048	1,718	5.9	83.9
Reeves—13,783	841	418	1.4	49.7
Scurry—16,921	2,684	1,460	5.8	54.5
Sterling—1,143	731	307	1.2	41.9
Terry—12,651	1,200	742	4.1	61.9
Tom Green—110,224	4,594	79	0.4	1.7
Upton—3,355	3,872	3,551	8.3	91.7
Val Verde—48,879	1,905	91	0.4	4.8
Ward—10,658	1,846	1,327	10.1	71.9

(Continued)

TABLE 2.7 (*Continued*)
Oil and Gas Property Taxes—FY 2012

County—Population	Total Tax Base (Mill. $)	Oil and Gas Tax Base (Mill. $)	Oil and Gas Tax Levy (Mill. $)	O&G% of Total Tax Base
Winkler—7,110	1,430	984	7.2	68.8
Yoakum—7,879	4,102	3,517	13.5	85.7
Total	*$105,579*	*$47,815*	*$192.80*	

Notes: Total tax base is the appraised value of all taxable property in each county. Oil and gas tax base is the appraised value of producing oil and gas properties. Oil and gas tax levy is the amount levied on producing oil and gas properties, and county populations are derived from the 2010 census.

producing mineral interests or their agents. Because the rights to receive production are owned by the royalty owner(s) and the working interest owner(s) based on the oil and gas lease that is the basis for the relationship between royalty and working interest owners, each such owner is also liable for the tax associated with the owned interest. Note that a royalty owner may have created a nonparticipating royalty interest that is liable for its share of the tax, and a working interest owner may have created interests out of its interest such as ORRI and PP, which interests are liable for their share of the tax. Based upon the royalty agreed to in the lease of each producing property, the ownership shares generally range from 1/8 (0.125) to 1/4 (0.25) for royalty owners and 7/8 (0.875) to 3/4 (0.75) for the working interest owners. Thus, when viewing the property tax amounts received by county table (Table 2.7), the *Oil and Gas Tax Levy* column reflects the aggregate paid by all owners of an interest in each producing property in each county, not just the working interest share.

In addition to these direct taxes on the industry, most locations have enacted hotel/motel taxes that, while not industry specific, certainly amount to a significant source of revenue to the local economies due to the significant amount of traveling done by individuals working in the industry.

Table 2.7 shows oil and gas tax amounts received by Texas Permian Basin counties for fiscal year 2012. Taxes received in fiscal year 2012 are those that were assessed on property in 2011. Tax bills are sent to the property owners in October of the assessed year (2011 in this report) and are payable any time between receipt of the bill and January 31 of the following year (2012) without penalty, interest, and collection fees. These figures represent taxes levied on producing oil and gas properties only and do not include other taxes assessed on pipelines, refineries, and other related properties. The source of these data is provided by the counties to the Texas Comptroller.

The oil and gas tax amounts include state taxes; however, the state receives no property taxes. Those taxes all are paid to the various taxing entities enumerated in the table (see Local Taxes) and in the amounts listed in Table 2.7 for each Texas Permian Basin county under the heading *Oil & Gas Tax Levy (Mill. $)*. The corresponding total for each county is divided between all the taxing entities in the county within the boundaries of which the production is located. For example, in Lubbock County for the year 2014, the tax rates are (as a percentage of $100s of valuation) Lubbock County—0.34534; Lubbock ISD—1.235; City of Lubbock—0.5044; Lubbock Hospital District—0.11844; and Lubbock Water District—0.008. Additionally, *Oil and Gas Tax Base* (third column) is the combined value of all types of hydrocarbon production even though there are different rates for oil and condensate (4.6% each) and gas (7.5%). Finally, the total for *Oil & Gas Tax Levy (Mill. $)* of $192.8 million includes only property (ad valorem) taxes in the Permian Basin counties.

2.11.2 New Mexico

Table 2.8 shows the tax distributions from oil and gas production in the five New Mexico Permian Basin Counties in 2012. The values in this table are compiled from the NM State Land Office, *Financial Monthly Reports* (http://www.nmstatelands.org), and from ONGARD Tax Type by County Distribution Period Report run on January 4, 2013. Data are for FY12. The table is reproduced, in part, from Meghan Downes (2012). County populations are derived from the 2010 census.

A thorough review and analysis that describes all taxes on the industry in New Mexico, their derivation, and the distribution of each is provided in "Economic Impact of New Mexico's Oil and Gas Industry—2012" (Downes, 2012) and includes a summary of the various sources and distributions of oil and gas taxes in New Mexico.

Consequently, the Texas property taxes of $192,800,000 and the New Mexico Ad Valorem and Land Grant Maintenance Fund distributions of $142,625,409 equal $335,425,409 in tax levies for the counties of the Permian Basin in 2012. This value does not include the multiple millions of dollars in previously discussed taxes and fees to the federal, state, and local government agencies and taxing entities annually.

TABLE 2.8
New Mexico Tax Distributions

County—Population	Ad Valorem Production	Ad Valorem Equipment	Total Ad Valorem Taxes Distributed to Counties (Ad Valorem Prod. and Equip.)	Land Grant and Maintenance Fund Distributions to Beneficiaries	Total County Level Tax Payments from Oil and Gas
Chaves—65,645	$2,256,046	$444,980	$2,701,026	$17,939,124	$20,640,150
Eddy—53,829	$41,634,030	$5,652,478	$47,286,508	$0	$47,286,508
Lea—64,727	$54,308,319	$8,687,532	$62,995,851	$0	62,995,851
Otero—63,797	$0	$0	$0	$10,608,391	$10,608,391
Roosevelt—19,846	$302,188	$48,779	$350,967	$743,542	$1,094,509
Total	$98,500,583	$14,833,769	$113,334,352	$29,291,057	$142,625,409

2.12 SUMMARY AND CONCLUDING REMARKS

The Permian Basin's oil and gas industry is an important driver of economic activity in the region and beyond. The industry's activities generate and sustain jobs, income, and output. The industry also provides substantially to the gross state products of both Texas and New Mexico. In addition, through various measures of taxation, the industry provides many localized benefits to the citizens of both New Mexico and Texas. Recent innovations and discoveries in both plays and technologies have given rise to increased production of oil and gas and have led to additional economic benefits that will likely impact the region for years to come. These benefits are in the form of sustainable growth and a more stable economy than has been experienced in the past.

Specifically, several potential and confirmed resource plays have been identified in the Permian Basin. Resource plays (shale plays being a subset) are the most active drilling areas in the United States and have a number of characteristics that make the economics favorable. First, these resource plays cover very large areas (multiple counties). Second, they resemble more of a manufacturing type of process where thousands of wells are drilled and enable producers to take advantage of economies of scale. Moreover, wells exhibit a repeatable statistical distribution, thus offer a predictable performance in a given geological subset adding to favorable economics. On the technology side, the Permian Basin has the greatest rig count of any basin/region in the world (27% of the United States and 56% of Texas). A rapidly increasing amount of the U.S. and Permian Basin wells are being drilled horizontally. As such, Permian Basin well productivity has improved dramatically since 2011 due to improved technology in horizontal drilling and hydraulic fracturing. Drilling efficiencies are being realized in all U.S. resource plays, and the Permian Basin is the least mature; thus, vast efficiency improvements are expected in the Permian Basin. Unquestionably, the Permian Basin's oil and gas industry is leading the way to new economic heights in Southeast New Mexico and West Texas.

ACKNOWLEDGMENTS

Funding for this research was provided by the Permian Basin Petroleum Association. The opinions expressed herein are those of the authors. The authors thank Tariq Ali, Roland O. Ezewu, Ibegbuna Ezisi, and Mark Thompson for research assistance and helpful comments.

APPENDIX A: COUNTIES INCLUDED IN THE PERMIAN BASIN IMPLAN MODELS

Texas counties (43)

Andrews
Borden
Cochran
Coke
Crane

Crosby
Culberson
Dawson
Dickens
Ector
Edwards
Floyd
Gaines
Garza
Glasscock
Hale
Hockley
Howard
Irion
Jeff Davis
Kent
Kimble
Lamb
Loving
Lubbock
Lynn
Martin
Midland
Mitchell
Motley
Nolan
Pecos
Reagan
Reeves
Scurry
Sterling
Terry
Tom Green
Upton
Val Verde
Ward
Winkler
Yoakum

New Mexico counties (5)

Chaves
Eddy
Lea
Otero
Roosevelt

APPENDIX B: SOCIETY OF PETROLEUM EVALUATION ENGINEERS (SPEE) PARAMETERS FOR EVALUATING RESOURCE PLAYS

The following *Tier 1* characteristics are nearly always observed in resource plays:

1. Wells exhibit a repeatable statistical distribution of estimated ultimate recoveries (EURs).
2. Offset well performance is not a reliable predictor of undeveloped location performance.
3. A continuous hydrocarbon system exists that is regional in extent.
4. Free hydrocarbons (nonsorbed) are not held in place by hydrodynamics.

If the reservoir being evaluated satisfies these four criteria, there is a very good chance the reservoir is a resource play. Conversely, if any one of these characteristics is absent, it is quite likely the reservoir is *not* a resource play. The Tier 1 criteria are listed in the order of significance, and both geological and engineering data must support these criteria.

The Tier 1 criteria possess aspects of engineering and geology, and determining whether a reservoir is a resource play requires consideration of both. The geological depositional model needs to describe a reservoir with regional extent, while the engineering data need to show statistically repeatable EURs over time. Obviously, sufficient time is required to arrive at these conclusions—time for historical data to accumulate, and time to analyze the data.

As a practical matter, it is anticipated that resource plays will encompass more than 100 completed wells in the reservoir. There are two rationales for this: first, developing a usable statistical model in a resource play typically requires about 100 wells, and second, a reservoir that has sufficient areal extent to be considered a resource play will likely encompass a minimum of 100 wells.

Although the following *Tier 2* reservoir characteristics are not required, these are commonly observed in resource plays:

1. Requires extensive stimulation to produce at economic rates.
2. Produces little in situ water (except for coalbed methane and tight oil reservoirs).
3. Does not exhibit an obvious seal or trap.
4. Low permeability (<0.1 md).

REFERENCES

2012 Annual reports of Anadarko Petroleum Corporation, Apache Corporation, Concho Resources, Devon Energy Corporation, Occidental Petroleum, Pioneer Natural Resources.

Baker Hughes. (2013) Available from Security and Exchange Commission EDGAR. http://gis.bakerhughesdirect.com/Reports/RigCountsReport.aspx.

Council of State Governments. http://www.csg.org.

DrillingInfo and DI Desktop™. http://www.drillinginfo.com/.

Energy Information Agency (EIA), Drilling Productivity Report. Engler, T. W. et al. (2011) Reasonable Foreseeable Development (RFD) Scenario Final report submitted to the U.S. Department of the Interior Bureau of Land Management, New Mexico at the Pecos District.

Federal Reserve Bank of St. Louis Economic Database. http://research.stlouisfed.org/fred2/.
Fueling the Texas Economy, The Texas Oil & Gas Industry, published by the Texas Oil & Gas Association.
IMPLAN Version 3.0 software and data sets for New Mexico and Texas.
Kaiser, M. (2010) Economic limit of field production in Louisiana. *Energy*, 35, 3399–3416.
Meghan Downes, C. (2012) Economic impact of New Mexico's Oil and Gas Industry—2012.
New Mexico Oil Conservation Division. http://www.emnrd.state.nm.us/OCD/.
New Mexico Oil and Gas Association. http://www.nmoga.org.
New Mexico Taxation and Revenue Department. http://www.tax.newmexico.gov/.
Oil and Gas in Texas, A Joint Association Message from the Texas Oil and Gas Industry produced by associations representing the Texas Oil and Gas industry.
Railroad Commission of Texas. (2013) http://www.rrc.state.tx.us/permianbasin/.
Texas Oil and Gas Association. Fueling the Texas Economy 2013.
Texas State Comptroller's Office. http://www.window.state.tx.us/.
The Texas Property Tax Code.
Tudor, Pickering, Holt & Co Research. http://www.tphco.com/reports.
U.S. Bureau of Economic Analysis. http://www.bea.gov/.
U.S. Bureau of Labor Statistics. http://www.bls.gov/.
U.S. Energy Administration. Drilling Productivity Report, various issues.
U.S. Energy Administration. Oil and gas lease equipment and operating costs report.
U.S. Energy Information Agency (EIA). (March 2015) Drilling productivity report: For key tight oil and shale gas regions. http://www.eia.gov/petroleum/drilling/pdf/dpr-full.pdf (accessed March 30, 2015).
U.S. Energy Information Administration. http://www.eia.gov/.
Watson, M. C. et al. (December 2010) Monograph III: Guidelines for the practical evaluation of undeveloped reserves in resource plays. Society of Petroleum Evaluation Engineers.

3 From Property Rights to Endangered Species

Legal Issues Surrounding Hydraulic Fracturing

William R. Keffer, J. Randall Miller, J. Berton Fisher, Taylor Stevenson, and Adrianne Waddell

CONTENTS

3.1 Overview 66
3.2 Introduction 66
3.3 The Fracks of Life (Or "The More Things Change, The More They Stay the Same") 66
3.3.1 Cover Your Base 67
3.3.2 Know Your Field of Play 67
3.3.3 Communicate with Your Neighbors 68
3.3.4 Find Out What the Locals Are Drinking 68
3.3.5 Get the Dirt...on How It's Being Used 68
3.3.6 Don't Be a Stranger 69
3.3.7 Find Out If Your Landowner Thinks He Looks Good in a Suit 69
3.3.8 All Politics Is Local 69
3.3.9 Share the Wealth 69
3.3.10 Knowledge Is Power 70
3.4 Subsurface Trespass at Common Law 70
3.5 Water Use in Hydraulic Fracturing: Perspective Is Reality 73
3.6 Role of the Endangered Species Act in the Oil and Gas Industry: A Focus on the Dunes Sagebrush Lizard and the Lesser Prairie Chicken 75
3.6.1 The Endangered Species Act 75
3.6.2 The Dunes Sagebrush Lizard 76
3.6.3 The Lesser Prairie Chicken 77
3.7 Summary and Concluding Remarks 78
References 78

3.1 OVERVIEW

The chapter discusses a variety of legal issues associated with hydraulic fracturing and provides practical guidance to landowners interested in leasing their lands for oil and gas exploration and mining.

3.2 INTRODUCTION

Living life in this world does not happen in a laboratory or an isolation chamber; it happens around people, property, and countless features we call the *environment*. As a result, exploring for, finding, and producing our natural resources as energy sources for our world, though necessary, never happen in a vacuum.

The development and full-fledged implementation of new and improved technology in the oil and gas industry has ushered in a renaissance of exploration and production in the United States. The combined use of 3-D seismic testing, horizontal drilling, and hydraulic fracturing has opened up vast new frontiers of reserves that until now had been uneconomic to recover or unknown to the industry. The sharp increase in shale-gas and shale-oil discoveries and development is literally changing the geopolitical balance regarding relative energy positions, and has made plausible the idea of energy independence for the United States. But as with most technological and economic advancements, unanticipated or unintended consequences often surface. Whether those ramifications become magnified or are properly managed depends on the industry's response in scope and timing.

The following four articles highlight some of the issues that have become magnified, in light of the hydraulic fracturing boom. Whether it is protecting people, property rights, water, or animals, there are statutes and constituencies in place designed to ensure that the proper balance is struck. Although those statutes and constituencies seem to be everywhere and all-consuming, those engaged in the business of finding, producing, and providing energy can quell opposition before it ever begins by respecting people, property, and the environment.

3.3 THE FRACKS OF LIFE (OR "THE MORE THINGS CHANGE, THE MORE THEY STAY THE SAME")

Our law practice has dealt with environmental impacts related to the oil and gas industry for more than 30 years: as in-house counsel for a major oil company, as outside counsel for several other major and independent oil companies, and as counsel for small and large landowners affected by these impacts. During that time, we have compiled an extensive depth and breadth of experience regarding practices that can minimize a company's exposure to claims arising from oil and gas operations. Hydraulic fracturing is an advancement with substantial economic benefits, and yet the process has developed a reputation as an industrial bogeyman, whose downside is far too risky and therefore, unacceptable. Whether companies subscribe to any of the alleged concerns is besides the point; the more important point is to realize that society's perception of this issue will become the reality, and litigation will likely follow.

Toward that end, we offer the following *lessons learned* and practical advice to companies seeking to be a part of the current revolution so that they might minimize the claims and allegations that will undoubtedly become part of their experience.

3.3.1 Cover Your Base

There is simply no substitute for baseline data. Ideally, before drilling and fracturing a well, you should survey and identify every available, actively used water well within a certain radius of the well to be drilled (e.g., at least twice the farthest calculated distance that the fracture will/can travel from the input point), then design and conduct a sampling exercise, in order to collect a sufficient amount of relevant data regarding area water quality. This data set will be your first and best line of defense against later claims of alleged frack-related groundwater contamination. The appropriate radius and list of test parameters will vary from location to location, so those decisions should be made with the advice and active participation of lawyers and experts who have had extensive experience in handling groundwater-pollution claims involving fracturing fluids or other oilfield-related operations.

3.3.2 Know Your Field of Play

Before drilling and fracturing your well, research the history of the field in which you are planning to drill. Usually, there will have been some amount of prior exploration, drilling, and production, as well as possible ongoing production. There might be wells that were abandoned and plugged decades ago with everything ranging from telephone poles to oilfield debris. There might be wells that have been temporarily abandoned for years, whose downhole condition is unknown. There might be producing wells that were drilled and completed at a time in history when steel was in short supply, so that perhaps those wells were *short-cased* and are not properly protective through the base of the deepest usable *fresh*water zone.

While you cannot always guard against—or even know about—every potential problem in a mature field, you can perform a certain level of due diligence that can inform and equip you with the ability to take precautionary steps that you might not otherwise have considered. For example, you can confer with field-administrative officials and both past and current landowners and lessees around the well site and area, and obtain a relatively inexpensive historical aerial photograph of an area, with at least a 2-mile perimeter around the well site. Factors such as surface and subsurface location conditions, state oil and gas well regulations, historical wellbore spacing, and current industry practice in the field can help determine the distance from your well site that should be examined. Being aware of secondary consequences that might result from your fracturing is another way to be armed with information that might protect you against claims for which landowners would seek to hold you primarily liable. As a general rule, exceeding (not merely meeting) administrative requirements for drill-site due diligence is a cost-effective practice that can enhance your credibility as a concerned, prudent operator.

3.3.3 Communicate with Your Neighbors

This is perhaps the simplest preventive step that can be taken, yet also the most often overlooked and underestimated. We can always go back into the company files and find that point in time, early in the history of the complaint, when a company representative chose poor, abusive, or no communication with a neighboring landowner, instead of proper, courteous, and respectful communication. What could have been resolved as a nominal complaint became a very expensive lawsuit. Always assume that neighboring landowners want to be treated exactly as you want to be treated.

It is an unfortunate reality that as an oil and gas company, you usually start off with three strikes against you when you show up on a neighboring landowner's doorstep: (1) you are an uninvited guest; (2) you are going to impact the way the neighboring landowner has been using and living on his property; and (3) you are likely to do something on his property that can be described as *contamination*. On top of all that, the neighboring landowner usually has absolutely no financial interest in the oil and gas to be produced. Instead, he has to tolerate your presence for the financial benefit of his neighbor. As a result, every communication—and every incentive to have communication—with that landowner should always be designed and carried out with those realities foremost in mind.

3.3.4 Find Out What the Locals Are Drinking

In addition to sampling area water wells to determine a baseline for quality, do a little reconnaissance to find out the source of drinking water in the area. Is everyone in the immediate area of your planned operation on private well water? Do they use their own filtration systems? For what particular water quality problem? Are they on a rural water system? How far from your landowner is the nearest municipal or rural water system, and are taps available? Where does the rural water system get its water? From other water wells in the area? Likewise, if there's a town or community nearby, where does the community system get its water? Collecting this information ahead of time can paint a picture regarding area water sources that could help you design a more precise strategy regarding the collection of baseline data.

3.3.5 Get the Dirt…on How It's Being Used

Take the time to learn about the current and historical use of the land. Has it been used to grow crops? If so, what kind and for how long? Has it been used for cattle grazing? Is there any kind of specialized use? For example, does the landowner take particular pride in his range management practices? Is his property used for hunting, fishing, wildlife observation, or wildflower tours? Knowing ahead of time how the landowner is using his property can not only provide helpful insight into the well placement and the size and location of related operations, but can also provide critical information for identifying certain parameters to look for when sampling area water wells.

3.3.6 Don't Be a Stranger

At the outset of a new project, there will often be a great deal of confidentiality and privacy for perfectly legitimate reasons. But to the landowner and everyone else in the community, it just looks like plain, ol' secrecy, and secrecy means you know something they don't know, which will almost always be interpreted as something detrimental to them and beneficial for you. Don't allow yourself to be cast in the role of foreigner or stranger. Instead, determine what kind of assistance you will need in conjunction with the local work, and see if you can include local talent on your team. When your landowner and other locals can see a familiar face involved with your activities, the level of stress and concern can be reduced substantially.

3.3.7 Find Out If Your Landowner Thinks He Looks Good in a Suit

Although easy to do, companies almost never research a landowner's litigation history at the local courthouse. Nothing predicts future behavior better than past behavior. If a landowner has made a habit of suing others, you can rest assured that there will come a time when he feels justified in suing you. This is another simple box to check under the heading of *due diligence* but is almost always passed over as an unnecessary expense, if even considered at all.

3.3.8 All Politics Is Local

Make it a point to learn about the elected officials that represent the area where you are planning to drill and fracture. Municipal and county officials are typically the most important folks to know. But it can also be enlightening to know the state legislators and congressional representatives. You never can tell when one of them, or one of their family members, might live or own property in the immediate area. The state or county might have enacted statutes or ordinances relating to oil and gas development. Learn ahead of time whether the elected officials have taken public positions on these issues.

3.3.9 Share the Wealth

Once you know you will be in the community for the long term, broaden your perspective to understand what all that means. You have decided that it will be profitable to continue operating in that community, so share some of that profit by investing back into it. First, remember that you are making money off of the resources that happen to be located in that community, so invest some of that money in things important to the community: the volunteer fire department; the local school; the community's infrastructure—roads, water, utilities, and so on. Second, and just as important, these decisions need to be made from an informed perspective. Tone-deaf, haphazard donations made just to check off a box on your to-do list can be more damaging than making no donation at all.

3.3.10 Knowledge Is Power

A truism borne from every pollution case we have ever handled is that *information is good*. If it's bad news (e.g., there is contamination present, a risk has been increased, etc.), then it's news that you would rather know now and have found out yourself, than to have operated in ignorance and have the bad news dumped on you later by hostile forces. If it's good—or innocuous—news, then all the better. *Head-in-the-sand* policies might be temporary avoidance measures, but they are ultimately recipes for disaster.

In this fast-paced, head-spinning, technologically advanced era in which we live, amazingly effective oilfield production tools, like hydraulic fracturing, make the future of U.S. oil and gas production exciting and seemingly unbounded. Yet, even with such positive developments, there are still responsibilities. Common courtesy and common sense are never out of date and can be the difference between peaceful coexistence and hostile, expensive, protracted litigation. These 10 tips cannot guarantee immunity from unpleasant and perhaps unfounded attacks, but they will go a long way toward minimizing your exposure and defusing those who are always looking for their next target.

3.4 SUBSURFACE TRESPASS AT COMMON LAW

One of the most critical questions presently facing the oil and gas industry is whether fluids injected into wellbores during hydraulic fracturing or wastewater disposal may constitute an actionable trespass if the fluids migrate to the subsurface property of a neighboring landowner. If claims for subsurface trespass were found to apply to various downhole oil and gas production methods, the ensuing flurry of plaintiff injunctions could critically stifle the U.S. energy industry. While no Supreme Court of Texas case has addressed this issue directly, two prior cases and one ongoing subsurface trespass action may shed some light on the available precedent and what future courts are likely to hold. Based on the relevant common law, it is likely that courts will find actionable subsurface trespass only when the subsurface invasion of plaintiff's property leads to actual harm to the property itself or to the plaintiff's use of their land.

In *Railroad Commission v. Manziel*, the Supreme Court of Texas held that when the Railroad Commission of Texas (RRC) permits a secondary recovery operation for the production of oil and gas, no actionable trespass exists when the injected, secondary recovery fluids move across lease lines and into the subsurface of an adjoining tract.[1] This 1962 case involved an RRC Rule 37-exception permit that authorized the defendant to drill an injection well located near the border of an adjacent lease owned by the plaintiffs.[2] The RRC permit authorized the defendant to water-flood the well as a part of a secondary oil and gas recovery effort, inevitably causing fluid to migrate across the lease line and into the plaintiffs' subsurface property. The plaintiffs claimed that the injection of water into their subsurface estate would "cause damage to their producing wells, and w[ould] result in loss and injury to their oil and gas interests due to premature flooding."[3] In response, the court noted that "[w]ater injected into an oil reservoir generally spreads out radially from the injection well bore, and it is impossible to restrict the advance of the water to lease lines."[4] It was this transient nature of injected fluids and the inability to stop them from migrating across lease lines that led the court to rule that secondary recovery by injection did not constitute an actionable subsurface trespass.[5] Furthermore, the court pointed to the concept of the *Negative Rule of Capture*, which, like the standard rule of capture, states that where a landowner may utilize a well to capture oil and gas that may migrate from adjoining tracts, so too may they utilize a well to inject substances into a reservoir, even though those substances may migrate to other adjoining tracts.[5]

The *Manziel* court essentially held, not that subsurface incursions may never constitute actionable trespass, but that in cases where the RRC authorized certain secondary recovery injections, the migration of fluids from one lease to the next does not constitute a subsurface trespass, nor are those operations "subject to an injunction on that basis."[1] Furthermore, the court found only one instance in which an injunction may be granted on the applied theory of subsurface trespass: when the injecting operator creates a "continuing, physical invasion of an adjoining mineral estate by drilling across lease lines."[6] Therefore, this court appears to suggest that if the RRC permits an injection, such as hydraulic fracturing, then that agency authorization trumps many, if not all, claims of subsurface trespass.

In the 2006 case, *Coastal Oil & Gas Corp. v. Garza Energy Trust*, the Supreme Court of Texas held that a plaintiff's claim for actionable trespass was precluded by the rule of capture.[7] In *Coastal*, the defendant (Coastal) leased acreage from the plaintiff (Garza) for the purpose of oil and gas production. Coastal also owned land directly adjacent to the property they leased from Garza and conducted hydraulic fracturing operations on both.[8] Garza argued that the wells Coastal fraced on the adjacent tract near the Garza property line led to drainage and created *substantial damage to their hydrocarbons* and that the migration of fluids constituted a subsurface trespass.[9] The court rejected Garza's reliance on the *ad coelum* doctrine, claiming this doctrine is a "maxim [that] has no place in the modern world" and that "[t]he law of trespass need no more be the same two miles below the surface than two miles above."[10] While the court did not directly address the question of whether the migration of fluids to a neighboring tract of land constituted a subsurface trespass, they stated that the rule of capture precluded actionable trespass in this case because "actionable trespass requires injury, and [the plaintiff's] only claim of injury – that Coastal's fracing operation made it possible for gas to flow from beneath [the plaintiff's property] to the [defendant's]... is precluded by the rule of capture."[7]

The *Coastal* court also differentiated deviated wells whose wellbore crosses a lease line from fracing operations by stating that produced oil and gas in a deviated well "d[o] not migrate to the wellbore from another's property; it [the deviated well] is already on another's property" and is therefore subject to trespass claims.[11] The court listed four policy reasons "not to change the rule of capture to allow one property owner to sue another for oil and gas drained by hydraulic fracturing that extends beyond lease lines."[11] First, there are multiple methods by which the owner of a tract may recover damages or prevent drainage. These methods of self-help include drilling an offset well, enforcing the implied covenant in the lease to protect against drainage, or pooling their tract either by contract or by force via the RRC.[11] Second, the RRC, *not* the judicial system, is the appropriate body to regulate oil and gas production activities such as hydraulic fracturing.[12] Third, the judicial system does not have the capability nor the expertise to evaluate the amount of oil and gas drained.[13] Finally, the oil and gas industry as a whole is opposed to extending liability for drainage to the field of hydraulic fracturing.[14] These four factors, coupled with previous precedent, led the *Coastal* court to conclude that, absent the evidence of an actual injury, the rule of capture precludes a claim for trespass when fluids migrate across lease lines during fracing operations.[15]

Currently, the Supreme Court of Texas is once again faced with the decision of whether the migration of injected fluids into the subsurface of another's property may constitute an actionable trespass.[16] In *FPL Farming Ltd. V. Environmental Processing Systems*, Environmental Processing Systems (EPS) received a permit from the Texas Commission on Environmental Quality (TCEQ) to drill and operate multiple, nonhazardous wastewater-injection wells less than a thousand feet from a neighboring rice farm owned by FPL Farming Ltd. (FPL). FPL sued EPS, claiming that the wastewater from the injection wells would likely migrate into their subsurface estate and contaminate their farm's water supply and, therefore, constituted an actionable subsurface trespass.[17] EPS argued that the injected wastewater caused no actual harm to FPL's farming operation. The Texas Court of Appeals shielded EPS from liability, stating that when the TCEQ authorizes and permits a drilling operation, there could be no damage claims for subsurface trespass.[18] The Supreme Court of Texas disagreed and held that "[a]s a general rule, a permit granted by an agency does not act to immunize the permit holder from civil tort liability from private parties for actions arising out of the use of the permit."[19]

Therefore, the court reversed and remanded, holding that just because there was a TCEQ permit did not mean that there could not have also been an actionable trespass.[20] During oral argument on January 7, 2014, the court pointed out that other jurisdictions required that plaintiffs claiming subsurface trespass prove actual damages to their property or hindrance of the use of their property in order to receive a judgment in their favor.[21] While the *FPL* case is ongoing, the oil and gas industry is waiting with bated breath, as an unfavorable ruling with respect to downhole wastewater injection could have major implications on hydraulic fracturing operations.

Whether oil and gas production methods that involve the downhole injection of fluids may constitute an actionable trespass if the fluids migrate to the subsurface estate of a neighboring landowner is an issue that will likely shape the future of the American oil and gas industry. The *Manziel* court stipulated that the migration of fluids across lease lines did not constitute actionable trespass if the operator held a Texas Railroad Commission permit for downhole fluid injection. The *Coastal* court held that where fluids migrated to the property of another, the rule of capture trumped claims of subsurface trespass. Finally, the court in *FPL* ruled that the simple fact one possesses an agency permit for wastewater injection does not mean that a plaintiff cannot make a claim for subsurface trespass. While the *FPL* case is ongoing, that court is poised to make a definitive ruling on the admissibility of subsurface-trespass claims, likely within the next year.

Based on the foregoing precedent, the stance courts should take with regard to subsurface trespass is summed up by University of Oklahoma College of Law Professor Owen Anderson, who stated that "[a]llowing injunctive relief or ejectment for subsurface trespass is particularly troubling and should be limited to situations where the harm to a neighboring landowner clearly outweighs the utility and practical necessity of the subsurface invasion."[22] This determination, he says, should be left to government regulatory agencies and not to the courts or the landowners involved.[23] While the debates over subsurface trespass are likely to continue, one thing is for certain—many in the oil and gas industry are waiting and hoping for a favorable ruling in *FPL Farming Ltd. V. Environmental Processing Systems* with ever-increasing anticipation.

3.5 WATER USE IN HYDRAULIC FRACTURING: PERSPECTIVE IS REALITY

Opponents of hydraulic fracturing claim that hydraulic fracturing uses vast quantities of water, raising the specter of water scarcity to frighten rural residents and others. At the core, these claims have no merit and instead rely on innumeracy, which is the inability to compare large numbers, and the unfamiliarity of most citizens with readily available, but sometimes arcane, sources of information. These claims can be persuasively countered by comparing water use in hydraulic fracturing to permitted water uses for domestic, public water supply, agricultural, and other existing purposes. Since water use is local, such comparisons are most persuasive when they are stated within the context of local water use.

Water is the primary carrier fluid in hydraulic fracturing and makes up in volume the bulk of hydraulic fracturing fluids. At the moment, most water used in hydraulic fracturing is fresh water taken from both surface and groundwater sources: the same sources used for domestic, public water supply, and agricultural purposes. Consequently, it is not surprising that a consistent theme in environmental criticisms of hydraulic fracturing is the charge that use of surface and shallow groundwater for hydraulic fracturing could lead to water scarcity. For example, the group "New Yorkers Against Fracking" specifically charges:

> Depletion and degradation of surface freshwater and shallow drinking water aquifers: Massive amounts of clean water are taken from lakes, ponds, streams and shallow aquifers for fracturing operations. The scale of this industrial drawdown, and subsequent contamination with fracking chemicals, will degrade water quality and could lead to water scarcity.[24]

The concern is even more acute in water-poor states. In Nevada, for example, Lander County and the Center for Biological Diversity[25] joined in filing formal administrative protests against a U.S. Bureau of Land Management oil and gas lease sale in central Nevada. The filing expresses concern about water use and questions how the water necessary for hydraulic fracturing will be obtained. Specifically, Lander County's intervention was on behalf of ranchers and farmers concerned that hydraulic fracturing could end up taking water away from them.[26] Given the ongoing dry conditions in the central and western United States, charges that a particular water use will result in water scarcity constitute a serious matter. In short, opponents of hydraulic fracturing consistently exploit rural residents' fear that their water supply will be compromised by hydraulic fracturing.

As with most emotionally charged conflicts, truth is an early casualty: here, it is occurring in discussions regarding the actual impact of hydraulic fracturing on water supplies. It is typical that concerns related to the volumes of water used for hydraulic fracturing are not separated from concerns related to the contamination of surface and groundwater. Some opponents of hydraulic fracturing attempt to bootstrap the issue of water volume into concerns about air quality, safety, and road repair:

> It has been estimated that the transportation of two to five million gallons of water (fresh or waste water) requires 1,400 truck trips. Thus, not only does water used for hydraulic fracturing deplete fresh water supplies and impact aquatic habitat, the transportation of so much water also creates localized air quality, safety and road repair issues.[27]

Although the specific number of truck trips cited sounds precise, a simple calculation demonstrates that the data in this paragraph are suspect. Large tank trucks typically have capacities ranging from 5,500 to 11,600 gal. The number of trips required to transport 2 million gallons of water would be between 364 and 172 trips, depending upon tank volume—not 1400 trips. The information presented earlier suggests that trucks used to transport the "two to five million gallons of water (fresh or waste water)" have a capacity of only between ~1400 and 3600 gal. The second issue is that the term "trips" is not defined; a round trip could be construed to be two trips—one full and one empty.

No one disputes that large volumes of water are used in hydraulic fracturing. Indeed, in a section entitled "Hydraulic Fracturing Water Usage," the FracFocus Chemical Disclosure Registry states: "...the multi-stage fracturing of a single horizontal shale gas well can use several million gallons of water..."[28] Hydraulic-fracturing opponents have hit upon the strategy of using the term "millions of gallons." A million gallons sounds like a lot of water, and multiple millions of gallons sounds even bigger. For example, the website "Dangers of Fracking" states hyperbolically that hydraulic fracturing requires "8 million gallons of water per fracking."[29] This claim is unsupported by any underlying data or citation. Indeed, an examination of the data in FracFocus[30] for 2010 through 2013, for records reporting water usage for individual well completions, shows that the median volume of water used in the hydraulic fracturing of an oil and gas well was ~1.6 million gallons (so half of the wells were fractured using less than ~1.6 million gallons of water), and that only 25% of all hydraulically fractured wells used more than 3.9 million gallons of water. This claim that each hydraulically fractured well used 8 million gallons of water is, to the say the least, incorrect.

To bring rationality to the discussion of water use and hydraulic fracturing, it is essential to put the amount of water used in hydraulic fracturing into the context of the amount of water used for other purposes. This is not a novel concept; the FracFocus website states:

> The amount of water used in hydraulic fracturing, particularly in shale gas formations, may appear substantial, but it is small when compared to other water uses such as agriculture, manufacturing and municipal water supply.[31]

It should be remembered that water use is local. Accordingly, the most meaningful comparison of water used for hydraulic fracturing with water used for other purposes will also be local and easily relatable to those uses whose scales are easily understood. FracFocus reported that the amount of water used in Oklahoma hydraulic fracturing for 2012 was just less than 5.5 billion gallons, with nearly 1.1 billion gallons of that water being used to hydraulically fracture wells in Canadian County alone. To put this in context, however, the volume of water used to hydraulically fracture wells in Canadian County comprised 0.1% of the permitted surface and groundwater withdrawals for all uses in Canadian County in 2012. This was a typical value for the share of water used by hydraulic fracturing in other Oklahoma counties where hydraulic fracturing was reported to FracFocus in 2012.[32]

In Nevada, Noble Energy, Inc. is actively pursuing an oil and gas play in the Elko formation near the town of Wells in Elko County.[33] This location averages precipitation just under 10 in. a year, so it is unsurprising that the citizens of Elko County are concerned about Noble Energy's water use. But an examination of actual water use in Elko County demonstrated that Noble's actual water use is projected to be much less than 1 million gallons per well, a miniscule amount. Even if Nobel used the apocryphal 8 million gallons/well, their proposed 20-well program would represent only 0.05% of the total water consumption of Elko County, which uses 933,041 acre-ft (304 billion gallons) of water each year.[34]

Given these examples, it becomes obvious that large numbers presented in a vacuum can become relatively small numbers when presented in context. The amount of water use may sound frightening when measured in gallons, but becomes less so when measured in acre feet. Opponents of fracturing often rely on misused or inaccurate numbers. In matters relating to hydraulic fracturing and water use, placing numbers in perspective can become persuasive.

3.6 ROLE OF THE ENDANGERED SPECIES ACT IN THE OIL AND GAS INDUSTRY: A FOCUS ON THE DUNES SAGEBRUSH LIZARD AND THE LESSER PRAIRIE CHICKEN

The growth of the oil and gas industry has led to an expansion of operations into habitats of potentially endangered or threatened species, triggering the Endangered Species Act. As a result, collaborative efforts began among federal, state, and private agencies to create conservation agreements that allow the oil and gas industry to continue expansion, while increasing the protection of endangered species. Among the latter are the dunes sagebrush lizard and the lesser prairie chicken, two species found in the middle of the overlap between industry and the Endangered Species Act. Each case provides a current example of how industry, state, and the federal government are working to keep oil and gas development moving forward within the bounds of the Endangered Species Act.

3.6.1 The Endangered Species Act

The Endangered Species Act aims to protect species in danger of extinction, along with their ecosystems.[35] The ESA is administered by the Interior Department's U.S.

Fish and Wildlife Service (FWS) and the Commerce Department's National Marine Fisheries Service (NMFS).[35] The FWS oversees terrestrial and freshwater organisms, while the NMFS focuses on marine wildlife.[35] Species are listed pursuant to Section 4 of the ESA according to five factors[35]: (1) damage to, or destruction of, a species' habitat; (2) overutilization of the species for commercial, recreational, scientific, or educational purposes; (3) diseases or predation; (4) inadequacy of existing protection; and (5) other natural or manmade factors that affect the continued existence of the species.[35] The FWS uses the best scientific information available to evaluate whether one or more of the five factors places a species in peril.[35] The FWS promotes the use of voluntary conservation agreements between landowners and other parties to reduce risks to species before they must be listed as threatened or endangered.[35] Such agreements have played a role in the conservation efforts of the dunes sagebrush lizard and lesser prairie chicken.

3.6.2 The Dunes Sagebrush Lizard

The dunes sagebrush lizard is a small, light-brown phrynosomatid lizard found in southeastern New Mexico and West Texas.[36] Oil and gas development contributes to the *greatest threat* to the dunes sagebrush lizard—the loss of its specialized habitat of shinnery oak (*Quercus havardii*).[37] In addition, each operational activity carries various risks to the dunes sagebrush lizard.[38]

The dunes sagebrush lizard is exposed to unnatural environmental stressors created by oil and gas development.[37] The use of caliche pads and roads makes it difficult for shinnery oak to emerge.[38] Studies have also shown increased lizard mortality related to the use of these caliche pads and roads, including collisions with vehicles, inaccessibility to habitat and mates, and exposure to predators.[39] The construction of pipelines contributes to the risks facing the species by removing vegetation, including shinnery oak, destabilizing the overall dune structures, and using large equipment capable of crushing nests and lizards hiding beneath the sand.[40] The open trenches formed by the pipelines can create pitfalls that—without escape ramps—trap small animals including the dunes sagebrush lizard.[39] Pipeline leaks can expose the lizard to toxins, and routine maintenance of the lines can increase its chances of being crushed by vehicles.[39] Power lines are another added stressor because they provide predators more perches, increasing predation.[40]

Seismic exploration is used on a limited basis, but trucks used in the process can crush lizards, and the actual process can directly harm dunes sagebrush lizards lying dormant beneath the surface in the winter months.[40] The species is also exposed to toxins associated with oil and gas wells, such as oil spills, hydrogen-sulfide gas emissions, and other chemicals.[41] The effects of long-term exposure to such toxins are unknown, but exposure is expected to result in "physiological dysfunction, impaired foraging abilities, increased mortality, and population declines."[42] Development also fragments the habitat into smaller parcels, where there is more edge habitat and less interior habitat.[37] The greater the fragmentation of a habitat, the greater chance a species may have of becoming extinct.[37] The risks associated with fragmentation are less severe, considering more than half of the dunes sagebrush lizard's habitat is not fragmented.[39]

In 2012, the proposal to list the dunes sagebrush lizard as an endangered species was withdrawn by the FWS.[43] The withdrawal was partly due to efforts under the Texas Conservation Plan.[43] The plan was developed with the help of the Texas Comptroller's Office, in conjunction with federal, state, and private partners.[44] The plan focuses on avoiding further degradation of the dunes sagebrush lizard's habitat, reclamation of habitat to reduce fragmentation, and removal of mesquite that encroaches into shinnery-oak dunes.[44] The plan allows for some loss of habitat in situations where it cannot be avoided, but where proper mitigation measures are used to minimize habitat loss.[44] Under its conservation provisions, the Texas Conservation Plan limits the amount of habitat lost to 1% in the first 3 years.[44] The other provisions are based on the Conservation Recovery Award System and habitat loss mitigation.[44] Under these provisions, 90% of the designated habitat must be avoided and up to 10% may be taken, but only if the same amount is created elsewhere by restoration measures or if the habitat is protected from mesquite encroachment.[44]

The FWS determined the Texas Conservation Plan will be effective at eliminating or reducing threats to habitat.[45] The decision is based on the plan's efforts to first avoid destroying the habitat and then, if development is necessary, limit the amount of habitat loss and strive to restore such loss.[45] A majority of the mapped Texas habitats of the dunes sagebrush lizard has already been enrolled in the plan.[45]

3.6.3 The Lesser Prairie Chicken

The lesser prairie chicken is a prairie grouse known for its colorful mating display.[46] The species can be found in parts of Texas, New Mexico, Oklahoma, Kansas, and Colorado.[46] In Texas, the lesser prairie chicken is found across 13 counties in the Permian Basin/Western Panhandle region and the Northeastern Panhandle region.[47] The species has faced many troubles over the past 15 years leading up to its listing as "threatened."[46] It is estimated that its habitat has been decreased by 84%, and a record low of 17,616 birds were recorded in 2013 (almost a 50% reduction from 2012 estimates).[46] The statewide 2012 population in Texas alone was estimated to be only 1800.[47]

Lesser prairie chickens have a relatively short lifespan of about 5 years and a high annual mortality rate.[48] Annual adult survival in Texas varies by habitat.[48] Threats to the species arise in the form of habitat fragmentation and severe drought in the Great Plains.[46] Lesser prairie chickens are sensitive to the size of habitat fragments regardless of habitat's quality, making fragmentation a great threat to the species.[49] Oil and gas development has led to fragmentation of the species' habitat through the building of access roads, drill pads, pipelines, and waste pits.[50] The birds' limited flight ability also makes them prone to collisions with fences, power lines, and vertical man-made structures.[49] Other oil and gas activities, such as surface exploration, exploratory drilling, field development, facilities construction, and operation and maintenance, also impact the species' habitat.[51]

The listing of the lesser prairie chicken as threatened also came with an unprecedented special rule that recognizes the ongoing conservation efforts among the five range states. This rule allows the oil and gas industry to escape regulations not included in the Western Association of Fish and Wildlife Agencies' (WAFWA) range-wide conservation plan.[46] The WAFWA plan encompasses all five states where the lesser prairie chicken is found and is a voluntary conservation measure that focuses on mitigation of unavoidable impacts on the lesser prairie chicken, along with financial incentives for landowners who manage their property to benefit the bird.[52] Another conservation effort includes the Rangewide Oil and Gas Industry Candidate Conservation Agreement with Assurances (CCAA) for the Lesser Prairie Chicken.[53] This plan incorporates measures to address impacts to the species from oil and gas activities on nonfederal lands throughout the species' range and provides coverage for 30 years.[53]

Notwithstanding these efforts, the FWS determined that because existing conservation efforts focus only on one or two of the threats facing the lesser prairie chicken, these efforts do not adequately target the significant threats the lesser prairie chicken faces from oil and gas development and wind and energy development.[54] As a result, the lesser prairie chicken is currently listed as "threatened" under the Endangered Species Act, with litigation likely to follow.[55] The Endangered Species Act has the potential to influence progress in the oil and gas industry, where operations expand into endangered or threatened species' habitats. Collaborative efforts to create conservation plans that avoid the necessity of regulations are not only endorsed by the FWS but seem to play a significant role in cases like the dunes sagebrush lizard and lesser prairie chicken.

3.7 SUMMARY AND CONCLUDING REMARKS

In light of recent exponential expansion in the oil and gas industry, the sections of this chapter examined issues tangential to and resulting from this boom. The manner in which an oil and gas company engages a community is as important as its considerations of wildlife and knowledge of property law. Increasing awareness of environmental impacts from the oil and gas industry could change the way the industry operates for years to come. The challenge becomes making sure that the benefits outweigh the costs. Oil and gas companies must maintain a balance between respect for life and property and making a profit. Additionally, they need to convey factual evidence to counteract hyperbolic statistics and engage in open dialogue with the community.

REFERENCES

1. *Railroad Commission v. Manziel*, 361 S.W.2d 560, 568–569 (Tex. 1962).
2. Chris Kulander and Robert Shaw, Toward uniformity in subsurface trespass jurisprudence-geophysical techniques, hydraulic fracturing, gas storage, and injection well disposal (2014), 12. https://www.utexas.edu/law/centers/energy/wp/wp-content/uploads/centers/energy/Hydraulic-Fracturing-and-the-Rule-of-Capture.pdf.
3. *Railroad Commission v. Manziel*, 361 S.W.2d 560, 566 (Tex. 1962).
4. *Railroad Commission v. Manziel*, 361 S.W.2d 563 (Tex. 1962).
5. Chris Kulander and Robert Shaw, Toward uniformity in subsurface trespass jurisprudence-geophysical techniques, hydraulic fracturing, gas storage, and injection well disposal (2014), 13. https://www.utexas.edu/law/centers/energy/wp/wp-content/uploads/centers/energy/Hydraulic-Fracturing-and-the-Rule-of-Capture.pdf.

6. *Railroad Commission v. Manziel*, 361 S.W.2d 567 (Tex. 1962).
7. *Coastal Oil & Gas Corp. v. Garza Energy Trust*, 268 S.W.3d 1, 12–13 (Tex. 2008).
8. Barclay Nicholson and Brian Albrecht, Subsurface trespass emerging issues (2012), 1. http://www.nortonrosefulbright.com/files/us/images/publications/20120201AOGRSubsurfaceTrespassNicholson.pdf.
9. Brannon Robertson, An important case on subsurface trespass, *King & Spalding Energy Newsletter* (New York, 2014), 2. http://www.kslaw.com/imageserver/KSPublic/library/publication/2014articles/5-14-14_law360.pdf.
10. *Coastal Oil & Gas Corp. v. Garza Energy Trust*, 268 S.W.3d 1, 11 (Tex. 2008).
11. *Coastal Oil & Gas Corp. v. Garza Energy Trust*, 268 S.W.3d 14 (Tex. 2008).
12. *Coastal Oil & Gas Corp. v. Garza Energy Trust*, 268 S.W.3d 14–15 (Tex. 2008).
13. *Coastal Oil & Gas Corp. v. Garza Energy Trust*, 268 S.W.3d 16 (Tex. 2008).
14. *Coastal Oil & Gas Corp. v. Garza Energy Trust*, 268 S.W.3d 1, 16–17 (Tex. 2008).
15. Brannon Robertson, Important case on subsurface trespass pending before the Texas Supreme Court, *King & Spalding Energy Newsletter*, May 2014, 1; Barclay Nicholson and Brian Albrecht, Subsurface trespass emerging issues, *The American Oil & Gas Reporter*, February 2012, 2.
16. John F. Sullivan, Anthony F. Newton, and Cleve J. Glenn, The eyes of Texas are upon a subsurface trespass case, *K&L Gates Legal Insight*, last modified January 13, 2014. http://m.klgates.com/files/Publication/bb722f3a-e3db-4f70-b866-b05426f8c2d4/Presentation/PublicationAttachment/c8419e71-4063-477e-8275-bf4d96388a15/Oil_Gas_Alert_13Jan14.pdf.
17. Brannon Robertson, Important case on subsurface trespass pending before the Texas Supreme Court, *King & Spalding Energy Newsletter*, last modified May 2014, 3; Barclay Nicholson and Brian Albrecht, Subsurface trespass emerging issues, *The American Oil & Gas Reporter*, February 2012, 1.
18. *FPL Farming Ltd. V. Envtl. Processing Sys., L.C.*, 351 S.W.3d 306, 309 (Tex. 2011).
19. *FPL Farming Ltd. V. Envtl. Processing Sys., L.C.*, 351 S.W.3d 310 (Tex. 2011).
20. *FPL Farming Ltd. V. Envtl. Processing Sys., L.C.*, 351 S.W.3d 314 (Tex. 2011).
21. John F. Sullivan, Anthony F. Newton, and Cleve J. Glenn, The eyes of Texas are upon a subsurface trespass case, *K&L Gates Legal Insight*, January 13, 2014, 4.
22. Owen L. Anderson, *Lord Coke, the Restatement, and Modern Subsurface Trespass Law*, 6 *Texas Journal of Oil Gas & Energy Law* 203, 206, 2010–2011.
23. Owen L. Anderson, *Lord Coke, the Restatement, and Modern Subsurface Trespass Law*, 6 *Texas Journal of Oil Gas & Energy Law* 206–207, 2010–2011.
24. Facts, *ny against fracking*, accessed June 26, 2014. http://nyagainstfracking.org/facts/
25. An NGO based in Nevada.
26. Martin Griffith, Rural county protests fracking in Nevada, *usatoday*, last modified June1,2014. http://www.usatoday.com/story/money/business/2014/06/01/rural-county-files-protest-over-fracking-in-nevada/9845941/.
27. Hydraulic fracturing 101, *Earthworksaction*, accessed June 26, 2014. http://www.earthworksaction.org/issues/detail/hydraulic_fracturing_101.
28. Hydraulic-fracturing-usage, *fracfocus*, accessed June 26, 2014. http://fracfocus.org/water-protection/hydraulic-fracturing-usage.
29. Hydraulic fracturing. *Dangersoffracking*, accessed June 26, 2014. http://www.dangersoffracking.com/.
30. The public view of FracFocus does not allow bulk downloads of data. However, there are organizations that have extracted data from the FracFocus site and have posted it for download. The discussions of bulk FracFocus data in this paper are based on FracFocus data available on the website Skytruth, http://frack.skytruth.org/fracking-chemical-database. I validated these data against a sampling of records from the FracFocus database. The data available from the Skytruth website are concordant with the FracFocus data.

31. Hydraulic fracturing water usage, *Fracfocus*, accessed June 26, 2014. http://fracfocus.org/water-protection/hydraulic-fracturing-usage.
32. Data for permitted water use in Oklahoma were obtained from the Oklahoma Water Resources Board at http://www.owrb.ok.gov/maps/pmg/owrbdata_WR.html; data were taken from the .dbf file component of Permitted Groundwater Wells and Dedicated Land for permitted groundwater withdrawals and from Permitted Surface Water Diversions and Areas of Use for permitted surface-water diversions.
33. Henry Brean, Oil/gas search: Fracking comes to Nevada, *reviewjournal*, last modified January 8, 2013. http://www.reviewjournal.com/business/energy/oilgas-search-fracking-comes-nevada.
34. Elko County Nevada Water Resource Management Plan 2007, available at http://www.leg.state.nv.us/74th/Interim_Agendas_Minutes_Exhibits/Exhibits/Lands/E060608F-1.pdf, personal communication, Lee Hinman, Noble Energy, 2014. Also, see Nevada Energy Forum, Hydraulic Fracturing 101 Presentation, Hydraulic Fracturing in Nevada, available at http://www.nevadaenergyforum.com/article/hydraulic-fracturing-101-presentation, by J.B. Fisher.
35. Endangered Species Act overview, *fws.gov*, last modified July 15, 2013. http://www.fws.gov/endangered/laws-policies/.
36. Endangered and threatened wildlife plants; withdrawal of the proposed rule to list dunes sagebrush lizard; Proposed Rule, *Federal Register* 77 (June 19, 2012): 36872–36873.
37. Endangered and threatened wildlife plants; withdrawal of the proposed rule to list dunes sagebrush lizard; Proposed Rule, *Federal Register* 77 (June 19, 2012): 36887.
38. Endangered and threatened wildlife plants; withdrawal of the proposed rule to list dunes sagebrush lizard; Proposed Rule, *Federal Register* 77 (June 19, 2012): 36888.
39. Endangered and threatened wildlife plants; withdrawal of the proposed rule to list dunes sagebrush lizard; Proposed Rule, *Federal Register* 77 (June 19, 2012): 36889.
40. Endangered and threatened wildlife plants; withdrawal of the proposed rule to list dunes sagebrush lizard; Proposed Rule, *Federal Register* 77 (June 19, 2012): 36890.
41. Endangered and threatened wildlife plants; withdrawal of the proposed rule to list dunes sagebrush lizard; Proposed Rule, *Federal Register* 77 (June 19, 2012): 36896.
42. Endangered and threatened wildlife plants; withdrawal of the proposed rule to list dunes sagebrush lizard; Proposed Rule, *Federal Register* 77 (June 19, 2012): 36897.
43. Endangered and threatened wildlife plants; withdrawal of the proposed rule to list dunes sagebrush lizard; Proposed Rule, *Federal Register* 77 (June 19, 2012): 36872.
44. Endangered and threatened wildlife plants; withdrawal of the proposed rule to list dunes sagebrush lizard; Proposed Rule, *Federal Register* 77 (June 19, 2012): 36885.
45. Daniel M. Ashe, Endangered and threatened wildlife plants; withdrawal of the proposed rule to list dunes sagebrush lizard, *Federal Register*, last modified June 19, 2012, 36886.
46. Claire Cassel, U.S. fish and wildlife service lists lesser prairie-chicken as threatened species and finalizes special rule endorsing Landmark State Conservation Plan, *fws*, last modified March 27, 2014. http://www.fws.gov/news/ShowNews.cfm?ID=04F68986-AE41-6EEE-5B07E1154C2FB2E7.
47. Daniel M. Ashe, Endangered and threatened wildlife plants; withdrawal of the proposed rule to list dunes sagebrush lizard, *Federal Register*, last modified June 19, 2012, 20015.
48. Daniel M. Ashe, Endangered and threatened wildlife plants; withdrawal of the proposed rule to list dunes sagebrush lizard, *Federal Register*, last modified June 19, 2012, 20004.
49. Daniel M. Ashe, Endangered and threatened wildlife plants; withdrawal of the proposed rule to list dunes sagebrush lizard, *Federal Register*, last modified June 19, 2012, 20001.
50. Daniel M. Ashe, Endangered and threatened wildlife plants; withdrawal of the proposed rule to list dunes sagebrush lizard, *Federal Register*, last modified June 19, 2012, 20018.
51. Daniel M. Ashe, Endangered and threatened wildlife plants; withdrawal of the proposed rule to list dunes sagebrush lizard, *Federal Register*, last modified June 19, 2012, 20052.

52. Daniel M. Ashe, Endangered and threatened wildlife plants; withdrawal of the proposed rule to list dunes sagebrush lizard, *Federal Register*, last modified June 19, 2012, 19990.
53. Daniel M. Ashe, Endangered and threatened wildlife plants; withdrawal of the proposed rule to list dunes sagebrush lizard, *Federal Register*, last modified June 19, 2012, 19993.
54. Daniel M. Ashe, Endangered and threatened wildlife plants; withdrawal of the proposed rule to list dunes sagebrush lizard, *Federal Register*, last modified June 19, 2012, 20065.
55. Thomas A. Campbell et al., 'Threatened' Chicken May Ruffle Permitting Feathers, *Law 360*, last modified April 28, 2014. http://www.law360.com/articles/531783/threatened-chicken-may-ruffle-permitting-feathersn.

4 Looking into the Crystal Ball

Potential New EPA Hydraulic Fracturing Rules Impacting Unconventional Oil and Gas Production from Shale Plays

Ron Truelove

CONTENTS

4.1 Overview 83
4.2 Introduction 84
4.3 Existing Regulations 84
4.4 Future Potential Requirements 85
4.5 April 15, 2014, White Papers 85
4.6 May 13, 2014, NGO Section 112 Petition 86
4.7 May 9, 2014, EPA ANPR for Fracturing Chemical Disclosure 87
4.8 June 5, 2014, ANPR for Indian Country Minor Source Permitting 88
4.9 Summary and Concluding Remarks 88
References 90

4.1 OVERVIEW

This chapter discusses the United States Environmental Protection Agency (USEPA) interpretations of New Source Performance Standard (NSPS) subpart OOOO that regulate flowback emissions from hydraulically fractured natural gas wells. White papers recently issued by the USEPA related to additional emission sources at oil and gas production facilities not currently regulated by NSPS Subpart OOOO will also be analyzed to predict additional, future regulations for hydraulically fractured oil wells. The chapter also discusses recent public action around hazardous air pollutants (HAP) regulations and USEPA action around hydraulic fracturing chemical use disclosures.

4.2 INTRODUCTION

Hydraulic fracturing is the process by which high-pressure water, proppant (typically sand), and other additives are pumped at a high pressure down a cased and perforated well bore to further fracture the reservoir rock to facilitate oil and/or gas flow to the well. Hydraulic fracturing is used when producing from relatively tight reservoir rock that would otherwise not yield enough oil and/or gas for the well to be commercial. Hydraulic fracturing has been utilized in conventional and unconventional reservoirs for many years and is not a new process.

The combination of hydraulic fracturing and horizontal drilling opened up the ability to commercially produce gas from the Barnett Shale formation in North Texas more than 10 years ago. The technology has since been utilized to commercially produce oil, condensate, and natural gas liquids from other shale reservoirs, such as the Eagle Ford Shale in South Texas.

After a well is hydraulically fractured, the reservoir rock near the well bore must be cleaned of water and excess sand that is not lodged within the fractured rock. This process is called *flowing back* a well. The pressure is released, and the water and sand flow back to the surface. The sand is removed from the water, and the water is stored in frac tanks prior to either reuse or disposal in permitted disposal wells. During this flowback period, reservoir fluids (i.e., liquid hydrocarbons and gas) become entrained in the fracturing water that was pumped down the well. When the combined reservoir fluids and flowback water reach the surface, air emissions consisting of methane, ethane, and other hydrocarbons considered volatile organic compounds (VOC) can be emitted into the atmosphere.

Due to growth of unconventional shale oil, condensate, and natural gas liquids development, U.S. drilling and hydraulic fracturing activities are at levels that have not been seen since the 1970s, making the United States one of the world's largest producers of oil and other hydrocarbon liquids. With the increase in drilling and other oilfield activity, regulatory and public concerns are also at an all-time high.

4.3 EXISTING REGULATIONS

There are essentially three types of existing regulations that affect hydraulic fracturing:

- State oil and gas commission (e.g., Texas Railroad Commission) permits to drill, specific rules governing well construction, and specific rules governing underground disposal of water[1]
- State oil and gas commission hydraulic fracturing chemical use disclosure requirements[2]
- USEPA rules governing air emissions during the flowback process from new, hydraulically fractured natural gas wells that were drilled after October 15, 2012.

Oil and gas companies have acted in accordance with state oil and gas commission and the Bureau of Land Management (BLM) rules for many years. Permits to drill, well construction requirements, and underground injection permits for produced water disposal are well-known requirements in all oil- and gas-producing states.

The requirement for disclosure of fracturing chemicals is a relatively new regulation. The industry established a voluntary chemical disclosure process through a website called the FracFocus Chemical Disclosure Registry.[3] As state oil and gas commissions have implemented chemical disclosure regulatory requirements, many, if not all, have required the use of FracFocus so that all data nationally are in a single, publicly accessible, web-based location. BLM has proposed chemical disclosure requirements and has also proposed the use of FracFocus as part of those requirements.

Clean Air Act NSPS Subpart OOOO was proposed on August 23, 2011, and was finalized and published in the Federal Register on August 16, 2012. The rule then was revised, with the revision published in the Federal Register on September 23, 2013. This rule regulates very specific new, modified, and reconstructed sources of VOC emissions from oil and gas operations. Regarding hydraulic fracturing, the rule regulates the VOC emissions generated during flowback from hydraulically fractured natural gas wells. Beginning October 15, 2012, oil and gas companies had to begin combusting emissions during flowback. Beginning January 1, 2015, the flowback emissions from some wells will have to be captured and put into the gas-gathering line for sale or otherwise recycled back into the production process and not vented or combusted. This is referred to in the rule as reduced emission completion (REC) or *green completions*. Wildcat, reservoir delineation, and low-pressure well flowback can still be controlled by combustion, but after January 1, 2015, all other wells will have to undergo REC. Again, this requirement to control emissions during flowback only applies to wells that are hydraulically fractured and are drilled principally for the production of natural gas (i.e., natural gas wells).

4.4 FUTURE POTENTIAL REQUIREMENTS

Looking into the oil and gas regulatory *crystal ball*, additional regulation may be on the horizon that could affect hydraulic fracturing operations, especially hydraulic fracturing of oil, condensate, and natural gas liquids wells. The remainder of this discussion provides some insight into those possible new requirements based on the following:

- April 15, 2014, EPA Oil and Gas White Papers[4]
- May 13, 2014, petition from nongovernment organizations (NGOs) to list oil and gas facilities in populated areas as a regulated area source under Section 112 of the Clean Air Act[5]
- May 9, 2014, EPA Advanced Notice of Public Rulemaking (ANPR) to propose requiring disclosure of fracturing chemicals federally.[6]

4.5 APRIL 15, 2014, WHITE PAPERS

Much of the anticipated new air emissions regulations are being driven by President Obama's March 28, 2014, Climate Action Plan. The cornerstone of this plan is the President's directive to the EPA to reduce methane emissions from oil and gas operations. Part of the strategy for reductions is to first identify sources and trends of methane emissions, then develop regulations to reduce those emissions. By 2020, the strategy calls for methane to be reduced from 2005 levels by 17%.

On April 15, 2014, the EPA issued five white papers identifying certain sources of methane emissions at oil and gas facilities and the EPA's understanding of the emissions from those sources. The EPA requested peer review and public comment of the white papers by June 16, 2014. These white papers specifically address the following:

- Oil well completions (i.e., flowback emissions) and associated gas venting and flaring
- Compressors
- Fugitive equipment leaks
- Liquids unloading
- Pneumatics (i.e., pneumatic controllers and pneumatic pumps)

Based on the information presented in the white paper on oil well completions and associated gas venting and flaring, one reasonable conclusion could be that the EPA will add regulations that flowback emissions from hydraulically fractured oil wells must be captured and either combusted or otherwise recovered through REC or green completions—after some future date. Basically, NSPS Subpart OOOO could be expanded to include emissions control during the flowback period for all wells that are hydraulically fractured, not just natural gas wells.

The EPA already has made an informal interpretation that NSPS Subpart OOOO is applicable to any well that is drilled and through which shale gas is brought to the surface, regardless of whether or not the well was drilled principally for the production of natural gas. This informal interpretation is based on statements in the preamble of the final rule that mentions *shale gas* as one type of natural gas. This informal interpretation opens up the potential for completion and flow back emission controls to be required by the EPA (under NSPS Subpart OOOO) for wells drilled principally for the production of shale oil and other shale liquids as well as wells where the shale gas is just a by-product of the liquids production.

Additionally, in some of the areas of recent shale oil development, gas-gathering infrastructure has not yet caught up with oil and condensate well drilling and completion activities, in which case any associated gas must either be vented or combusted. From the associated white paper on the subject, NSPS Subpart OOOO could be expanded to require recovery of the resource instead of venting or combusting the associated gas. Such provisions are currently in NSPS Subpart OOOO for hydraulically fractured natural gas wells but not oil or condensate wells.

4.6 MAY 13, 2014, NGO SECTION 112 PETITION

On May 13, 2014, more than 60 environmental and community NGOs petitioned the EPA to set federal HAP limits on oil and gas facilities. The petition listed benzene, formaldehyde, and naphthalene as HAPs of concern. The petition also stated that the petitioners believe that the EPA does not effectively regulate these emissions due to the growth of hydraulic fracturing and of the number of oil and gas facilities and associated HAP emissions. The petitioners claimed that technology exists to reduce these HAP emissions.

Specifically, the NGOs petitioned the EPA to create an area source category consisting of oil and gas production wells (and associated equipment) located in metropolitan statistical areas and consolidated metropolitan statistical areas with a population in excess of 1 million. After creating a source category, the petition calls for the EPA to set Maximum Achievable Control Technology (MACT) standards in the National Emissions Standards for Hazardous Air Pollutants (NESHAP) located at 40 Code of Federal Regulations (CFR) Part 63 for oil and gas production wells (and associated equipment).

To clarify, the federal NESHAP program does regulate specific oil and gas HAP emissions. The industry is not unregulated for HAP emissions; for instance, NESHAP Subpart ZZZZ[7] regulates formaldehyde emissions from engines, and NESHAP Subparts HH[8] and HHH[9] regulate benzene, toluene, ethyl benzene, and xylene (BTEX) emissions from glycol dehydration units, storage tanks with the potential for flash emissions, and equipment fugitive emission leaks for equipment in volatile organic HAP service.

If the EPA acts on the May 13, 2014 petition, other HAP emission sources could become regulated under the NESHAP program. Many of the white paper subjects discussed earlier in the context of methane and VOC emissions could also be sources of HAP emissions.[10] If these emissions sources are addressed in the NSPS (e.g., NSPS Subpart OOOO), the requirements would only impact new, modified, and reconstructed sources. If these emission sources are addressed in the NESHAP as called for in the petition, existing emission sources could also become regulated under federal requirements.

4.7 MAY 9, 2014, EPA ANPR FOR FRACTURING CHEMICAL DISCLOSURE[11]

On May 9, 2014, the EPA announced as a part of an ANPR that it will seek public comment over a 90-day period on what hydraulic fracturing chemical information could be reported and disclosed. The EPA seeks comments on the possible approaches that could be used for obtaining fracturing chemical information, including nonregulatory approaches. The EPA also solicited input on possible incentive and recognition programs that could help drive the development and use of *safer* hydraulic fracturing chemicals.

As mentioned earlier, the oil and gas exploration and production industry created the website and associated database called FracFocus. Many key industry companies voluntarily use the website to report fracturing chemical use. The website is also utilized by state oil and gas commissions that require chemical disclosure as a part of their regulatory program. The data reported into FracFocus are publicly accessible. The EPA mentions FracFocus in the ANPR as an option. Many oil and gas companies support FracFocus as an appropriate option for any mandatory federal reporting due to its existing extensive use by state oil and gas commission regulations and its voluntary use by the industry. Use of another system would result in a duplication of work by the industry, especially in those states with mandatory fracturing chemical reporting requirements.

4.8 JUNE 5, 2014, ANPR FOR INDIAN COUNTRY MINOR SOURCE PERMITTING

On June 5, 2014, the EPA announced that it would create the Indian Country Minor New Source review program in the oil and natural gas production sector. Deadlines are fast approaching for existing minor sources and new minor sources to submit air quality permit applications for operations in Indian Country pursuant to the Indian Country minor source permitting rule published in the Federal Register on July 1, 2011. The EPA asked for comment on three possible approaches to minor source permitting for oil and gas sources:

- General permit (GP)
- Federal implementation plan
- Permit by rule (PBR)

Determining the air permitting mechanism and establishing permit application forms is a necessary step for the EPA to implement the Indian Country minor source permitting rule. However, this process could also be used by the EPA to pursue additional emission controls for the industry.

In the ANPR, the EPA also stated that it believes that managing emissions from existing oil and natural gas sources in Indian Country could lead to instability with requirements from surrounding states. Additionally, the EPA believes it is vital to address existing sources due to the substantial activity connected with the oil and gas production sector in Indian Country. These statements set the stage for consistent and possibly stringent emission controls in any type of streamlined permitting process, such as a GP or PBR process. The companies operating in Indian Country may be forced to choose from either a streamlined permitting process requiring NSPS- or NESHAP-level (or beyond) emission controls or a case-by-case permit, which could take several months for approval. In a business environment where oil and gas companies must be very responsive to opportunities, the streamlined permits could result in tighter emission controls for new sources of emissions.

4.9 SUMMARY AND CONCLUDING REMARKS

Oil and gas companies currently must now comply with federal regulations that they were not concerned with 10 years ago. The regulatory *crystal ball* seems to indicate that additional regulation for methane, VOC, and HAP may be forthcoming in the very near future. Additional equipment will likely become regulated through new NSPS and NESHAP as well as new permitting regulations of emissions in Indian Country. Further regulation could also be on the horizon for hydraulic fracturing chemical use disclosure. In addition to these federal regulatory activities, several state environmental agencies have recently implemented more onerous oil and gas emission limits through regulation changes and permits. Colorado recently changed its Regulation 7 to regulate methane emissions in addition to VOC by expanding NSPS Subpart OOOO storage tank requirements to existing sources, and leak detection and repair (LDAR) requirements to gathering compressor stations and exploration and production facilities. This is a first for state agencies. Pennsylvania, Ohio, and Wyoming have begun to implement exploration and production site LDAR requirements through streamlined permitting actions. To qualify for the streamlined permit that is quicker to get administratively, oil and gas companies must agree to more rigorous emission controls. The more oil and gas companies agree to these requirements to get their air quality permits to construct and operate equipment, the more these controls become required through the case-by-case permit process. Additional state agencies in oil- and gas-producing states are expected to follow suit with more rigorous requirements.

In the current product pricing environment, shale oil drilling, hydraulic fracturing, and the construction of oil and gas production facilities continue to grow in the United States. Oil and gas companies can invest resources trying to battle the additional regulations through the development of sound science on which to base regulations, through an active commenting role, and/or through the courts should ineffective requirements be required. Oil and gas companies can also invest resources in planning for the very likely future of increased emission controls, recordkeeping, and reporting. Even if only a portion of what has been discussed here comes to pass, the increased burden on oil and gas companies will be significant.

As such, oil and gas companies should consider the following to best position themselves for the future:

- Implementation of an environmental (and possibly safety) management information system (EMIS) where requirements can be housed, records kept, reports generated, and responsibilities identified and managed automatically through e-mail reminders and sign-offs
- Consistent equipment design and best practices that address current regulatory requirements and enable relatively easy retrofits in case of anticipated, future requirements
- Internal and/or external compliance assessments to ensure compliance and to establish compliance plans for anticipated, future requirements
- Budgeting in 2015 and 2016 for potential equipment changes due to anticipated new regulatory requirements
- Involvement in the federal and state regulatory development process, including sharing data with the EPA and the states so that any new regulation can be based on sound science

The *crystal ball* says more oil and gas environmental regulation is coming. The *crystal ball* also says that the most prepared oil and gas companies will be able to navigate new requirements in the most cost-effective ways. The *crystal ball* says that the future is here.

REFERENCES

1. BLM. Oil and gas; hydraulic fracturing on federal and Indian lands, 2012. http://www.blm.gov/pgdata/etc/medialib/blm/wo/Communications_Directorate/public_affairs/hydraulicfracturing.Par.91723.File.tmp/HydFrac_SupProposal.pdf, accessed February 4, 2015.
2. BLM. Comments on BLM's proposed rule regarding well stimulation on public and Indian lands, 2012. http://www.energyxxi.org/comments-blms-proposed-rule-regarding-well-stimulation-public-and-indian-lands, accessed February 4, 2015.
3. FracFocus. Fracfocus chemical disclosure registry, 2014. http://fracfocus.org/, accessed February 4, 2015.
4. EPA. White Papers. EPA, 2014. http://www.epa.gov/airquality/oilandgas/whitepapers.html.
5. EPA. Petition to the United States Environmental Protection Agency. Earthjustice, 2014. http://earthjustice.org/sites/default/files/files/OilGasToxicWellsPetition51314.pdf.
6. EPA. Hydraulic fracturing chemicals and mixtures; advance notice of proposed rule-making [Rin 2070-Aj93; Frl-9909-13], 2014. http://www.epa.gov/oppt/chemtest/pubs/prepub_hf_anpr_14t-0069_2014-05-09.pdf.
7. EPA. 40 CFR Part 60, Subpart ZZZZ. EPA, http://www.ecfr.gov/cgi-bin/text-idx?c=ecfr;rgn=div6;view=text;node=40%3A14.0.1.1.1.1;idno=40;sid=e94dcfde4a04b27290c445a56e635e58;cc=ecfr. Updated 2014.
8. EPA. 40 CFR Part 60, Subpart HH. EPA. http://www.ecfr.gov/cgi-bin/text-idx?node=sp40.11.63.hh. Updated 2014.
9. EPA. 40 CFR Part 60, Subpart HHH. EPA. http://www.ecfr.gov/cgi-bin/text-idx?node=sp40.13.63.hhhh. Updated 2014.
10. The White House. Climate action plan—Strategy to reduce methane emissions, 2014. http://democrats.energycommerce.house.gov/sites/default/files/documents/White-House-Methane-Climate-Strategy-2014-3-28.pdf.
11. EPA. Hydraulic fracturing chemicals and mixtures, federal register. EPA, 2014 http://www.gpo.gov/fdsys/pkg/FR-2014-05-19/pdf/2014-11501.pdf.

5 Commentary on Health and Environmental Risks from Hydraulic Fracturing

David Klein, Jennifer Knaack, and Audra Morse

CONTENTS

5.1 Overview 91
5.2 Introduction 91
5.3 Industrial and Public Risk Posed by Hydraulic Fracturing 93
5.4 Impacts of Chemical Exposure Occurring during Hydraulic Fracturing 94
5.5 Summary and Concluding Remarks 97
References 97

5.1 OVERVIEW

This chapter highlights human health and environmental risks associated with hydraulic fracturing operations.

5.2 INTRODUCTION

Petrochemical production has been happening in the United States since the 1900s.[1] In the United States, total oil production had been declining for decades as shown in Figure 5.1.

With the rise in oil prices, unconventional oil is now being recovered in ever-increasing quantities. The International Energy Agency has defined unconventional oil as hydrocarbons that must be obtained using nontraditional technologies, which include hydraulic fracturing. Figure 5.2 presents the thousands of barrels produced per day from January 2000 through May 2011, highlighting the increase in tight oil production in the United States in recent years.

Hydraulic fracturing was first invented in 1947.[2] Although Halliburton had been employing hydraulic fracturing for many years, it was not in significant use until 2003 when hydraulic fracturing was coupled with horizontal drilling. Tight oil, also known as shale oil or light tight oil, is responsible for the increase of oil production in the United States.

The Energy Policy Act passed by the U.S. Congress in 2005 is a document that runs 551 pages.[3] In the entire document, there is only a single mention of hydraulic fracturing in Section 322. It reads:

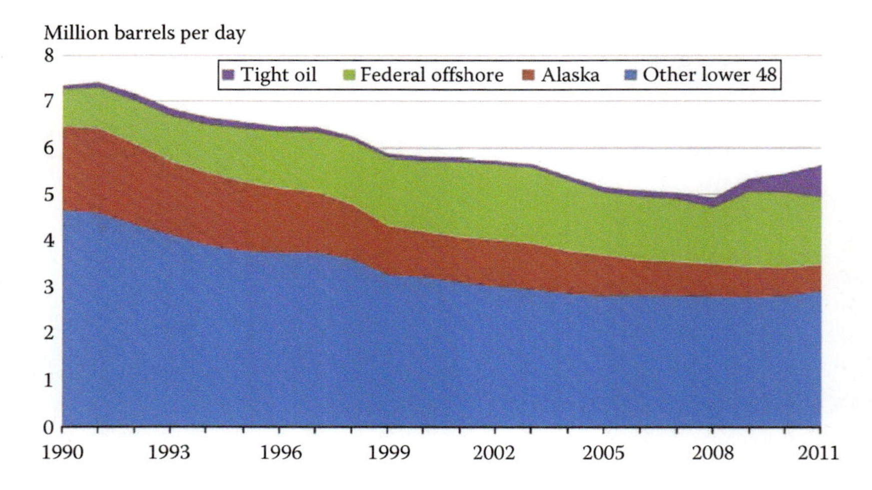

FIGURE 5.1 Total oil production in the United States from 1990 to 2011. (From the U.S. Energy Information Administration, Washington, DC; based on HPDI, LLC 2011 is through November.)

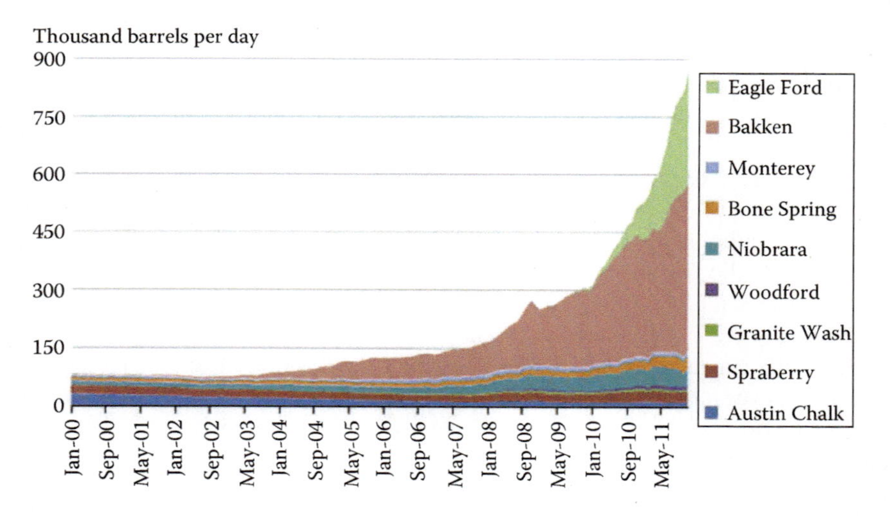

FIGURE 5.2 Increase in tight oil production in the United States. (From the U.S. Energy Information Administration, Washington, DC; based on HPDI, LLC; 2011 is through November.)

SEC. 322. HYDRAULIC FRACTURING.

Paragraph (1) of section 1421(d) of the Safe Drinking Water Act (42 U.S.C. 300h(d)) is amended to read as follows:

"(1) UNDERGROUND INJECTION.— The term 'underground injection'—
"(A) means the subsurface emplacement of fluids by well injection; and
"(B) excludes—
"(i) the underground injection of natural gas for purposes of storage; and
"(ii) the underground injection of fluids or propping agents (other than diesel fuels) pursuant to hydraulic fracturing operations related to oil, gas, or geothermal production activities."

The Energy Policy Act provided an environmental basis to allow hydraulic fracturing to proceed because the Safe Drinking Water Act would not apply to the hydraulic fracturing process. Additionally, the Energy Policy Act effectively prevented regulatory oversight or litigation with regard to any effects of drinking water that might be attributed to the hydraulic fracturing process.

There are two general questions to address in this chapter. First, are there increased health risks from hydraulic fracturing? Second, with the increase in density of hydraulic fracturing are risks further increased? In any industrial process, there are hazards and risks. As a society we weight the cost/benefit ratio constantly. The answer is that of course there are risks associated with hydraulic fracturing but are they acceptable? This is where there are two widely divided groups and we hope to address this issue in a fair, unbiased manner. No modern industrial process can proceed risk-free; it is the relative risk that we will address in this chapter.

5.3 INDUSTRIAL AND PUBLIC RISK POSED BY HYDRAULIC FRACTURING

In the case of hydraulic fracturing, there are risks associated with the construction of the well and drilling that occurs. These risks are borne by the workers in the oil and gas industry. In one example in 2012, the Occupational Safety and Health Administration (OSHA) produced a notice warning of inhalation injury from silica that arises from exposure to a large amount of silica sand used in the hydraulic fracturing process.[4] Injuries and deaths directly tied to hydraulic fracturing are not easy to determine because injury and death data is grouped with all mining processes. As shown in Figure 5.3, there is information available on risks associated with the oil and gas extraction process.

There are also the risks faced by the public due to processes of hydraulic fracturing. Both air and water can be polluted by drilling and hydraulic fracturing processes, as covered in detail in Chapter 7. Only under the tightest regulation and control do industrial processes proceed without error, and releases are often a component of the error process. As such, industrial releases would provide higher levels of exposure to chemicals compared to other environmental exposures. However, much of the focus on health risk has been placed on potential leaks from the well bore.[5] It has been reported the leaking of old well heads may be a potential source

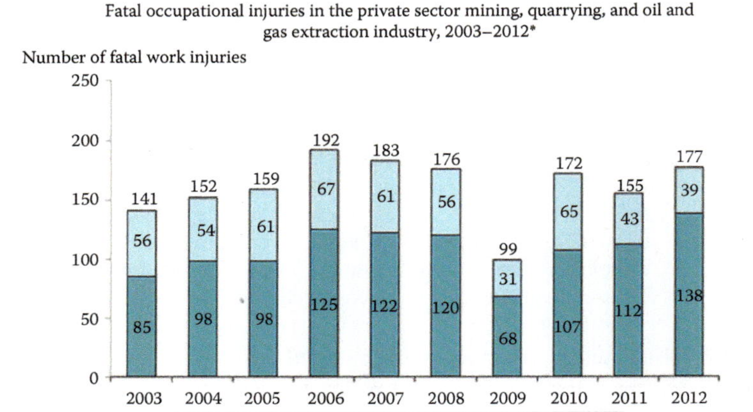

FIGURE 5.3 Fatal occupational injuries in oil and gas extraction industries versus all other mining, 2003–2012. Fatal work injuries in the private mining industry increased by 14% in 2012. Fatalities in the oil and gas extraction industries accounted for 78% of the fatal work injuries in the mining sector in 2012. *Note:* Oil and gas extraction industries include oil and gas extraction (NAICS 21111), drilling oil and gas wells (NAICS 213111), and support activities for oil and gas operations (NAICS 213112). (From the U.S. Bureau of Labor Statistics, U.S. Department of Labor, Washington, DC.)

of concern, as chemicals used the oil and gas extraction process remaining in the wells may release chemicals into the air, soil, and water surrounding the well as the bore casing continues to age with time.[6] In any case, a leaking well head can allow petroleum products to escape into the biosphere.[7]

Another area that has been addressed but not well publicized is the risk involved with the rapidly expanding hydraulic fracturing industry and surface transportation. The Associated Press has conducted a study of traffic fatalities in Texas counties with intensive hydraulic fracturing in comparison to other Texas counties.[8] It has been reported in Texas that traffic deaths have declined 20% while in counties with hydraulic fracturing activity, traffic fatalities rose 18%. Transportation accidents affect the general public as well as the workers of the oil and gas industry.

5.4 IMPACTS OF CHEMICAL EXPOSURE OCCURRING DURING HYDRAULIC FRACTURING

The use of toxic chemicals in hydraulic fracturing poses a significant risk to human health. Exposure to these chemicals can occur from a variety of sources including contaminated air and water around natural gas wells. During the hydraulic fracturing process, sand, water, and large quantities of chemicals, many of them either known, or suspected to be toxic, are pumped deep into the ground using high pressure to fracture shale and release stores of natural gas. During the flowback process, waste products from hydraulic fracturing rise to the surface of the well and contain the toxic compounds along with sand and brine water. At least four well worker deaths have been attributed to chemical exposure, possibly to hydrocarbons, during flowback operations.[9] Both underground and surface drinking water sources can become contaminated during hydraulic fracturing and flowback. The air around natural gas wells can become contaminated with petroleum hydrocarbons and methane released from the shale and pose a significant inhalation hazard. Additionally, the wastewater produced by hydraulic fracturing can cause further contamination of agricultural land and food sources if not handled or disposed properly.

Published studies have demonstrated that a wide variety of toxic chemicals are present in the air and water surrounding natural gas wells where hydraulic fracturing has occurred. In one study, air samples taken from rural home sites and near well pads were analyzed for the presence of 56 different hydrocarbons.[10] Among the compounds tested for, trimethylbenzenes, aliphatic hydrocarbons, and xylenes were frequently found in air samples and provided the greatest contribution to the hazard index for residents near natural gas wells. Benzene, toluene, and other aromatic hydrocarbons were also found in most air samples collected from both home sites and near well pads. In another study, silica levels in air around wells were shown to often exceed the maximum use concentration for half-mask, and sometimes full-face, air-purifying respirators.[10] A study of surface and ground water near drilling sites has identified contamination with endocrine-disrupting chemicals.[11] Spills at wells can also cause groundwater contamination with benzene, toluene, ethylbenzene, and xylene.[12] Methane gas has also been shown to contaminate groundwater around drilling sites.[8]

Despite the wealth of evidence that chemicals used or produced during hydraulic fracturing can pose exposure risks to local residents through air and water contamination, no large-scale exposure studies have been conducted to date. Some anecdotal evidence for exposures to these chemicals can be found across the Internet.[13] In these cases, blood or urine analysis has indicated that exposure to a chemical commonly used in hydraulic fracturing has occurred. However, the limited sample size in each case prevents any larger conclusions from being made, and it is unclear whether the chemical exposure is a result of hydraulic fracturing processes or exposure from elsewhere. For example, benzene can be found in flowback waste and also in tobacco smoke, motor vehicle exhaust, air surrounding gas stations, and products such as glue and paint.[14] Likewise, phenol can be found in flowback waste and in foods such as fried chicken and smoked summer sausage, tobacco smoke, and automobile exhaust.[15] Without a large-scale epidemiological study, preferably with matching urine and blood samples taken before and after drilling begins, isolating hydraulic fracturing as the cause for chemical exposures will be difficult.

Standard analytical methods exist to analyze blood and urine for a large number of toxic chemicals that have been identified in air and water samples around hydraulic fracturing sites. Most methods rely on extraction of the chemicals or metabolites from urine followed by high performance liquid chromatography (HPLC) or gas chromatography and detection by mass spectrometry (MS) or ultraviolet–visible detectors.

Benzene is one of the major contaminating chemicals associated with hydraulic fracturing that can be found in contaminated ground and surface water as well as in polluted air around drilling sites. Once in the body, benzene can metabolize into ringed structures including hydroquinone and phenol. Alternatively, benzene can metabolize into two nonring structures: *trans*, *trans* muconic acid (*t,t*MA), and *S*-phenyl mercapturic acid (SPMA). Unmetabolized benzene and benzene adducted to plasma proteins can be measured in blood while unmetabolized benzene and benzene metabolites are measured in urine.[16] Unchanged benzene and benzene metabolites can be found in blood and urine specimens from petrochemical workers, but these biomarkers may not be reliable at low exposure levels.[17]

Toluene, ethylbenzene, and xylene are other major contaminating chemicals associated with hydraulic fracturing. Like benzene, these chemicals can be found in both polluted air and contaminated water sources. Upon exposure, these chemicals can remain unchanged in blood or be excreted unchanged or as metabolites in urine. A rapid, high-throughput HPLC–MS method for measuring the urinary metabolites from benzene, toluene, ethylbenzene, and xylene has been developed.[18] Detection limits for each metabolite are in the low ng/mL range and may represent subtoxic levels of exposure.

Trimethylbenzenes are aromatic hydrocarbons and are composed of three isomers: 1,2,3-trimethylbenzene, 1,2,4-trimethylbenzene, and 1,3,5-trimethylbenzene. All three forms have been found in air samples taken near wells[10] and the 1,2,4- and 1,3,5-isoforms have been measured in deep groundwater wells.[19] In the body, trimethylbenzenes are metabolized into dimethylbenzoic acids and further metabolized to dimethylhippuric acids. The dimethylhippuric acids are excreted into urine and all six isoforms can be measured after liquid–liquid extraction using HPLC with ultraviolet detection.[20]

During hydraulic fracturing, crystalline silica, in the form of sand, is pumped into shale under high pressure to induce factures and to hold the fractures open, thus allowing natural gas to escape into the well. The use of silica during the fracturing process can result in aerosolization of silica particles aboveground, which presents a serious inhalation health hazard and can result in silicosis, chronic obstructive pulmonary disease, chronic renal disease, and an increased risk of tuberculosis.[21] Unlike small chemical molecules, crystalline silica does not metabolize once it enters the body. Instead, the dust particles become embedded in the lung and are ingested by macrophages, ultimately resulting in inflammation of the lungs.

Diagnosis of exposure to inhaled crystalline silica is made based on exposure history, x-ray examination of the lungs that reveals inflammation, and the absence of any other cause. However, recent efforts have identified malondialdehyde and neopterin as potential biomarkers of exposure to crystalline silica.[22] Silica particles may cause oxidative stress resulting in lipid peroxidation.[23] In blood serum, malondialdehyde serves as a biomarker of lipid peroxidation and neopterin serves as a urinary biomarker of oxidative stress caused by activation of the immune system.[23] Both malondialdehyde and neopterin can be measured using HPLC with an ultraviolet detector and were shown to strongly correlate with exposure to crystalline silica in glass sandblasters.[23]

In addition to known chemical toxins used during hydraulic fracturing process, there are many unknown chemicals that are considered proprietary by the companies producing fracturing fluids. Determining the toxicities or developing methods to diagnose exposure to these chemicals will not be possible until the chemicals are elucidated. Many companies have begun listing the chemicals used during the hydraulic fracturing process on FracFocus.org,[24] which will aid in the work of evaluating worker and societal risk to the components used in hydraulic fracturing. Due to public pressure, many companies are exploring chemicals that are nontoxic to humans and the environment, such as the use of UV disinfection instead of chemical disinfectants to treat the hydraulic fracturing fluid. Furthermore, companies are encouraged to continue modifications to the hydraulic fracturing fluid so that the fluid components are nontoxic.

5.5 SUMMARY AND CONCLUDING REMARKS

The process of hydraulic fracturing has led to a great employment boom in many areas of the country and a boom in energy production for the United States. The products of hydraulic fracturing may even become chemical exports, ensuring hydraulic fracturing activity in the United States for years to come. This chapter provides description of some of the problems that have been documented, indicating a need for increased research on the affects to human health.

For human health, an area of considerable concern is the ability to have clinical tests to document exposure to different petrochemicals. Clinical tests should include both workers in the industry and residents in proximity to the hydraulically fractured wells. Much more work is needed to develop tests for specific chemicals.

The chapter ends on the same question it began with: What are the risks from hydraulic fracturing and are these risks acceptable? The answer is that there have been documented environmental and human health effects in and around hydraulic fracturing sites. It is therefore important that we continue monitoring and studying the effects in order to find suitable mitigation strategies.

REFERENCES

1. Bommer, P. *A Primer of Oilwell Drilling*, 7th edn. Austin, TX: The University of Texas. 2008.
2. Montgomery, C. T.; Smith, M. B. Hydraulic fracturing: History of an enduring technology. http://www.jptonline.org/index.php?id=481 (accessed July 1, 2014).
3. Congress, U., Energy Policy Act of 2005. http://energy.gov/sites/prod/files/2013/10/f3/epact_2005.pdf (accessed July 1, 2014).
4. OSHA, Hazard Alert. 2012. https://www.osha.gov/dts/hazardalerts/hydraulic_frac_hazard_alert.html (accessed July 1, 2014).
5. Claudio, B. et al., From mud to cement-building gas wells. *Oil Field Review* 2003, Fall, 62–76.
6. Skogdalen, J. E.; Vinnem, J. E. Quantitative risk analysis of oil and gas drilling, using deepwater horizon as case study. *Reliability Engineering & System Safety* 2012, 100, 58–66.

7. McKenzie, L. M.; Witter, R. Z.; Newman, L. S.; Adgate, J. L. Human health risk assessment of air emissions from development of unconventional natural gas resources. *The Science of the Total Environment* 2012, 424, 79–87.
8. Osborn, S. G.; Vengosh, A.; Warner, N. R.; Jackson, R. B. Methane contamination of drinking water accompanying gas-well drilling and hydraulic fracturing. *Proceedings of the National Academy of Sciences of the United States of America* 2011, 108(20), 8172–8176.
9. Snawder, J.; Esswein, E.; King, B.; Breitenstein, M.; Alexander-Scott, M.; Retzer, K.; Kiefer, M.; Hill, R. Reports of worker fatalities during flowback operations, May 19, 2014, Centers for Disease Control, NIOSH Blog. http://blogs.cdc.gov/niosh-science-blog/2014/05/19/flowback (accessed July 1, 2014).
10. Esswein, E. J.; Breitenstein, M.; Snawder, J.; Kiefer, M.; Sieber, W. K. Occupational exposures to respirable crystalline silica during hydraulic fracturing. *Journal of Occupational and Environmental Hygiene* 2013, 10(7), 347–356.
11. Kassotis, C. D.; Tillitt, D. E.; Davis, J. W.; Hormann, A. M.; Nagel, S. C. Estrogen and androgen receptor activities of hydraulic fracturing chemicals and surface and ground water in a drilling-dense region. *Endocrinology* 2014, 155(3), 897–907.
12. Gross, S. A.; Avens, H. J.; Banducci, A. M.; Sahmel, J.; Panko, J. M.; Tvermoes, B. E. Analysis of BTEX groundwater concentrations from surface spills associated with hydraulic fracturing operations. *Journal of the Air & Waste Management Association* 2013, 63(4), 424–432.
13. Murphy, D.; Murphy, J. Fracking destroys the environment and poisons the air we breathe. http://www.globalresearch.ca/fracking-destroys-the-environment-and-poisons-the-air-we-breathe/5371475 (accessed June 29, 2014); Phillips, S. A link between heavy drilling and illness? Doctors search for solid answers StateImpact Pennsylvania [Online], 2012. http://stateimpact.npr.org/pennsylvania/2012/04/27/doctors-in-shale-country-search-for-answers-but-come-up-short/ (accessed June 29, 2014); Tibbetts, P. Phase II: BTEXs metabolites results 2014. http://fromthestyx.wordpress.com/2014/04/23/phase-ii-btexs-metabolites-results/ (accessed June 29, 2014); List of the harmed. http://pennsylvaniaallianceforcleanwaterandair.wordpress.com/the-list/ (accessed June 29, 2014).
14. Prevention, C. f. D. C. a. Facts about benzene. http://www.bt.cdc.gov/agent/benzene/basics/facts.asp (accessed June 29, 2014).
15. Registry, A. f. T. S. a. D. Public Health Statement for phenol toxic substances portal—Phenol [Online], 2008. http://www.atsdr.cdc.gov/PHS/PHS.asp?id=146&tid=27 (accessed June 29, 2014).
16. Weisel, C. P. Benzene exposure: An overview of monitoring methods and their findings. *Chemico-Biological Interactions* 2010, 184 (1–2), 58–66.
17. Hoet, P.; De Smedt, E.; Ferrari, M.; Imbriani, M.; Maestri, L.; Negri, S.; De Wilde, P.; Lison, D.; Haufroid, V. Evaluation of urinary biomarkers of exposure to benzene: Correlation with blood benzene and influence of confounding factors. *International Archives of Occupational and Environmental Health* 2009, 82(8), 985–995.
18. Alwis, K. U.; Blount, B. C.; Britt, A. S.; Patel, D.; Ashley, D. L. Simultaneous analysis of 28 urinary VOC metabolites using ultra high performance liquid chromatography coupled with electrospray ionization tandem mass spectrometry (UPLC-ESI/MSMS). *Analytica Chimica Acta* 2012, 750, 152–160.
19. Agency, E. P. Ground-water quality investigation—Pavillion, Wyoming 2011. http://www2.epa.gov/sites/production/files/documents/Nov8-2011_MethodsGraphicsAndDataTablesSummary.pdf (accessed June 29, 2014).
20. Jarnberg, J.; Stahlbon, B.; Johanson, G.; Lof, A. Urinary excretion of dimethylhippuric acids in humans after exposure to trimethylbenzenes. *International Archives of Occupational and Environmental Health* 1997, 69(6), 491–497.

21. Prevention, C. f. D. C. a. Worker exposure to crystalline silica during hydraulic fracturing 2012. http://blogs.cdc.gov/niosh-science-blog/2012/05/23/silica-fracking/ (accessed June 29, 2014).
22. Azari, M. R.; Ramazani, B.; Mosavian, M. A.; Movahadi, M.; Salehpour, S. Serum malondialdehyde and urinary neopterin levels in glass sandblasters exposed to crystalline silica aerosols. *International Journal of Occupational Hygiene* 2011, 3(1), 29–32.
23. Kamal, A. A.; Gomaa, A.; el Khafif, M.; Hammad, A. S. Plasma lipid peroxides among workers exposed to silica or asbestos dusts. *Environmental Research* 1989, 49(2), 173–180.
24. FracFocus Chemical Disclose Registry. http://fracfocus.org/ (accessed October 12, 2014).

6 Health and Environment Risks from Oil and Gas Development

Anne C. Epstein

CONTENTS

6.1 Overview 101
6.2 Introduction 102
6.3 Toxic Air Emissions: Benzene and Other Volatile Organic Compounds 102
6.4 Direct Measurements of Benzene Air Emissions during Oil and Gas Activity 102
6.5 Benzene Air Emissions Due to Leaks 103
6.6 Volatile Organic Compounds Emitted at Oil and Gas Sites 104
6.7 Evidence of Human Illness Associated with Toxic Air Emissions from Oil and Gas Development 104
6.8 Groundwater Contamination and the Risk to Private Water Wells 105
6.9 Groundwater Contamination Reports 106
6.10 Groundwater and Air Contamination Risks from Open Drilling Reserve Pits Used in West Texas 106
6.11 Improper Well Construction and the Risk of Subsurface Migration 107
6.12 Evidence for Water Well Contamination Due to Underground Migration of Gas 107
6.13 Evidence of Groundwater Contamination in Texas 108
6.14 Presumption of Groundwater Contamination due to Oil and Gas Drilling Activity in State Law 108
6.15 Risk to Human Health: What Level of Proof Do We Need? 109
6.16 Summary and Concluding Remarks 109
References 110

6.1 OVERVIEW

This chapter discusses health and environmental risks from oil and gas development and reviews results from various epidemiological studies. The chapter highlights the need for better monitoring of emissions and exposure in order to assess environmental risks from oil and gas development.

6.2 INTRODUCTION

The major risks to public health from oil and gas activities in West Texas come from toxic air emissions and the potential to contaminate drinking water with hazardous pollutants. Toxic air emissions include benzene, a known carcinogen, as well as other known hazardous air pollutants that are released routinely as part of both oil and gas activities. Emissions may rise to hazardous levels in proximity to the well site and cause human illness. Leaking equipment may produce especially dangerous levels. Children and developing fetuses are especially vulnerable to the toxic effects of air pollutants, and may become ill at lower ambient air levels than is considered safe for healthy adults.

Drinking water in private water wells may become contaminated due to spills, leaks, accidents, and improper well construction. Open drilling reserve pits increase the risk of toxic spills and also contribute to air pollution. Water contaminants have included arsenic as well as benzene and heavy metals and have been found in private water wells close to drilling sites at levels that pose significant risks to human health. Protecting public health requires understanding the risks posed to human health from oil and gas development and taking precautions in order to reduce those risks.

6.3 TOXIC AIR EMISSIONS: BENZENE AND OTHER VOLATILE ORGANIC COMPOUNDS

Oil and gas development releases a number of hazardous air pollutants that endanger people who live and work in close proximity to oil and gas activities. Chief among these air hazards is benzene. Benzene is a group 1 carcinogen, a known cause of cancer. It occurs naturally in crude petroleum, and on exposure to the air it quickly evaporates and contaminates the air. The World Health Organization has determined that there is no safe level for benzene in the air,[1] but the Texas Commission on Environmental Quality (TCEQ) has set guidelines for acceptable air levels.[2] Acute exposure to high levels of benzene can cause headaches, dizziness, drowsiness, confusion, tremors, and loss of consciousness. Chronic exposure to lower levels of benzene causes leukemia and lymphoma. Benzene also causes noncancer poisoning of the blood cells, which produces a serious form of anemia.[1]

6.4 DIRECT MEASUREMENTS OF BENZENE AIR EMISSIONS DURING OIL AND GAS ACTIVITY

Esswein and colleagues at the National Institute for Occupational Safety and Health measured occupational exposure to volatile organic compounds (VOCs) at six different oil and gas completion sites in Colorado and Wyoming in 2014.[3] Volunteer workers wore personal air monitors, and VOC levels were determined by real-time, direct-reading instruments. Fifteen of 17 benzene samples taken from workers gauging flowback or production tanks exceeded the recommended 8 h exposure limits. The researchers also found that benzene exposures were variable, unpredictable, and sometimes very high, with several readings exceeding 200 ppm close to open tank hatches, compared to a recommended 15 min exposure limit of 1 ppm. On the basis of these findings, the researchers concluded that airborne levels of benzene potentially pose health risks to oil and gas workers during flowback operations.

McKenzie and colleagues measured airborne levels of a variety of VOCs including benzene at ground level 45–152 m (150–500 ft) from the well pad center during well completions in Colorado, in an area largely devoid of other sources of benzene. For comparison, they measured VOCs in nearby areas of active natural gas development located 762 m (2500 ft) away from the nearest gas well. They discovered elevated levels of benzene close to gas wells, some of which exceeded recommended long-term exposure limits. However, 762 m (2500 ft) away from the drilling site, levels were much lower. Other VOC levels were also significantly elevated close to wells compared to further away.[4]

Using an elevated ground-based monitoring station, Colborn and colleagues measured levels of benzene and other VOCs over a 1-year period during natural gas drilling and fracturing activities at a site 1.1 km (0.7 miles) away from natural gas wells in Colorado. They found that benzene was emitted during drilling activities even prior to well stimulation; although at this distance, benzene levels were not elevated to the dangerous range.[5]

Petron and colleagues, in collaboration with the Center for Disease Control (CDC) National Institute for Occupational Safety and Health, did a "top-down" assessment of benzene levels in the upper atmosphere at an area of intense oil and gas development in Northeastern Colorado using airplane measurements during 12 different flights. Again, this site was largely devoid of other benzene sources, and the researchers correlated their findings with atmospheric data that might have affected the results. They found markedly elevated levels of benzene that were seven times that estimated by the state of Colorado inventory.[6]

6.5 BENZENE AIR EMISSIONS DUE TO LEAKS

One source of benzene is gas leaks. Leaks occur at many points in oil and gas production, including oil wells that co-produce natural gas, as well as leaking equipment including seals, valves, and multiple types of connecting points and moving parts.[7]

The Texas Commission on Environmental Quality (TCEQ), which is the state agency responsible for air quality in Texas, used infrared, mobile cameras to survey 94 natural gas sites in the Dallas-Fort Worth area in 2009. Air samples were collected at 73 of those sites, and at 21 of those sites, in 12 different areas, benzene levels exceeded the U.S. Environmental Protection Agency (EPA) level for long-term health effects, and 2 sites required immediate action for levels of benzene so high as to pose an immediate threat to health and safety: one was at a storage tank and one was at a wellhead. Interestingly, many of the elevated benzene levels could be attributed to equipment malfunctions and were brought into compliance after repairs were made, highlighting the need for air monitoring and enforcement.[8]

6.6 VOLATILE ORGANIC COMPOUNDS EMITTED AT OIL AND GAS SITES

Oil and gas development is associated with air emissions of multiple other VOCs known to be hazardous to human health, such as toluene, xylene, and ethyl benzene, which may cause headaches; dizziness; nausea; eye, nose and throat irritation; and shortness of breath.[9–11] Other VOCs include formaldehyde and ethylene glycol. Some of these VOCs occur naturally in petroleum products or in the target geological formation, some are injected into hydraulically fractured wells, and some are used on drill rigs for other purposes such as drill wash. Emissions of VOCs occur during all phases of oil and gas activities.[12]

The average uncontrolled emissions resulting from a shale single oil well completion is 18,143 kg (20 tons) of VOCs per well during a 7-day flow-back period, or 5,805 kg (6.4 tons) based on a 3-day flow-back period.[13] For comparison, the EPA defines a "major" source of air pollution as one that emits 9,072 kg (10 tons) per year of any of the listed toxic air pollutants, or 22,680 kg (25 tons) of a mixture of air toxins.[14]

The EPA investigated well production activities at the Fort Brethold Indian reservation and found that VOCs were the most common air pollutants emitted during the long production phase of oil and gas development. The EPA has published federally enforceable guidelines to control VOC emissions from oil well production.[13] Storage tanks emit VOCs including benzene, and are subject to the EPA's 2012 New Source Performance Standards for VOCs if they have the potential to emit 5443 kg or more (6 or more tons) of VOCs per year.[15]

6.7 EVIDENCE OF HUMAN ILLNESS ASSOCIATED WITH TOXIC AIR EMISSIONS FROM OIL AND GAS DEVELOPMENT

Congenital heart defects: McKenzie and colleagues performed an exhaustive study comparing birth records to the location of active drilling sites in rural Colorado, an area of very intense oil and gas activity, but few other sources of air contamination. They found a highly statistically significant increase in the risk of congenital heart disease for people who lived within 16 km (10 miles) of gas wells, and risk increased in proportion to the degree of well concentrations. The statistical significance remained robust even after controlling for multiple confounding factors.[16]

Health status and proximity to well sites: Rabinowitz and colleagues administered a survey about health complaints to residents in an area of intense natural gas development in rural Pennsylvania in 2012.[17] They chose a computer-generated random sample of 180 households with private water wells that were located at various distances from natural gas wells, and did not inform the subjects about the purpose of the study. The researchers found that people who lived less than 1 km (0.6 miles) from the nearest gas well reported significantly more health complaints than people living more than 2 km (1.2 miles) away: Thirty-nine percent of people living in close proximity to wells reported upper respiratory symptoms compared to 18% of people who lived further away. These results were statistically significant after adjustment for multiple risk factors.

Elevated hazard index in proximity to well sites: McKenzie and her colleagues in 2012 calculated the risk of human illness resulting from the elevated level of air toxins that they measured within 152 m (500 ft) of drilling sites in Colorado compared to the much lower levels that were present 762 m (2500 ft) away, in the study mentioned earlier. They determined that people living 152 m (500 ft) or closer to well sites in an area of intense oil and gas activity had a five times increased risk of noncancer illness compared to those living 2500 ft away.[4]

Breast cancer: In response to public concerns about a possible cancer cluster in Flower Mound, Texas related to intense oil and gas development, the Texas Department of State Health Services performed a cancer cluster investigation in 2014.[19] They found a statistically significant increase in the expected incidence of female breast cancer occurring in the Flower Mound area during the years 2002–2011. Although the state health officials did not indicate a potential link to oil and gas activities, the Institute of Medicine has determined that benzene is a possible cause of breast cancer.[20]

Children are at greater risk: Paulson has pointed out that air contamination standards are determined by a standard for healthy adults; however, developing fetuses, babies, and children will have a greater risk of illness and cancer from the same level of exposure.[21] Also, the elderly and those with underlying medical illnesses will be at a greater risk of noncancer illnesses.

Occupational fatalities: The EPA has begun collecting data on injuries to oil and gas workers due to toxic exposures to VOCs, and has compiled four case reports of worker death due to acute exposure to VOC since 2010.[22]

One case of direct evidence linking symptoms to toxic air emissions: In the first successful nuisance suit of its kind against an oil and gas operator, an individual in the Barnett Shale area alleged that her multiple medical symptoms were due to exposure to air toxins released by gas wells, and provided her own medical tests showing that her blood stream contained high levels of hazardous chemicals. In April 2014, the court awarded her $2.9 million in damages.[23,24]

Risks from air emissions may be episodic and are inadequately monitored: Brown and colleagues have shown the health risks from benzene and other air pollutants may be inadequately monitored by current protocols, which rely on average values over days to weeks. They showed that benzene and other toxic air pollutants may be released in spikes and bursts that go undetected unless continuous monitoring devices are in place, with peaks occurring up to 30% of the time in a given residential monitoring site. Since the toxicity of a compound depends on the intensity and duration of exposure, when peaks are missed the health risks of exposure may be underestimated. Thus, studies that show low levels of toxins in ambient air samples in an area of oil and gas development may be failing to detect the intermittent occurrence of high levels that have the potential to make people sick.[25]

6.8 GROUNDWATER CONTAMINATION AND THE RISK TO PRIVATE WATER WELLS

When oil and gas activity contaminates groundwater, the main threat to public health occurs when that groundwater is used as a drinking water source in private water wells. Fifteen million Americans obtain their drinking water from private water wells.[26] Wells that serve fewer than 15 people are not regulated under the U.S. Safe Drinking Water Act. There is no legal requirement for water safety testing; testing for water quality is at the discretion of the well owner.[26] Oil and gas development poses risks to underground water resources through surface spills and leaks of hazardous fluids, and through casing or cementing failures that allow stray gas to leak out of improperly constructed wells.[27]

6.9 GROUNDWATER CONTAMINATION REPORTS

Based on reports from drilling operators, complaints, and routine investigations, the Texas Railroad Commission documented 532 cases of groundwater contamination due to oil and gas activity in 2013, 72 of which were new. Contamination incidents involved crude oil, hydrocarbons, benzene, toluene, ethyl benzene, xylene, and other toxins.[28] In August 2014, the Pennsylvania Department of Environmental Protection (DEP) reported 243 "cases where DEP determined that a private water supply was impacted by oil and gas activities" during the years 2008–2014.[29] Gross et al. published a peer-reviewed study of cases of drinking water contamination due to surface spills in Colorado: 77 spills of oil or produced water were reported from July 2010 through July 2011. Benzene levels exceeded maximum contaminant levels in groundwater in 90% of the spills. The average spill was 3.78 m^3 (1000 gal). As of May 2012, 12% of the spills had not yet been remediated.[30]

In 2012, the EPA documented hazardous substances in four private water wells in Northeastern Pennsylvania, in an area that had been drilled by Cabot since 2008. Spills from oil and gas activities as well as high methane levels in private water wells had already been documented in the area. The EPA toxicologist found dangerously elevated levels of arsenic, Bis (2-ethylhexyl) phthalate (DEHP) manganese, and phenol, all of which are classified as hazardous substances. Arsenic is a known human carcinogen, and DEHP is a probable human carcinogen. Other agents known to be associated with drilling activities were also found.[31]

6.10 GROUNDWATER AND AIR CONTAMINATION RISKS FROM OPEN DRILLING RESERVE PITS USED IN WEST TEXAS

Evaporation from open pits results in VOCs and hazardous air pollutants being released into the air.[32] Reserve pits can and do leak, even when lined.[33] When pits leak, the toxins in them can contaminate groundwater and drinking water sources.[34] In addition, pits can overflow due to rain, flooding, rise in water levels, and rupture of pit berms.[32] Drilling reserve pits contain a number of hazardous substances that can cause human illness, especially substances classified as hazardous air pollutants and VOCs, when released into the ground or into the air. Drilling mud may contain benzene as well as arsenic, barium, chromium, lead, and other metals.[32] The Bureau of Land Management states, "In the absence of site-specific information, Exploration and Production (E&P) waste pits should be considered to contain potentially hazardous waste harmful to human health…"[35]

The EPA found 103 sites of leaking pits in 2003 in Region 8, which encompasses Colorado, Utah, Wyoming, Montana, North Dakota, and South Dakota.[36]

To mitigate health and environmental risks associated with open pits and impoundments, the Railroad Commission of Texas recommends closed loop drilling,[37] as does the Bureau of Land Management.[38] The State of New Mexico functionally requires closed loop drilling,[39] as well as several city ordinances in Texas, including those of Fort Worth,[40] Arlington,[41] and Flower Mound[42] in the Barnett Shale area.

6.11 IMPROPER WELL CONSTRUCTION AND THE RISK OF SUBSURFACE MIGRATION

During oil and gas development, layers of casing and cementing inside the wellbore are designed to protect groundwater from contamination. Casing and cementing technology, like all technologies, has a failure rate. In the most extensive and complete study so far, Ingraffea and colleagues studied over 41,000 oil and gas wells drilled in Pennsylvania from 2000 to 2012. Most of the wells drilled after 2009 were unconventional wells. Using state inspection reports, the researchers were able to determine that the overall structural failure rate of unconventional wells in Pennsylvania was 1.9%; however, there was considerable variation depending on when the wells were drilled and where they were drilled. Unconventional wells drilled since 2009 in northeastern Pennsylvania, a drilling boom area, had a failure rate of 9.18%, even after what might have been an initial learning curve. Overall, unconventional wells had a 2.7% higher casing and cementing failure rate than conventional wells. Low inspection rates may have contributed to an underestimation of the failure rate. The researchers found that "each additional inspection in the pre-2009 stratum increases the risk of identifying a cement or casing problem by 17.7%."[43]

Olawoyin and colleagues examined DEP reports of all oil and gas violations in Pennsylvania from 2008 to 2010. There were a total of 2601 violations, 67% of which were considered serious. Of the 65 operators with violations, only 27 contributed to significant environmental risk. The authors pointed to the need to identify particular operators who posed a potential public health hazard.[44]

6.12 EVIDENCE FOR WATER WELL CONTAMINATION DUE TO UNDERGROUND MIGRATION OF GAS

Osborn and colleagues found a marked increase in methane, the major constituent of natural gas, in the Marcellus Shale drilling area of Pennsylvania and New York within 1 km (0.6 miles) of oil and gas wells. They were able to show that the methane had its origin deep underground and did not come from other sources. At a distance greater than 1 km (0.6 miles) from gas wells, methane contamination fell off rapidly.[45] Jackson and colleagues extended Osborne's findings in 2013, showing groundwater contamination with propane and ethane, both components of natural gas, within 1 km (0.6 miles) of a subset of natural gas wells in NE Pennsylvania.[46]

More recently, Darrah and colleagues used a highly sensitive technique to identify the sources of the methane that had contaminated domestic water wells in both

Northeastern Pennsylvania and in the Barnett Shale area in Texas.[47] They documented eight distinct clusters of domestic water wells that had been contaminated by fugitive gasses from leaking oil and gas wells. Sources included leaking casing and cement, and one case of underground well failure. Their analysis did not reveal any cases of contamination due to hydraulic fracturing.

Residents of Northeastern Pennsylvania, the same area in which Ingraffea and colleagues found high rates of well failure, filed nearly 1000 complaints alleging drilling contamination of private water wells from 2008 to 2012, according to records from the Pennsylvania Department of Environmental Protection that were obtained by *The Sunday Times*.[48] There were 90 documented cases of private water wells damaged by underground migration of gas from leaky natural gas wells.

6.13 EVIDENCE OF GROUNDWATER CONTAMINATION IN TEXAS

Fontenot and his colleagues from UT Arlington studied toxins found in water wells in the Barnett Shale area of Texas, and compared them to water wells in both non-drilling areas and in historical controls. In active drilling areas, arsenic levels were elevated above maximum contaminant limits (MCLs) in 32% of wells tested, and selenium was elevated above MCLs in 20%. Neither historical controls nor wells in nonactive drilling areas had significantly elevated levels of either compound. Total dissolved solids (TDS) were also elevated in Barnett Shale wells compared to control wells.[49]

Health risks from arsenic: Arsenic is odorless and tasteless, and when it contaminates groundwater it may be undetected for many years. Arsenic causes skin rashes, skin cancer, lung cancer, and bladder cancer, as well as cardiovascular disease, diabetes, injury to the brain and nervous system, and birth defects. Arsenic is one of the World Health Organization's 10 chemicals of major public health concern. Arsenic is found in native drinking water in many areas even in the absence of oil and gas drilling, which makes it imperative to establish baseline levels of arsenic and other heavy metals in the water supply prior to drilling.[50] In Hood County, Texas, in the Barnett Shale region, the EPA determined that two gas wells developed by Range Resources in 2009 contaminated domestic water wells with methane, benzene, toluene, as well as other components of natural gas, at levels that endangered health.[51]

6.14 PRESUMPTION OF GROUNDWATER CONTAMINATION DUE TO OIL AND GAS DRILLING ACTIVITY IN STATE LAW

In response to litigation and the difficulty in establishing proof of responsibility for well water contamination, largely due to the absence of baseline water quality data, the state of Pennsylvania enacted a law stipulating that if any water well becomes contaminated within 305 m (1000 ft) of a conventional oil and gas well, and if that oil or gas well had been completed within the previous 6 months, then the oil or gas well operator is presumed responsible for the water well contamination unless he or she can prove otherwise. For unconventional wells, the distance is 762 m (2500 ft) and the time is 12 months. Similarly, the state of West Virginia passed a law stipulating that if a water supply located within 457 m (1500 ft) of the vertical section of a horizontal well becomes contaminated, it is presumed that the oil and gas operator caused the contamination.[52]

6.15 RISK TO HUMAN HEALTH: WHAT LEVEL OF PROOF DO WE NEED?

Oil and gas development activities are associated with risk to public health from exposure to hazardous air pollutants and toxic chemicals in groundwater. The magnitude of those risks and the causes of variation in those risks is not yet fully known. An emerging body of evidence links oil and gas development directly to human illness. Although not conclusive, the studies demonstrate patterns that would be expected from the toxic profiles of the hazardous pollutants that are known to be emitted into the air or released into groundwater at oil and gas sites.

The American Public Health Association endorsed the Precautionary Principle in 2001, which states that "Where there are threats of serious or irreversible damage, lack of full scientific certainty shall not be used as a reason for postponing cost-effective measures to prevent environmental degradation." This principle was endorsed by the United States in signing the Rio Declaration on Environment and Development in 1992. The American Public Health Association's resolution stated, "Recognizing that fetuses and children are often more susceptible to environmental contaminants than adults, and that agency policies and decisions often fail to reflect this unique susceptibility; and recognizing that proof of cause and effect relationships is often difficult to establish because of non-specificity of health effects, long latent periods, subtle changes in functions that are difficult to detect without resource-intensive studies, and complex interactions of variables that contribute to adverse health effects...; [The American Public Health Association] reaffirms its explicit endorsement of the precautionary principle as a cornerstone of preventive public health policy and practice..."[53]

From a medical and a public health perspective, the question is not, "Can we prove beyond a reasonable doubt that oil and gas development causes illness?," but rather, "Is there enough evidence to suggest that oil and gas development poses a risk to public health?" The practice of medicine does not adhere to the standard of "full scientific certainty," but instead adopts standards of evidence according to our best medical and scientific judgments and the need to care for patients at the time. There is no reason that medical and public health judgments should be held to a different standard when it comes to energy.

6.16 SUMMARY AND CONCLUDING REMARKS

Much more research is needed in order to fully understand the health risks from oil and gas activity; however, in my opinion we have enough information at present to recommend measures to protect public health while awaiting further data. Extensive monitoring for groundwater and air pollution in proximity to oil and gas development is needed. Known practices that reduce air emissions and reduce the risk of groundwater contamination should be extensively employed at all sites. Public policymakers and industry representatives should recognize that a certain rate of accidents, equipment failures, malfunctions, spills, and leaks is to be expected, and they should engage detection, remediation, and safety practices to minimize them and take precautions to protect members of the public from harm when such mishaps inevitably occur. One key precaution is to increase the distance between the sources of toxins and the public by increasing setback distances.

REFERENCES

1. World Health Organization (WHO). (2010). Exposure to benzene: A major public health concern. Retrieved from http://www.who.int/ipcs/features/benzene.pdf, accessed May 10, 2014.
2. Haney, J.T. (2014). Texas Commission on Environmental Quality, "Benzene: CAS Registry Number: 71-43-2. 24-Hour Ambient Air Monitoring Comparison Value," Proposed February 2014.
3. Essewein, E.J. et al. (2014). Evaluation of some potential chemical exposure risks during flowback operations in unconventional oil and gas extraction: Preliminary results, *Journal of Occupational and Environmental Hygiene 11*: D174–D184. Retrieved from http://www.tandfonline.com/doi/pdf/10.1080/15459624.2014.933960, accessed January 14, 2015.
4. McKenzie, L.M., Witter, R.Z., Newman, L.S., and Adgate, J.L. (2012). Human health risk assessment of air emissions from development of unconventional natural gas resources, *Science of the Total Environment*. Retrieved from http://cogcc.state.co.us/library/setbackstakeholdergroup/Presentations/Health%20Risk%20Assessment%20of%20Air%20Emissions%20From%20Unconventional%20Natural%20Gas%20-%20HMcKenzie2012.pdf, accessed September 1, 2014.
5. Colborn, T., Schultz, K., Lucille, H., and Kwiatkowski, C. (2012). An exploratory study of air quality near natural gas operations, *Human and Ecological Risk Assessment 20*(1): 86–105. Retrieved from http://endocrinedisruption.org/assets/media/documents/HERA12-137NGAirQualityManuscriptforwebwithfigures.pdf, accessed July 1, 2014.
6. Petron, G. et al. (2014). A new look at methane and non-methane hydrocarbon emissions from oil and natural gas operations in the Colorado Denver-Julesburg Basin, *Journal of Geophysical Research: Atmospheres 119*(11): 6836–6852. Retrieved from http://onlinelibrary.wiley.com/doi/10.1002/2013JD021272/abstract, accessed May 14, 2014.
7. Environmental Protection Agency (EPA). (2014). Oil and natural gas sector leaks. Retrieved from http://www.epa.gov/airquality/oilandgas/pdfs/20140415leaks.pdf, accessed June 22, 2014.
8. Memo from Shannon Ethridge, MS, Toxicology Division, Chief Engineer's Office to Mark R. Vickery, P.G., Executive Director, re: Health Effects Review of Barnett Shale Formation Area Monitoring Projects. Texas Commission on Environmental Quality Interoffice Memorandum January 27, 2010, retrieved from http://www.tceq.state.tx.us/assets/public/implementation/barnett_shale/2010.01.27-healthEffects-BarnettShale.pdf, accessed on July 8, 2013.
9. Sadlier, J. and Honeycutt, M. (2010). "Barnett Shale Study" Presentation, Texas Commission on Environmental Quality 2010. Retrieved from http://www.tceq.state.tx.us/assets/public/implementation/barnett_shale/presentations/2010.01.27-sadlier-bshaleUpdate.pdf, accessed May 15, 2014.
10. Environmental Protection Agency (EPA). (2012). Toluene. Technology transfer network-air toxics web site. Retrieved from http://www.epa.gov/airtoxics/hlthef/toluene.html, accessed October 14, 2014.

11. Agency for Toxic Substances and Disease Registry (ATSDR). (2007). Toxicological profile for xylene. Retrieved from http://www.atsdr.cdc.gov/toxprofiles/tp71.pdf, accessed July 10, 2014.
12. Agency for Toxic Substances and Disease Registry (ATSDR). (2007). Toxicological profile for ethylbenzene. Retrieved from http://www.atsdr.cdc.gov/toxprofiles/tp110.pdf, accessed July 10, 2014.
13. Moore, C.W. (2014). Air impacts of increased natural gas acquisition, processing, and use: A critical review, *Environmental Science and Technology*, *48*(15), 8349–8359.
14. EPA. (2014, April). Oil and natural gas sector hydraulically fractured oil well completions and associated gas during ongoing production: Report for oil and natural gas sector oil well completions and associated gas during ongoing production review panel. U.S. EPA Office of Air Quality Planning and Standards (OAQPS).
15. EPA. (2013). Technology transfer network-air toxics web site: Pollutants and sources, last update September 18, 2013. Retrieved from http://www.epa.gov/airtoxics/pollsour.html, accessed October 14, 2014.
16. EPA's air rules for the oil and natural gas industry, Final updates to requirements for storage tanks used in oil and natural gas production and transmission. Updates and Clarifications to 2012 VOC New Source Performance Standards, U.S. EPA December 19, 2014. Retrieved from http://www.epa.gov/airquality/oilandgas/pdfs/20141219fs.pdf, accessed February 11, 2015.
17. McKenzie, L. et al. (2014). Birth outcomes and maternal residence proximity to natural gas development in rural Colorado, *Environmental Health Perspectives*, *122*(4): 412–417.
18. Rabinowitz, P.M. et al. (2014, September 10). Proximity to natural gas wells and reported health status: Results of a household survey in Washington County, Pennsylvania, *Environmental Health Perspectives*. Retrieved from http://dx.doi.org/10.1289ehp.1307732.
19. Texas Department of State Health Services (2014, July 30). Updated summary report: Occurrence of cancer in zip codes 75022 & 75028, Flower Mound, Denton County, Texas, Time Period: 2002–2011, Texas Department of State Health Services, Austin, TX.
20. Committee on Breast Cancer and the Environment: The Scientific Evidence, Research Methodology, and Future Directions; Institute of Medicine. (2011, December 7). *Breast Cancer and the Environment: A Life Course Approach*, The National Academies Press, Washington, DC. Retrieved from http://www.iom.edu/Reports/2011/Breast-Cancer-and-the-Environment-A-Life-Course-Approach.aspx, accessed October 13, 2014
21. Paulson, M.D., Jerome. (2014, February 4). Testing of chemicals and data reporting of information under Toxic Substances Control Act (TSCA) Sections 4 and 8, Testimony on behalf of the American Academy of Pediatrics before the House Energy and Commerce Subcommittee on the Environment and the Economy.
22. Snawder, J. et al. (2014, May 19). Reports of worker fatalities during flowback operations, NIOSH Science Blog.
23. Lisa Parr and Robert Parr, Plaintiffs vs Aruba Petroleum and Encana Oil & Gas, Defendants, Cause No. 11-01650-E, Filed 17 September 2013 in County Court No.5 of Dallas County, Texas, Plaintiffs' Eleventh Amended Petition.
24. Deam, J. (2014, April 23). Jury awards Texas family nearly $3 million in fracking case, *Los Angeles Times*. Retrieved from http://articles.latimes.com/2014/apr/23/nation/la-na-fracking-lawsuit-20140424, accessed April 23, 2014.
25. Brown, D. et al. (2014). Understanding exposure from natural gas drilling puts current air standards to the test, *Reviews on Environmental Health*, *29*(4), 277–292.
26. EPA. Drinking water from household wells EPA 816-K-02-003, January 2002.

26. Vengosh, A. et al. (2014). A critical review of the risks to water resources from unconventional shale gas development and hydraulic fracturing in the United States, *Environmental Science and Technology*, *48*(15), 8334–8348.
27. Texas Groundwater Protection Committee. Joint groundwater monitoring and contamination Report 2013, April 2014.
28. Pennsylvania Department of Environmental Protection. Water supply determination letters. Retrieved from http://files.dep.state.pa.us/OilGas/BOGM/BOGMPortalFiles/OilGasReports/Determination_Letters/Regional_Determination_Letters.pdfm, accessed August 29, 2014.
29. Gross, S.A. et al. (2013). Analysis of BTEX groundwater contamination form surface spills associated with hydraulic fracturing operations, *Journal of the Air & Wastewater Management*, *63*(4), 424–432.
30. U.S. EPA Region III, Action Memorandum-Request for Funding for a Removal Action at the Dimock Residential Groundwater Site, Dimock Township, Susquehanna County, PA, From Richard Fetzer, Received January 19, 2012.
31. EPA report to Congress 1987: Management of wastes from the exploration, development, and production of crude oil, natural gas, and geothermal energy, Volume 1 of 3: Oil and Gas. U.S.EPA, Office of Solid Waste and Emergency Response, EPA/530-SW-88-003, December 1987.
32. The Expert Panel on Harnessing Science and Technology to Understand the Environmental Impacts of Shale Gas Extraction Environmental impact of shale gas extraction in Canada, Council of Canadian Academies, Ottawa, ON, Canada, 2014.
33. Ramirez, P. Reserve pit management, risks to migratory birds, U.S. fish and Wildlife Service Region 6 Environmental Contaminants Program, September 2009.
34. United States Department of the Interior, Bureau of Land Management, Wyoming State Office, Memo to District Managers From Deputy State Director, Minerals and Lands, Subject Management of Oil and Gas Exploration and Production Pits, November 15, 2011.
35. EPA. Report of the United States EPA Region 8 oil and gas environmental assessment effort 1996–2002, January 2003.
36. Railroad Commission of Texas. Annotated bibliography of waste minimization technology for crude oil and natural gas exploration, production, and pipeline transportation operations, published by Oil and Gas Division, July 2004.
37. U.S. Department of the Interior, Bureau of Land Management, Wyoming State Office, Memo to District Managers from Deputy State Director, Minerals and Lands, Instruction Memorandum No. WY-2012-007, November 15, 2011.
38. State of New Mexico Energy, Minerals and Natural Resources Department Oil and Conservation Commission, Order of the Conservation Commission, Case No. 14015, Order No. R-12939, May 9, 2008.
39. Fort Worth City Ordinance No. 18449-02-2009. Amendment to Article II of Chapter 15,"Gas" entitled "Gas Drilling and Production" of the Code of Ordinances of Fort Worth, Texas, February 3, 2009.
40. Ordinances governing gas drilling and production in the City of Arlington Texas, Ordinance No 11-068, "Gas Drilling and Production" (December 6, 2011).
41. Town of Flower Mound, Ordinance No. 29-11. Repealing existing Article VII, "Oil and Natural Gas Well Drilling and Operations" of Chapter 34 of the Code of Ordinances of the Town of Flower Mound, Texas, and Adopting a New Article VII, Entitled "Oil and Natural Gas Well drilling and Operations", July 2011.
42. Ingraffea, A. et al. Assessment and risk analysis of casing and cement impairment in oil and gas wells in Pennsylvania, 2000–2012, *PNAS* Early Edition, May 30, 2014.
43. Olawoyin, R. et al. (2013, June). Environmental Safety Assessment of Drilling Operations in the Marcellus-Shale Gas Development, *SPE Drilling & Completion*, *18* (2).

44. Osborn, S. et al. Methane contamination of drinking water accompanying gas-well drilling and hydraulic fracturing, *PNAS* Early Edition, April 14, 2011.
45. Jackson, R. et al. Increased stray gas abundance in a subset of drinking water wells near Marcellus shale gas extraction, *PNAS* Early Edition, June 3, 2013.
46. Darrah, T.H. Noble gases identify the mechanisms of fugitive gas contamination in drinking-water wells overlying the Marcellus and Barnett Shales, *PNAS* Early Edition, August 12, 2014. Retrieved from http://www.pnas.org/content/111/39/14076, accessed September 15, 2014.
47. Legere, L. (2013, May 19). Sunday Times review of DEP drilling records reveals water damage, murky testing methods, *The Times-Tribune*. Retrieved from http://thetimes-tribune.com/news/sunday-times-review-of-dep-drilling-records-reveals-water-damage-murky-testing-methods-1.1491547, accessed on October 14, 2014.
48. Fontenot, B. et al. (2013). An evaluation of water quality in private drinking water wells near natural gas extraction sites in the Barnett shale formation, *Environmental Science and Technology*, *47* (17), 10032–10040.
49. World Health Organization. Arsenic: Fact Sheet No. 372, December 2012.
50. EPA Region VI Emergency Administration Order in the Matter of: Range Resources Corporation and Range Production Company, Docket Number SDWA-06-2011-1208.
51. Hall, K. (2013). Hydraulic fracturing contamination claims: Problems of proof. *Ohio State Law Journal Furthermore* 74, 71–85.
52. American Public Health Association. (2001, March).The precautionary principle and children's health. *American Journal of Public Health 91*(3): 495–496.

7 Addressing Concerns about Impacts from Unconventional Drilling Using Advanced Analytical Chemistry

Doug D. Carlton Jr., Zacariah L. Hildenbrand, Brian E. Fontenot, and Kevin A. Schug

CONTENTS

7.1 Overview 115
7.2 Introduction 115
7.3 Target Analytes Related to Unconventional Drilling 116
7.4 Dissolved Gases 117
7.5 Organic Constituents 119
7.6 Ions and Isotopes 124
7.7 Summary and Concluding Remarks 128
References 129

7.1 OVERVIEW

This chapter discusses the role of analytical chemistry and recent advancements in characterizing potential contaminants arising from oil and gas production activities.

7.2 INTRODUCTION

Natural gas has recently emerged as an abundant, economic, and clean source of energy. It has helped to reduce CO_2 emissions and curtail the production of industrial chemicals by the power sector.[1] Advanced drilling techniques such as hydraulic fracturing and shale acidization have made the extraction of natural gas from previously inaccessible deep shale formations economically advantageous. Hydraulic fracturing uses a mixture of water, proppants, and chemical additives injected at high pressures to create and maintain fissures or fractures in the shale formation to release trapped gases. Shale acidization uses copious amounts of hydrochloric and/or hydrofluoric acid under low pressures to dissolve sediments of the shale formation to increase the formation permeability for trapped gases. Even with the proven effectiveness of these approaches to liberate sequestered hydrocarbons, they are not without unique environmental risks. Concerns have been raised over their potential impacts on surrounding ground and surface water. Apprehensions over environmental responsibility, alongside the aspiration of using natural gas to lead the United States to attaining energy independence, have driven the need for numerous investigations of relationships between unconventional drilling and groundwater quality.

To understand the relationship between unconventional drilling and groundwater quality, a multifaceted approach must be implemented that involves basic water quality measurements as well as detailed chemical speciation. In cases of traditional groundwater measurements, pH, salinity, conductivity, and total dissolved solids (TDS) are the most common parameters measured in studies of groundwater quality. These measurements provide a superficial look at water quality and can be taken quickly in the field to detect the presence of possible abnormalities, such as broad fluctuations in pH that may be indicative of anthropogenic contamination. Researchers have also investigated the viability of biological assays[2] and total organic carbon/total nitrogen (TOC/TN) content as bulk water measurements for detecting perturbations in groundwater systems. While these measurements are considered common practice in the field, they are not as informative as more advanced analytical measurements as they do not identify specific chemical compounds or those that might more clearly enable the assignment of sources of perturbation.

In this chapter, we will discuss the analytes chosen by researchers as possible indicators of groundwater contamination from unconventional drilling processes. We have broadly categorized these analytes as dissolved gases, organic compounds, ions, and isotopes. We will also provide an overview of the chosen analytical methods for the determination of these target compounds. Analytical techniques include gas chromatography (GC), ion chromatography (IC), elemental analysis, and mass spectrometry. While significant guidance continues to be taken from established standard methods (e.g., American Society for Testing and Materials [ASTM] and United States Environmental Protection Agency [EPA]), the methods described herein have sometimes been tailored to the chemical compounds specifically related to unconventional drilling processes. Furthermore, these methods can continue to be improved by the addition of new target compounds or new technologies to improve current performance metrics. Provided is a snapshot of the current state-of-the-art and a means with which to help understand the potential impacts of an important industrial process on the environment.

7.3 TARGET ANALYTES RELATED TO UNCONVENTIONAL DRILLING

Ideally, the analytical approaches for monitoring groundwater should either be specific for compounds related to unconventional drilling or complemented by a broad spectrum of analyte measurements to better discern possible contamination sources.

To date, the EPA, ASTM, and AOAC International have not developed standard methods to monitor for specific suites of chemical compounds that may induce groundwater contamination specifically from unconventional drilling processes. Past investigations[3,4] conducted through regulatory agencies have used established methods for general volatile and semi-volatile organic compounds, ions, and metals to survey potentially contaminated water and then cross-referenced these data with available information about industrial activity, to putatively identify specific sources of contamination. A shortcoming of using standardized regulatory methods to investigate this variable and proprietary process is that they are only suited to identify a set list of compounds with little emphasis toward additional compounds that may be detected but not identified. Academic researchers have targeted larger numbers of industrial byproducts, artificially elevated groundwater constituents, and specific compounds or ions thought to be indicative of contamination through unconventional drilling.[5–8] Linking contamination events to industrial sources requires sampling of proprietary industrial chemical mixtures and flowback or produced waters that are not typically available to researchers. As such, this broad approach does not allow for definitive conclusions about the source of contaminants as these studies necessarily rely on correlative (e.g., geospatial[9]) analyses to link contamination events with their respective sources.

7.4 DISSOLVED GASES

Dissolved gases such as chlorofluorocarbons,[4,10,11] sulfur hexafluoride,[4,12,13] tritium,[4,11,13] helium,[4,11,13] neon,[4,13] argon,[4,13] krypton,[4,11] and xenon[4] have been used as environmental tracers to age and source groundwater. However, the majority of investigations into the effects of unconventional oil and gas (UOG) activities on groundwater have quantified light hydrocarbons, (e.g., methane, ethane, and propane) within areas of UOG extraction.[6,14,15]

The vast majority of natural gas is methane with lesser amounts of ethane and propane. Methane in the subsurface can be found naturally in two forms—biogenic and thermogenic methane. Biogenic methane results as a byproduct of bacterial metabolism at shallow depths.[16] Thermogenic methane, a key objective of UOG exploration, is produced through geological processes occurring at depths greater than 1000 m. Over an extended time period, high temperatures and pressures transform decomposing organic materials into methane gas.[16]

Scenarios that could plausibly result in introduction of thermogenic methane to overlying aquifers include the failure of borehole casings or hydraulic fractures intersecting abandoned oil wells or unmapped faults.[17,18] It has been suggested that stray gas from shallow hydrocarbon formations is able to permeate upward through gas well annuli (i.e., voids between concentric piping or casings),[19] and gas from deep target shale formations can be introduced through poorly constructed or failing well casings.[18]

With each study where methane in groundwater is detected, the origin (biogenic or thermogenic) must be identified. Methane sources can be attributed based on measured methane to ethane and propane ratios or through measured isotopic abundances of carbon-13 (^{13}C) and deuterium (^{2}H) isotopes.[14] Such measurements can even be used to distinguish thermogenic methane produced by different geological formations. Compared to biogenic methane, thermogenic methane mixtures have a greater proportion of ethane to pentane (C_2–C_5) hydrocarbons relative to methane (C_1). Biogenic methane mixtures are predominately methane and carbon dioxide (CO_2).[20] Attempts have been made to use these signatures to delineate the origin of methane and identify the formation from which detected stray gas may have migrated into the groundwater.[6,15] This can be valuable information for sourcing contamination if, for example, UOG drilling is present in close proximity to the point where the water sample was taken and this is in a location near to multiple geological formations.

Quantification of light hydrocarbons is most often performed by separation using GC and detection using a flame ionization detector (FID). Some studies have used a mass spectrometer (MS) or thermal conductivity detector (TCD) to have the additional capabilities of quantifying N_2, O_2, CO_2, H_2S, and H_2O.[14,21] Dissolved gases of interest are liberated from groundwater by either purging the sample with an inert gas and trapping it on a sorbent (i.e., purge and trap[22]), or by creating a headspace in the vial then agitating and/or heating the sample prior to withdrawing the vapor phase for analysis (i.e., headspace analysis[23]). Separation of light hydrocarbons and fixed gases can be achieved with a PLOT (porous layered open tubular) column, especially one with a divinylbenzene Q stationary phase.[22,24] Separations of light hydrocarbons can be performed with 3.175 mm stainless-steel packed columns,[25] but typically require different column fittings and pressure controllers on the GC compared to those used for capillary columns. Mega-bore packed capillary columns (0.53 mm ID) are becoming more typical in today's market since these are more compatible with typical capillary GC systems. While multiple columns may separate the selected compounds of interest, sample throughput needs to be considered in column selection. A particular column may be excellent at separating permanent gases, but it may also have a very high affinity for short-chain hydrocarbons,[26] which can increase analysis time and sacrifice efficiency for late eluting peaks. Along with the necessary hardware for methane analysis, a meticulous sample collection procedure is important to reduce the degassing of methane and other volatile organic compounds as the water is pumped from the groundwater well. Suggested sampling procedures include sampling the water from a purged well at a low flow rate and under constant pressure. This is best done through valved, non-gas permeable tubing into an evacuated bladder, which has been preloaded with a biocide to reduce bacterial gas degradation and/or generation during the 14-day maximum holding time.[27] Delineation of this procedure is shown in Figure 7.1.

Substantial surveying of methane in groundwater in areas of UOG activity has taken place in the Marcellus Shale formation of Pennsylvania.[6,15] An early report[6] determined methane concentrations present in private groundwater to be greatest in areas within 1 km of UOG wells. The majority of methane detected was characteristic of deep, thermogenic methane. Methane was detected in approximately 80% of the collected samples. The root cause of these methane contamination events remains to be judged, although it was proposed that the creation of fractures by UOG drilling activities into abandoned legacy oil and gas wells could allow connectivity for the thermogenic methane to migrate into private

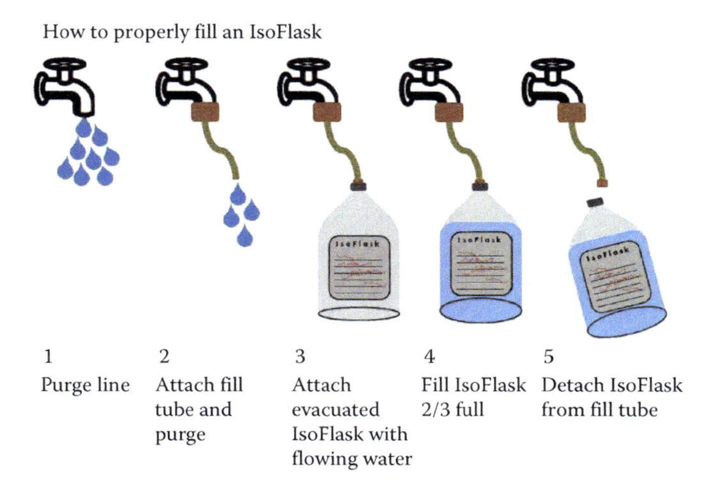

FIGURE 7.1 Sample collection for dissolved gas sampling. Dissolved gas sampling schematic by Isotech Laboratories. (From Isotech Laboratories, Collection of groundwater samples from domestic and municipal water wells for dissolved gas analysis, http://www.isotechlabs.com/customersupport/samplingprocedures/IsoBagSM.pdf, accessed August 6, 2014.)

water wells.[18] In Pennsylvania, approximately 350,000 legacy oil and gas wells have been drilled. Precise locations for ~100,000 of these wells are still unknown.[1] Additionally, stray gas and chemical contamination can result from gas well casing failures, an occurrence in approximately 3% of new gas well operations.[1] Some believe that thermogenic methane is naturally present in drinking waters of northeastern Pennsylvania due to the proximity of hydrocarbon-rich Catskill and Lock Haven formations, which may also include the presence of coal stringers that pass through the aquifers.[28]

7.5 ORGANIC CONSTITUENTS

Hydraulic fracturing fluids pumped into UOG wells and produced waters that return out of UOG wells are potential sources of anthropogenic organic compounds that could potentially interface with and contaminate groundwater systems. Stray gas can migrate vertically through improper or failing well construction or incipient faults and fractures, but hydraulic fracturing fluids would require significantly more intricate routes to travel to overlying aquifers,[17,18] assuming that hydraulic fracturing is performed thousands of meters below an aquifer. Upward migration of drilling fluids and produced waters would be greatly hampered due to their higher densities and viscosities compared to that of stray gas.[29,30] In contrast, a model considering the physical strain on the geological formation due to UOG drilling has posited an accelerated rate of upward brine migration, as soon as 6 years, and mixing with the aquifer.[31] Another hypothesized avenue for fluid to reach the overlying aquifers would be through induced fractures stretching thousands of feet upward to the groundwater.[31] However, others believe pressure from the overlying aquifer and various geological layers present between the shale formation and the aquifer would limit and even deflect the induced fractures originating from the borehole.[32,33]

Another mechanism for introducing hydraulic fracturing fluids into groundwater would be casing failures, which occur in the well at corresponding depths with the aquifer.[34] Despite being protected by multiple layers of piping and cement, the UOG drilling process does penetrate through the plane of the aquifer, and some recent findings indicate that structural failure of this protective casing can occur at a rate as high as 12% within the first year.[35] Groundwater can also become contaminated from other avenues, as shown in Figure 7.2, such as surface spills of chemicals or the improper containment of the flowback or produced waters, which may result in the leaching of chemicals into groundwater.[8]

While hydraulic fracturing fluids can be the source of organic compounds of concern, these fluids are actually about 95%–99% water. The small percentage of additives to the fluids typically includes a solid proppant and chemicals acting as lubricants, biocides, scale inhibitors, cross-linkers, surfactants, and corrosion inhibitors. The U.S. Congress has compiled a list of additives, volunteered from the 14 leading oil and gas service companies, revealing more than 2500 products used

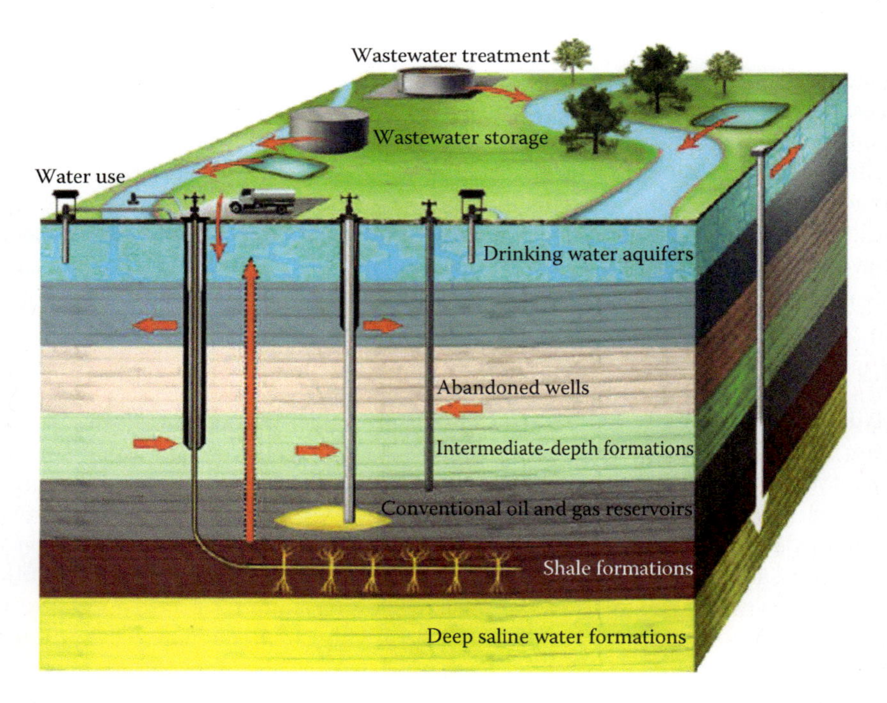

FIGURE 7.2 Depiction of possible pathways of contaminating drinking water aquifers. (Reprinted with permission from Vengosh, A., Jackson, R.B., Warner, N.R., Darrah, T.H., and Kondash, A., A critical review of the risks to water resources from unconventional shale gas development and hydraulic fracturing in the United States, *Environ. Sci. Technol.*, 48, 8334–8348. Copyright [2014] American Chemical Society.)

that contain over 750 chemical compounds from 2005 to 2009. In that time period, over 780 million gal of fracturing products, not including water, were used at registered UOG well sites.[36] A more recent compilation of chemical compounds used from the voluntary industry self-reporting website FracFocus (www.fracfocus.org) and the environmental monitoring organization SkyTruth (www.skytruth.org) lists 90 compounds that are publically disclosed.[37,38]

The analytical instruments that have been used to detect and identify potential contaminants in groundwater have predominately been GC–MS instruments. GC was chosen because the majority of the organic species are volatile or semi-volatile. Additionally, GC has been used extensively in the petroleum industry for decades. The MS detector has become almost a requirement when performing this research because of the unknown nature of possible contaminants and required confidence in compound identification. Many established industrial quality control and regulatory methods are written to utilize a flame ionization detection (FID) for the list of predetermined compounds of interest, but have demonstrated false-positives in contaminated groundwater related to UOG drilling because of the lack of qualitative information.[3]

The electron ionization (EI) source present in a typical GC–MS is a great tool for the initial process of identifying unknown compounds. The hard ionization of the source produces diagnostic fragments of each compound in a systematic and reproducible manner. The fingerprints can then be used for spectral matching across public libraries generated by the National Institute of Standards and Technology (NIST), National Institutes of Health (NIH), and EPA, among others. To complement the patterns of the resulting EI-generated spectra, a softer ionization technique such as chemical ionization (CI) can be used to identify the molecular ion of the compound. The negative chemical ionization (NCI) source can also be the source of choice for those wanting selective ionization of only halogenated compounds. NCI–MS allows for spectral confirmation of halogenated compounds while maintaining comparable picogram sensitivity with the electron capture detector (ECD) used for most quality control and regulated methods of halogenated species.

The type of MS detector (MSD) can also add capabilities and sensitivity to the analyses being performed. A single quadrupole (Q-MSD) can be classified as the simplest and most common option available. This MSD operates in scan or selected-ion monitoring (SIM) mode and offers unit resolution with no fragmentation capabilities. Unit resolution is the standard for which spectral databases have been created, so this is not a shortcoming for spectral matching in well-resolved mixtures. SIM with the Q-MSD improves sensitivity compared to scan operations, but the spectral information is limited. An ion trap (IT) or triple-quadrupole (QqQ) MSD allows for tandem mass spectrometry (MS/MS) operation, which means that generated ions in the gas phase can be fragmented to gain further confirmation of a chemical. The IT-MSD has a limited dynamic range; so it is generally not used extensively for quantitative purposes, but it can be operated in full scan mode without loss of sensitivity. The QqQ-MSD can handle many MS/MS events per second, making its speed valuable for achieving sufficient data points across a peak with faster chromatography. It is also very valuable for operation in selected or multiple reaction monitoring (SRM or MRM) modes, where limits of detection and specificity can be improved through the settings of optimal isolation, fragmentation, and detection windows for known analytes or analyte classes.

The time-of-flight (TOF) MSD can have the fastest scan speeds of the mentioned MSD. The rapid duty cycle of the TOF-MSD makes it suitable for the high efficiency peaks generated at high speed using two-dimensional GC (GC × GC). Confidence in library spectral matching of known and unknown compounds is improved by the increased chromatographic resolution of peaks from highly complex mixtures. A TOF-MSD does not allow for conventional MS/MS, but is capable of high resolution, mass accurate measurements up to four decimal places with less than 5 ppm error in mass accuracy. Such measurements allow for the generation of possible molecular formulae for ions of interest, since there are limited ways that carbon, nitrogen, oxygen, and sulfur, among other elements, can be combined to generate theoretical masses and isotopic abundances that match the accurately measured values. Newer versions of public spectral libraries are beginning to include accurate mass spectra for compounds when available. Many other tools to aid in this process are available commercially or on the web. A hybrid MSD such as a Q-TOF or IT-TOF can combine the benefits of MS/MS with high resolution spectra, but these are more commonly coupled with a liquid chromatography (LC)–MS system, rather than GC–MS systems.

The most elaborate investigation of organic compounds in an instance of possible contamination from UOG extraction was conducted by the EPA and the United States Geological Survey (USGS) in Pavillion, Wyoming.[3,4] The collected groundwater from private water wells and monitor wells was analyzed for volatile, semivolatile, gasoline, and diesel range organics,[39] and anionic surfactants according to multiple EPA methods are noted in Table 7.1.[3,4] The 8000-series methods are general screening methods that suggest multiple extraction and injection techniques for classes of analytes. The injection techniques for these 8000-series methods span a wide range with direct injection, static headspace, and purge-and-trap, with cryofocusing suggested for certain compounds. Suggested capillary columns are nonpolar 100% dimethyl polysiloxane, 95% dimethyl-, 5% phenyl polysiloxane phases, and a

TABLE 7.1
Summary of EPA Methods Used in Pavillion, WY

EPA Method	Analyte Class	Instrument
8260B	Volatiles	GC–MS
8270D	Semi-volatiles	GC–MS
8015D	Gasoline and diesel range organics	GC–FID
425.1	Methylene blue active substances	UV-VIS

Sources: DiGiulio, D.C. et al., Investigation of ground water contamination near Pavillion, Wyoming. U.S. Environmental Protection Agency Report, 2011; Wright, P.R. et al., Groundwater-quality and quality-control data for two monitoring wells near Pavillion, Wyoming, U.S. Geological Survey, Data Series 718, 2012.

mid-polarity 6% cyanopropyl phenyl, 94% dimethyl polysiloxane phase. The column dimensions are either 30 or 60 m lengths, standard- to mega-bore (0.25, 0.32 mm and 0.53, 0.75 mm ID, respectively), and film thicknesses ranging from 1 to 3 μm, all selected based upon a function of analyte selectivity, sample capacity, and needed resolution. Using these methodologies, the EPA reported detection of petroleum-based additives including gasoline and diesel range organics, BTEX (benzene, toluene, ethylbenzene, and *o*-, *m*-, and *p*-xylene), naphthalenes, and trimethylbenzenes.[3] Follow-up work performed by the USGS in 2012 could not confirm any of the compounds initially reported by the EPA.[4]

A verification study[40] of a draft standard operating procedure (SOP) developed by EPA was recently published online for the detection of 2-methoxy ethanol (2-ME), 2-butoxyethanol (2-BE), diethylene glycol (Di-EG), triethylene glycol (Tri-EG), and tetraethylene glycol (Tetra-EG) in drinking water using LC–MS/MS. The column used was a Waters Atlantis dC18 (Milford, MA; 2.1 × 150 mm, 3 μm), which is able to withstand 100% aqueous separation conditions for polar compounds such as the five chosen. Some laboratories in the multi-laboratory study had difficulty ionizing and detecting 2-ME and 2-BE. The report also includes a page of recommendations of analytical parameters that need to be further investigated to improve the robustness and accuracy of data acquisition and analysis.

A survey of groundwater across the Barnett Shale[5] used a semi-targeted GC–MS method for compounds commonly found in hydraulic fracturing fluids (according to the previously mentioned Congressional Report[36]). The compounds consisted of alcohols, aromatics, ethers, and other hydrocarbons. This method utilized SIM on a Q-MSD for the base peak of analytes of interest and a scan throughout the method to obtain spectral information for SIM peaks and to detect potential non-targets also present in the sample. Attention was paid to allow enough cycle time for the scan event and still provide sufficient data points across the chromatographic peak. This work was performed with a 95% dimethyl-, 5% diphenyl polysiloxane column. This column was not sufficient for retaining and resolving light alcohols and solvents for quantification, so a second, more selective method was also developed.

Light solvents in groundwater were sampled by an automated headspace sampler and separated and detected with a ZB-BAC2 column (Phenomenex; Torrance, CA; 30 m × 0.32 × 1.2 μ) originally designed for use in blood alcohol content analyses.[41] An FID was chosen as the detector over the MS in this application based on detector noise because the light alcohol ions would fall below an *m/z* of 50, a spectral region in which it is difficult to quantitate with an MSD due to interferences from constant atmospheric gas ions. Through this method, methanol and ethanol were quantified in groundwater. These alcohols are multifaceted in UOG fluids, commonly used for anticorrosive, cross-linking, stabilizing, and surfactant properties.[37] The alcohol concentrations did not correlate to UOG well proximity in this study.[5]

Trihalomethanes (THMs) are another class of compounds that may be found in groundwater containing high dissolved solids and stray methane gas.[8] THMs are compounds with three halogen atoms, most commonly chlorine or bromine, substituted for three hydrogens in the methane molecule. These chemical compounds are typically encountered as disinfection by-products of chlorination treatment in drinking water.[42] Chlorination of drinking water has reduced the mortality rate from infectious diseases, but THMs that can be generated from the process have been proven to be carcinogenic.[43] The formation water is a brine solution that contains significant amounts of bromide, which can facilitate formation of bromoform ($CHBr_3$), a compound with worse health effects than chlorinated analogs.[43]

Endocrine-disrupting chemicals have also been associated with groundwater in proximity to UOG processes in Garfield, CO. In this area, groundwater contamination had already been identified in the form of stray gas and salinization[44] before the effects of estrogenic compounds were detected through reporter gene assays.[45] No further efforts were made to identify the specific endocrine disrupting chemicals in that study; however, many metabolites of volatile and semi-volatile aromatic compounds (such as BTEX) are known to form highly endocrine disrupting compounds (e.g., phenols) during metabolism and environmental degradation.[46]

7.6 IONS AND ISOTOPES

Modern techniques to recover previously inaccessible hydrocarbon can result in the introduction of highly ionic and/or metallic compounds or solutions into groundwater. This can occur through both mechanical processes and as a byproduct of the geological formations that are affected by drilling activity. The drilling muds used during wellbore completion contain a large proportion of barite, a mineral of barium sulfate, which is used as a weighting agent for the mud to help control the subsurface pressure. Hydraulic fracturing and well completion fluids contain various salts, acids, and bases for pH control.[37] Multiple geological formations may be encountered on the way to the target shale formation, each with their own properties and ion or metal compositions that usually differ from that of overlying aquifers. Intrusion of sealants and cements from the wellbore can introduce clays with varying metallic properties depending on pH. Once hydraulic fracturing has commenced, the flowback water that resurfaces during hydraulic fracturing is a mixture of the original fracturing fluid and the shale formation water, which is often quite saline. The formation brines can contain high levels of chloride, bromide, lithium, sodium, barium, and strontium ionic species.[47]

Elemental analysis is a popular technique for assessing the possible introduction of drilling fluids or formation brines into groundwater because of the drastic concentration differences between the aquifer and foreign solutions. Typically, more than one element of interest is measured when surveying groundwater, whether quantifying naturally occurring major water ions or those thought to be related to UOG activities. To increase throughput, simultaneous measurement techniques such as inductively coupled plasma–optical emission spectroscopy (ICP–OES) or mass spectrometry (ICP–MS) are implemented as opposed to single-element methods like flame atomic absorption (FAA), graphite furnace atomic absorption (GFAA), or ion selective electrodes.[48] The ICP source can be affected by various forms of physical and chemical matrix effects that can lower quantification accuracy.[49,50] The two detectors each come with their own considerations. The mass spectrometry is susceptible to interference from isobaric ions, for example, $^{40}Ar\ ^{35}Cl^+$ and $^{75}As^+$ or $^{40}Ar\ ^{16}O^+$ and $^{56}Fe^+$, and OES can have spectral overlap between separate emission lines. Expected analyte concentration can also dictate detector selection, with MS

TABLE 7.2
Instrument Selection Matrix

	FAA	GFAA	ICP–OES	ICP–MS
How many elements?				
Single, >ppm	✓			
Single, <ppm		✓		
Many, >ppm			✓	
Many, <ppm			✓	✓
What sensitivity?				
ppm	✓		✓	
ppb		✓	✓	✓
ppt				✓

Source: Adapted from PerkinElmer, Inc., Atomic spectroscopy—A guide to selecting the appropriate technique and system, http://www.perkinelmer.com/CMSResources/Images/44-74482BRO_WorldLeaderAAICPMSICPMS.pdf, accessed September 10, 2013.

generally being the more sensitive one; however, OES detectors can provide up to 10 orders of magnitude in analytical working range.[51] The greater sensitivity of mass spectrometry can be a drawback in cases, as OES is much more robust to samples with high solute concentrations. Table 7.2 is a decision matrix to assist in proper instrument selection based upon experimental needs.

Even with the advantages of increased data and throughput with simultaneous techniques, there is a market for metal- and ion-specific assays, primarily in field-test kits, which need to be portable, rugged, and easy to use. Drilling operators can use quick measurements for single elements such as barium, chloride, and iron to monitor potential contamination events in real time due to their understanding of when, where, and how much of these constituents may be expected to be involved in any particular phase of UOG drilling activity. Hach, a manufacturer of analytical equipment for water quality measurements, published the *Hydraulic Fracturing Water Analysis Handbook*, a compilation of colorimetric single-analyte measurements relevant to water used in various stages of the hydraulic fracturing process.[52] Each method has listed interferents from other ions in the solution, and their accuracy becomes more limited as samples become more complex. Elements and ions considered to be the most relevant for such a monitoring strategy are barium, boron, chloride, iron, sulfate, and sulfide.

Elemental analysis has proven useful not only for identifying groundwater contamination from target elements, but also for providing valuable information about the nature of contamination events when correlated with measurements of other contaminants of concern. Water wells in Colorado were identified to have increased chloride concentrations which were correlated with a concordant increase in dissolved thermogenic methane as UOG activity increased in the area over time.[44] To date, groundwater affected by stray gas in Pennsylvania has not shown similar signs of salinization as seen in Colorado, thus illustrating how region-specific industrial practices and geological conditions can result in different contamination scenarios in different regions.[8]

Some contamination events have also been attributed to byproducts from UOG activities and waste disposal errors.[1] Groundwater in Wyoming contained elevated levels of potassium and chloride, as well as an alkaline pH.[3,4] Well completion and MSDS reports for products used in UOG activities in this area indicate that KOH was used as a cross-linker and solvent. Additionally, 6% KCl solutions were used in foam jobs and up to 60% potassium metaborate was used for cross-linking products.[3] Other metals that have been found at considerably elevated levels in flowback waters from the Marcellus Shale formation are boron, bromide, arsenic, selenium, iron, lithium, strontium, sodium, calcium, magnesium, barium, radium, and manganese.[53,54] This wastewater can enter groundwater if not recovered at the wellhead and/or properly confined until disposal.[7]

Selenium and arsenic are two metals present in many geological formations. Both selenium and arsenic were detected at elevated concentrations in groundwater from areas of active UOG extraction in the Barnett Shale formation of Texas.[5] The most likely mechanism for these occurrences was mechanical or vibrational disturbance of poorly maintained private water wells that have accumulated rust scale which can form complexes with arsenic and selenium.[55] Liberating this scale would mobilize the iron oxide and complexed metals, increasing the turbidity and metal content of the groundwater. Strontium and barium also form sulfate or carbonate scale along the inner casings of these same wells and could also be mobilized through the same mechanism. Arsenic, selenium, strontium, and TDS each had a proximity correlation with UOG wells. Figure 7.3 depicts the location and concentration level of arsenic that was present in groundwater in the Barnett Shale region in 2011.

The ratio and bivariate plots of particular metals can also be used to speciate various formation waters as well as foreign compounds. Shale formation water mixed with shallow aquifers will have a higher bromide to chloride (Br/Cl) ratio (>0.001).[56] Other elemental ratios that have been used to compare groundwater to formation waters are Na/Cl, Sr/Cl, Ba/Cl, and Li/Cl.[56] Linear plots of bromide, sodium, strontium, lithium, and barium with chloride in groundwater and chloride in other potential sources (e.g., brine, road salt, septic effluent, or animal waste) can also indicate the origin of elevated chloride concentrations in groundwater.[7,47]

Elemental isotopes can also be used to characterize the source(s) of any intrusions into the groundwater. Isotopic measurements must be made with an isotope ratio MS, which possesses a resolution and stability sufficient to precisely measure isotopic abundances. For example, the ratio of strontium-87 to strontium-86 ($^{87}Sr/^{86}Sr$) is specific to particular geological formations. The difference between the Sr ratio of unaffected waters and the Sr ratio of contaminated waters can vary by as little as only a few ten thousandths,[56] but the isotope ratio MS can distinguish this variation. Isotopic abundances of ^{13}C, ^{18}O, and ^{2}H may also be used to identify the origin of water, methane, and dissolved inorganic carbon.[8] The use of ^{13}C and ^{2}H

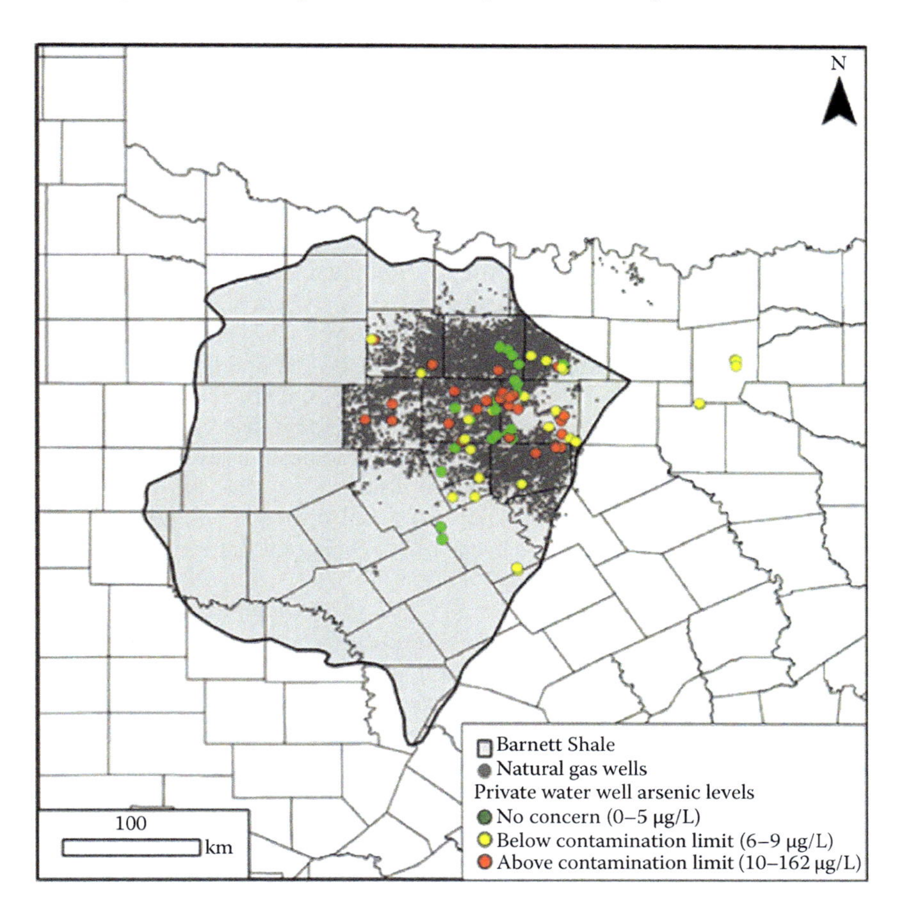

FIGURE 7.3 Location and concentration of arsenic in the Barnett Shale in 2011. (From Fontenot, B.E. et al., *Environ. Sci. Technol.*, 47, 10032, 2013.)

isotopes to distinguish between methane gas generated through microbial processes and that generated from carbonaceous material (i.e., biogenic versus thermogenic methane) has already been discussed earlier. The hydrogen isotope tritium, ^{3}H, and the ratio of $^{3}He/^{4}He$ can also be used as environmental tracers of formation water intrusion.[4] Additionally, the isotope ^{11}B can be used to characterize the source of dissolved boron constituents in the groundwater.[6]

While isotopic measurements have been used for identifying subsurface mixing through geochemical fingerprinting, the detection and measuring of isotopic ratios of naturally occurring radioactive material (NORM) is also of tremendous value. This is due to the human health implications when radioactive materials are introduced into aquifers that provide drinking and irrigation water in rural and suburban areas. The radioactive isotopes of radium (^{226}Ra and ^{228}Ra), the natural decay products of uranium-rich geological formations, are present in flowback water generated in the Marcellus Shale and possibly other shale formations.[7] Levels of NORMs in produced water reported in the scientific literature are not high enough to induce acute radiotoxicity,[57] but sediments downstream of wastewater treatment plants that have treated flowback water can exhibit ^{226}Ra levels approximately 200 times greater than background sediments.[7] The abundance ratio of ^{228}Ra to ^{226}Ra can also be used for geochemical fingerprinting to identify the likely source of radium.[8]

The wet chemical methods developed by the EPA for radium isotopes (methods 903.0 and 904.0) were to test drinking water and are inadequate for the high ionic strength of UOG flowback waters. Five radium quantification methods including barium sulfate co-precipitation (EPA method 903.3), radium extraction disks, emanation of ^{222}Rn decay product, manganese dioxide pre-concentration, and gamma spectroscopy were compared for their efficacy.[58] EPA method 903.3 using barium sulfate co-precipitation yielded a recovery <1% of ^{226}Ra. The techniques of MnO_2 concentration, ^{222}Rn measuring, and gamma spectroscopy had recoveries >90%.[58] If one is unable to speciate radium isotopes, bulk radioactivity of the groundwater can still indicate brine mixing. Progress has been made toward an improved EPA method for gross alpha and gross beta activity and flowback and produced water from hydraulic fracturing operations by measuring bulk radioactivity from uranium, thorium, radium, lead, and polonium.[59]

Measurement of metal oxidation states and polyatomic forms, for example, As(III)/As(IV) and Br^-/BrO_3^-, is sometimes used to further characterize the groundwater system. IC with conductivity detection is the most amenable route for speciating these ions. Instrumentation has recently been marketed for measurement of in-line total conductivity with automatic dilution, suitable for handling brine and flowback waters of unknown conductivities.[60] Some researchers have suggested that chloride concentration measurements are a good proxy for possible groundwater contamination from UOG activity because of the low solution mobility and elevated concentrations of chloride in produced waters.[61]

7.7 SUMMARY AND CONCLUDING REMARKS

Significant progress has been made by researchers in recent years toward developing analytical methods and furthering our understanding of the potential impacts that the rapid and recent shale gas boom may have on our groundwater resources. However, this field of research is still in its infancy and would benefit greatly from collaborative efforts between researchers in fields such as hydrogeology, petroleum and chemical engineering, human health research, ecology, and analytical chemistry. Collaboration between industry and academia would also provide exceptionally useful opportunities to advance knowledge. Continued research will not only bring a stronger understanding of the relationship between UOG drilling and groundwater quality, but it may also influence current and/or pending legislation on the issue. California, Colorado, Illinois, New York, Pennsylvania, and several countries in Europe, Asia, and South America already have or are currently in the process of creating legislation to protect groundwater resources in UOG drilling areas. These types of legislation may suggest best management practices to prevent environmental impacts and/or measurements of various groundwater constituents to monitor for contamination events, but further research is needed to ensure that these approaches will encompass all potential contaminants and provide the highest level of confidence that UOG activities can be carried out safely.[62]

REFERENCES

1. Vidic, R. D.; Brantley, S. L.; Vandenbossche, J. M.; Yoxtheimer, D.; Abad, J. D. Impact of shale gas development on regional water quality. *Science* 2013, *340*, 1235009.
2. Hildenbrand, Z. L.; Osorio, A.; Carlton Jr., D. D.; Fontenot, B. E.; Walton, J. L.; Hunt, L. R.; Oka, H.; Hopkins, D.; Bjorndal, B.; Schug, K. A. Rapid analysis of eukaryotic bioluminescence to detect instances of groundwater contamination. *J. Chem.* Submitted.
3. DiGiulio, D. C.; Wilkin, R. T.; Miller, C.; Oberley, G. Investigation of ground water contamination near Pavillion, Wyoming. US EPA 2011. http://www2.epa.gov/sites/production/files/documents/EPA_ReportOnPavillion_Dec-8-2011.pdf (accessed March 14, 2014).
4. Wright, P. R.; McMahon, P. B.; Mueller, D. K.; Clark, M. L. Groundwater-quality and quality-control data for two monitoring wells near Pavillion, Wyoming. U.S. Geological Survey 2012, Data Series 718. http://pubs.usgs.gov/ds/718/ (accessed March 14, 2014).
5. Fontenot, B. E.; Hunt, L. R.; Hildenbrand, Z. L.; Carlton, D. D., Jr.; Oka, H.; Walton, J. L.; Hopkins, D. et al. An evaluation of water quality in private drinking water wells near natural gas extraction sites in the Barnett Shale formation. *Environ. Sci. Technol.* 2013, *47*, 10032–10040.
6. Osborn, S. G.; Vengosh, A.; Warner, N. R.; Jackson, R. B. Methane contamination of drinking water accompanying gas-well drilling and hydraulic fracturing. *Proc. Natl. Acad. Sci. U.S.A.* 2011, *108*, 8172–8176.
7. Warner, N. R.; Christie, C. A.; Jackson, R. B.; Vengosh, A. Impacts of shale gas wastewater disposal on water quality in western Pennsylvania. *Environ. Sci. Technol.* 2013, *47*, 11849–11857.
8. Vengosh, A.; Jackson, R. B.; Warner, N. R.; Darrah, T. H.; Kondash, A. A critical review of the risks to water resources from unconventional shale gas development and hydraulic fracturing in the United States. *Environ. Sci. Technol.* 2014, *48*, 8334–8348.
9. Meng, Q.; Ashby, S. Distance: A critical aspect for environmental impact assessment of hydraulic fracking. *Extr. Ind. Soc.* 2014, *1*, 124–126.
10. Dunkle, S. A.; Plummer, L. N.; Busenberg, E.; Phillips, P. J.; Denver, J. M.; Hamilton, P. A.; Michel, R. L.; Coplen, T. B. Chlorofluorocarbons (CCl_3F and CCl_2F_2) as dating tools and hydrologic tracers in shallow groundwater of the Delmarva Peninsula, Atlantic Coastal Plain, United States. *Water Resour. Res.* 1993, *29*, 3837–3860.
11. Ekwurzel, B.; Schlosser, P.; Smethie, W. M.; Plummer, L. N.; Busenberg, E.; Michel, R. L.; Weppernig, R.; Stute, M. Dating shallow groundwater: Comparison of the transient tracers $^{3}H/^{3}He$, chlorofluorocarbons, and ^{85}Kr. *Water Resour. Res.* 1994, *30*, 1693–1708.
12. Busenberg, E.; Plummer, L. N. Dating young groundwater with sulfur hexafluoride—Natural and anthropogenic sources of sulfur hexafluoride. *Water Resour. Res.* 2000, *36*, 3011–3030.
13. Plummer, L. N.; Bexfield, L. M.; Anderholm, S. K.; Sanford, W. E.; Busenberg, E. Geochemical characterization of groundwater flow in the Santa Fe Group aquifer system, Middle Rio Grande Basin, New Mexico. Geological Survey Water-Resources Investigations Report 03-4131, 2004, p. 395.
14. Osborn, S. G.; McIntosh, J. C. Chemical and isotopic tracers of the contribution of microbial gas in Devonian organic-rich shales and reservoir sandstones, northern Appalachian Basin. *Appl. Geochem.* 2010, *25*, 456–471.

15. Jackson, R. B.; Vengosh, A.; Darrah, T. H.; Warner, N. R.; Down, A.; Poreda, R. J.; Osborn, S. G.; Zhao, K.; Karr, J. D. Increased stray gas abundance in a subset of drinking water wells near Marcellus shale gas extraction. *Proc. Natl. Acad. Sci. U.S.A.* 2013, *110*, 11250–11255.
16. Stolper, D. A.; Lawson, M.; Davis, C. L.; Fereira, A. A.; Santos Neto, E. V.; Ellis, G. S.; Lewan, M. D. et al. Formation temperature of thermogenic and biogenic methane. *Science* 2014, *344*, 1500–1503.
17. Kissinger, A.; Helmig, R.; Ebigbo, A.; Class, H.; Lange, T.; Sauter, M.; Heitfeld, M.; Klünker, J.; Jahnke, W. Hydraulic fracturing in unconventional gas reservoirs: Risks in the geological system, part 2. *Environ. Earth Sci.* 2013, *70*, 3855–3873.
18. Jackson, R. E.; Gorody, A. W.; Mayer, B.; Roy, J. W.; Ryan, M. C.; Van Stempvoort, D. R. Groundwater protection and unconventional gas extraction: The critical need for field-based hydrogeological research. *Ground Water* 2013, *51*, 488–510.
19. Schlumberger Oilfield Glossary: Annulus. http://www.glossary.oilfield.slb.com/en/Terms.aspx?LookIn=term%20name&filter=annulus (accessed July 17, 2014).
20. Schoell, M. The hydrogen and carbon isotopic composition of methane from natural gases of various origins. *Geochim. Cosmochim. Acta* 1980, *44*, 649–661.
21. ASTM Standard D1945-03. Standard test method for analysis of natural gas by gas chromatography. American Society for Testing and Materials, Conshohocken, PA, 2010.
22. Chambers, L. Purge-and-trap GC analysis of methane in water samples associated with hydraulic fracturing. Application Note 37920312, 2012, http://pdf.directindustry.com/pdf/oi-analytical/purge-and-trap-gc-analysis-water-samples-associated-wtih-hydraulic-fracturing-natural-gas-extraction/28296-400161.html (accessed April 11, 2012).
23. Hudson, F. Sample preparation and calculations for dissolved gas analysis in water samples using a GC headspace equilibrium technique.RSK SOP-175, 2004, http://www.epa.gov/region1/info/testmethods/pdfs/RSKsop175v2.pdf (accessed March 15, 2014).
24. Ji, Z. GC/TCD analysis of a natural gas sample on a single HP-PLOT Q column. Application Note 228-387, 2000, http://www.chem.agilent.com/Library/applications/HPI_Compendium%20-%20Crude%20Oil%20and%20Natural%20Gas.pdf (accessed March 15, 2014).
25. Burger, B. Res-Sil™ C bonded GC packings for analyses of light hydrocarbons. http://www.restek.com/Technical-Resources/Technical-Library/Petroleum-Petrochemical/petro_A001 (accessed August 24, 2014).
26. Restek Corporation. ShinCarbon ST micropacked GC columns. Lit. Cat.# PCTS1472-UNV, 2011, http://www.restek.com/pdfs/PCTS1472-UNV.pdf (accessed March 25, 2014).
27. Isotech Laboratories. Collection of groundwater samples from domestic and municipal water wells for dissolved gas analysis. http://www.isotechlabs.com/customersupport/samplingprocedures/IsoBagSM.pdf (accessed August 6, 2014).
28. Wilson, B. Geologic and baseline groundwater evidence for naturally occurring, shallowly sourced, thermogenic gas in northeastern Pennsylvania. *Am. Assoc. Petrol. Geol. Bull.* 2014, *98*, 373–394.
29. Flewelling, S. A.; Sharma, M. Constraints on upward migration of hydraulic fracturing fluid and brine. *Ground Water* 2013, *52*, 9–19.
30. Harrison, S. S. Contamination of aquifers by overpressuring the annulus of oil and gas wells. *Ground Water* 1985, *23*, 317–324.
31. Myers, T. Potential contaminant pathways from hydraulically fractured shale to aquifers. *Ground Water* 2012, *50*, 872–882.
32. Flewelling, S. A.; Tymchak, M. P.; Warpinski, N. Hydraulic fracture height limits and fault interactions in tight oil and gas formations. *Geophys. Res. Lett.* 2013, *40*, 3602–3606.

33. Fisher, K.; Warpinski, N. Hydraulic fracture-height growth: Real data. *SPE Annual Technical Conference*, Denver, CO 2011, SPE 145949, pp. 1–18.
34. Bureau of Oil & Gas Regulation. Well permit issuance for horizontal drilling and high-volume hydraulic fracturing to develop the Marcellus Shale and other low-permeability gas reservoirs. http://www.dec.ny.gov/energy/58440.html (accessed April 17, 2012).
35. Ingraffea, A. R.; Wells, M. T.; Santoro, R. L.; Shonkoff, S. B. C. Assessment and risk analysis of casing and cement impairment in oil and gas wells in Pennsylvania, 2000–2012. *Proc. Natl. Acad. Sci. U.S.A.* 2014, *111*, 10955–10960.
36. US House of Representatives Committee on Energy and Commerce. Chemicals used in hydraulic fracturing. 2011, http://democrats.energycommerce.house.gov/sites/default/files/documents/Hydraulic-Fracturing-Chemicals-2011-4-18.pdf (accessed October 11, 2011).
37. Stringfellow, W. T.; Domen, J. K.; Camarillo, M. K.; Sandelin, W. L.; Borglin, S. Physical, chemical, and biological characteristics of compounds used in hydraulic fracturing. *J. Hazard Mater.* 2014, *275*, 37–54.
38. Ritter, S. K. Mixed review for fracking chemicals. *C&EN* 2014, *92,* 7.
39. Environmental Integrity Project. Fracking beyond the law. http://environmentalintegrity.org/wp-content/uploads/Fracking-Beyond-the-Law.pdf, pp. 1–21 (accessed August 20, 2014).
40. Schumacher, B. A.; Zintek, L. The verification of a method for detecting and quanti-fying diethylene glycol, triethylene glycol, tetraethylene glycol, 2-butoxyethanol and 2-methoxyethanol in ground and surface waters. 2014, EPA/600/R-14/008, http://www2.epa.gov/sites/production/files/2014-07/documents/schumacherzintecverification_glycols_method.pdf (accessed August 24, 2014).
41. Fernandez, C.; Kelly, K.; Countryman, S. Improved separation of blood alcohols using Zebron ZB-BAC1 and BAC2. TN-2036 Applications, 2009, http://az621941.vo.msecnd.net/documents/66f5bc58-c01f-45a8-9544-1584351dbf9f.pdf (accessed March 31, 2014).
42. Richardson, S. D. Environmental mass spectrometry: Emerging contaminants and current issues. *Anal. Chem.* 2012, *84*, 747–778.
43. Boorman, G. A.; Dellarco, V.; Dunnick, J. K.; Chapin, R. E.; Hunter, S.; Hauchman, F.; Gardner, H.; Cox, M.; Sills, R. C. Drinking water disinfection byproducts: Review and approach to toxicity evaluation. *Environ. Health Perspect.* 1999, *107*, 207–217.
44. Thyne, T. Review of phase II hydrogeologic study (Prepared for Garfield County). 2008, http://s3.amazonaws.com/propublica/assets/methane/thyne_review.pdf (accessed March 15, 2014).
45. Kassotis, C. D.; Tillitt, D. E.; Davis, W.; Hormann, A. M.; Nagal, S. C. Estrogen and androgen receptor activities of hydraulic fracturing chemicals and surface and ground water in a drilling-dense region. *Endocrinology* 2014, *155*, 897–907.
46. Bahadar, H.; Mostafalou, S.; Abdollahi, M. Current understanding and perspective on non-cancer health effects of benzene: A global concern. *Toxicol. Appl. Pharm.* 2014, *276*, 83–94.
47. Lautz, L. K.; Hoke, G. D.; Lu, Z.; Siegel, D. I.; Christian, K.; Kessler, J. D.; Teale, N. G. Using discriminant analysis to determine sources of salinity in shallow groundwater prior to hydraulic fracturing. *Environ. Sci. Technol.* 2014, *48*(16), 9061–9069.
48. Olesik, J. Elemental analysis using ICP-OES and ICP/MS: An evaluation and assessment of remaining problems. *Anal. Chem.* 1991, *63*, 12A–21A.
49. Agatemor, C.; Beauchemin, D. Matrix effects in inductively coupled plasma mass spectrometry: A review. *Anal. Chim. Acta* 2011, *706*, 66–83.

50. Carlton Jr., D. D.; Fontenot, B. E.; Hildenbrand, Z. L.; Davis, T. M.; Walton, J. L.; Schug, K. A. Varying matrix effects for elemental analysis identified from groundwater in the Barnett Shale. *Int. J. Environ. Sci. Technol.* In review.
51. PerkinElmer, Inc. Atomic spectroscopy—A guide to selecting the appropriate technique and system. http://www.perkinelmer.com/CMSResources/Images/44-74482BRO_WorldLeaderAAICPMSICPMS.pdf (accessed September 10, 2013).
52. Hach. *Hydraulic Fracturing Water Analysis Handbook.* 2013, DOC022.53.80225, http://www.hach.com/fracingDownloads (accessed July 29, 2014).
53. Haluszczak, L. O.; Rose, A. W.; Kump, L. R. Geochemical evaluation of flowback brine from Marcellus gas wells in Pennsylvania, USA. *Appl. Geochem.* 2013, *28*, 55–61.
54. Jack, R.; Kondratyuk, T.; Stolz, J. *In Analysis of Flowback Water from Marcellus Unconventional Gas Extraction Using IC and ICP-OES.* Pittcon: Chicago, IL, 2014.
55. Groat, C. G.; Grimshaw, T. W. Fact-based regulation for environmental protection in shale gas development. 2006, http://www.velaw.com/UploadedFiles/VEsite/Resources/ei_shale_gas_reg_summary1202[1].pdf (accessed April 5, 2014).
56. Warner, N. R.; Jackson, R. B.; Darrah, T. H.; Osborn, S. G.; Down, A.; Zhao, K.; White, A.; Vengosh, A. Geochemical evidence for possible natural migration of Marcellus Formation brine to shallow aquifers in Pennsylvania. *Proc. Natl. Acad. Sci. U.S.A.* 2012, *109*, 11961–11966.
57. Rowan, E.; Engle, M.; Kirby, C.; Kraemer, T. Radium content of oil- and gas-field produced waters in the Northern Appalachian basin (USA): Summary and discussion of data. *Sci. Invest. Rep. (U.S. Geol. Surv.)* 2011, *5135*, 31.
58. Nelson, A. W.; May, D.; Knight, A. W.; Eitrheim, E. S.; Mehrhoff, M.; Shannon, R.; Litman, R.; Schultz, M. K. Matrix complications in the determination of radium levels in hydraulic fracturing flowback water from Marcellus Shale. *Environ. Sci. Technol. Lett.* 2014, *1*, 204–208.
59. Schumacher, B. A.; Griggs, J.; Askren, D.; Litman, B.; Shannon, B.; Mehrhoff, M.; Nelson, A.; Schulz, M. K. Development of rapid radiochemical method for gross alpha and gross beta activity concentration in flowback and produced waters from hydraulic fracturing operations. 2014, EPA/600/R-14/107, http://www2.epa.gov/hfstudy/development-rapid-radiochemical-method-gross-alpha-and-gross-beta-activity-concentration (accessed August 24, 2014).
60. Fisher, C.; Lopez, L. Analysis of fracking flowback water from Marcellus Shale using in-line conductivity, automated dilution, and ion chromatography. http://apps.thermoscientific.com/media/cmd/hypersite-events/Pittcon-2014/posters/PN70994_PC2014.pdf (accessed March 21, 2014).
61. Kresse, T. M.; Warner, N. R.; Hays, P. D.; Down, A.; Vengosh, A.; Jackson, R. B. Shallow groundwater quality and geochemistry in the Fayetteville shale gas-production area, North-Central Arkansas, 2011. 2012, Scientific Investigations Report 2012-5273, http://www.aogc.state.ar.us/notices/USGS%20Final%20Report.pdf (accessed March 17, 2014).
62. Simon, J. A. Editor's perspective—An update on the hydraulic fracturing groundwater contamination debate. *Remed. J.* 2014, *24*(2), 1–9.

8 Water Availability in the Permian Basin Region of West Texas

Venkatesh Uddameri and Danny Reible

CONTENTS

8.1 Overview 133
8.2 Introduction 133
8.3 Overview of Water Resources in the Study Area 135
8.4 Overall Groundwater Use within the Study Area 139
8.5 How Much Water Is Used for Unconventional Oil and Gas Production in West Texas? 142
8.6 Factors Influencing Water Use for Unconventional Oil and Gas Production and Associated Water Use 145
8.7 Does Increased Hydraulic Fracturing Imply Greater Freshwater Production in the Permian Basin? 146
8.8 Groundwater Production for Fracturing—Local-Scale Effects 150
8.9 Groundwater Production for Fracturing and Competition with Agriculture 153
8.10 Summary and Closing Remarks 155
Acknowledgment 156
Note on Units 156
References 156

8.1 OVERVIEW

This chapter discusses three broad strategies, namely, brackish groundwater use, reuse and recycle of produced water, and agriculture water conservation for reducing the freshwater footprint of hydraulic fracturing with an emphasis on the Permian Basin region of Texas.

8.2 INTRODUCTION

The Permian Basin is considered to be the richest hydrocarbon basin in the United States with 14 identified oil- and gas-bearing geologic plays. It is estimated to contain 106 billion barrels of oil or roughly one-fourth of all the oil discovered in the United States (Ruppel, 2009). Advances in drilling technologies, particularly

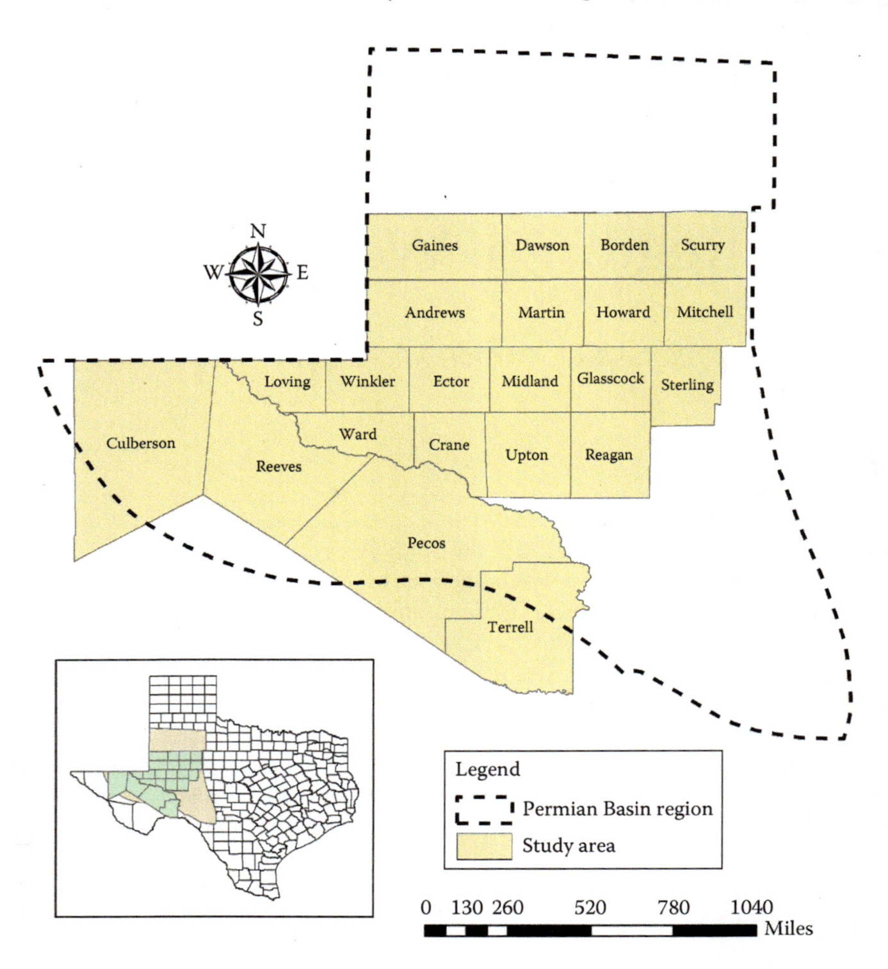

FIGURE 8.1 Permian basin region of Texas and the 22-county study area.

hydraulic fracturing and horizontal drilling, are likely to increase these estimates. The Permian Basin in Texas occupies an area of nearly 130,000 km^2 (50,000 sq mi). Figure 8.1 depicts its geographic extent and highlights a 22-county study area where oil and gas production has increased considerably in the last few years. The increase is primarily due to the use of advanced horizontal drilling and hydraulic fracturing technologies in shale plays such as the Barnett–Woodford and Avalon–Bone Spring. Increased oil and gas production in the region has contributed to the nation's decreased reliance on foreign oil (EIA, 2014) and boosted the regional economy (Ewing et al., 2014). Oil and gas production in this region is projected to increase significantly in years to come.

Water is an important input for both conventional and unconventional oil and gas production. In the case of the Permian Basin region of Texas, a semiarid region that is experiencing a prolonged drought in recent years, water is a scarce resource.

Irrigation agriculture is practiced extensively in this predominantly rural area. Conventional oil and gas production in the Permian Basin region peaked during the 1970s and tapered off considerably over the next three decades. As such, the water use associated with the recent surge in oil and gas production can be viewed as a new stress on the region's water resources. In particular, unconventional oil and gas production techniques involving hydraulic fracturing and horizontal wells require more freshwater than conventional oil production techniques. The projected proliferation of unconventional oil and gas activities and the associated water use is perceived by some as the reason for water shortages and a direct threat to the very existence of rural communities and their way of life in this region (Goldenberg, 2013). Others point to agricultural irrigation and prolonged droughts as reasons for water deficits (e.g., Blackmon, 2013). Water produced for oil and gas exploration and production is exempt from permitting by statute in Texas, while water use for municipal, agricultural, and other industrial activities may be regulated through local groundwater conservation districts. This policy choice also fuels the concerns of unfettered and potentially large-scale water withdrawals by oil and gas operators in many rural areas.

Water is indeed the lifeblood for the economic well-being of communities in arid and semiarid areas. Sustaining this resource is clearly important for long-term vitality of the Permian Basin region. Therefore, it is critical to understand (1) how much water is used by oil and gas operations, particularly through unconventional methods, (2) what factors control the water use by this industry, (3) where does the water come from, (4) do water withdrawals for fracturing pose competition with other water users in the region, and (5) what can be done to mitigate any potential water stress caused by unconventional oil and gas production and to reduce the overall water footprint in the region. The answers to these questions form the focus of the present study and are essential to ensure scarce water resources of the region are managed sustainably with minimal impacts on the economy and the environment. We begin by presenting a broad overview of water resources within the study area and then discuss the questions related to water use, particularly as they pertain to hydraulic fracturing.

8.3 OVERVIEW OF WATER RESOURCES IN THE STUDY AREA

Figure 8.2a–d depicts the major sources of water in the region. Precipitation in the region progressively decreases westward, from an annual average of approximately 55–25 cm (22 in. to about 10 in.). Most of the region has an average precipitation of 38–45 cm (15–18 in.). Rainfall patterns tend to be highly erratic, and most of the rainfall volume occurs in a few storm events. Such high-intensity, short-duration events result in considerable surface runoff, which is subsequently evaporated. These rainfall events are not conducive to infiltration and eventual aquifer recharge. In addition to noted spatial heterogeneity, the rainfall also exhibits considerable inter-annual variability with extremely wet years followed by unusually dry years (see Figure 8.3 for an example from Martin County, TX).

Surface water bodies occupy less than 1% of the total land area in the Permian Basin region (Figure 8.2). The Pecos River is the largest river, along with tributaries

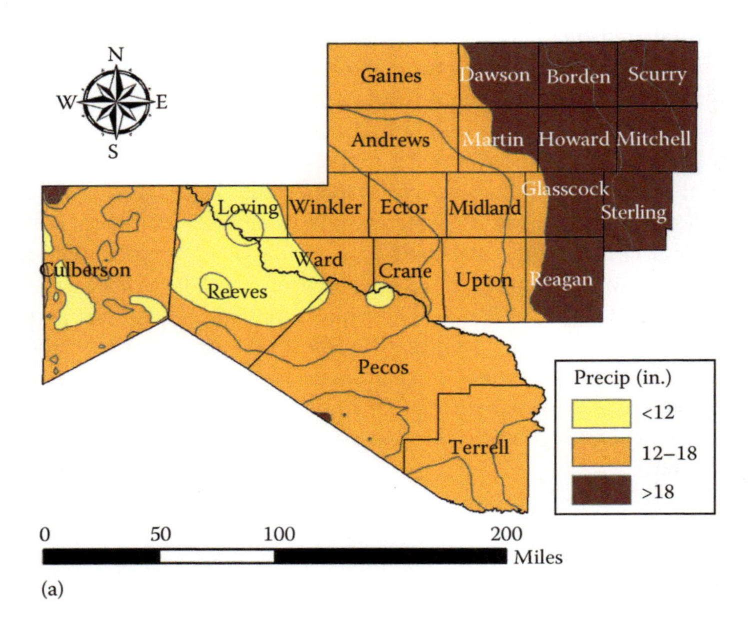

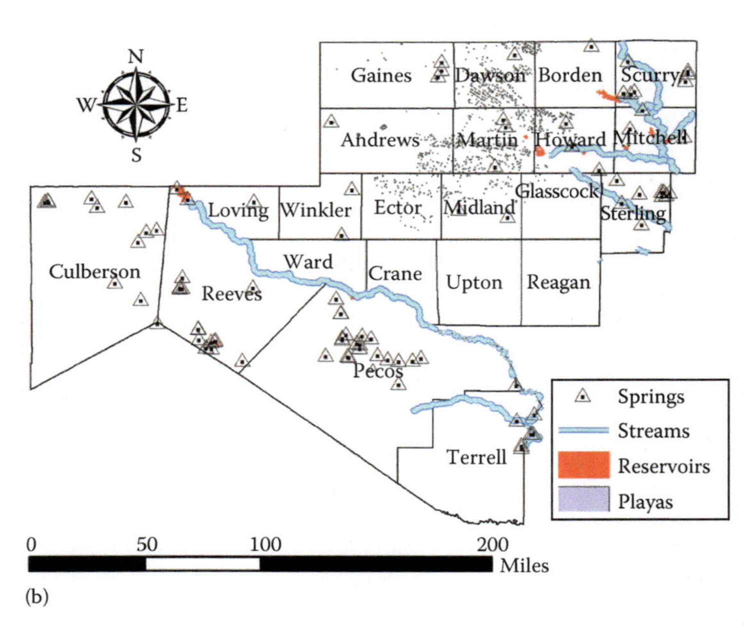

FIGURE 8.2 (a) Historical annual precipitation (inches), (b) surface water resources including springs and playas. *(Continued)*

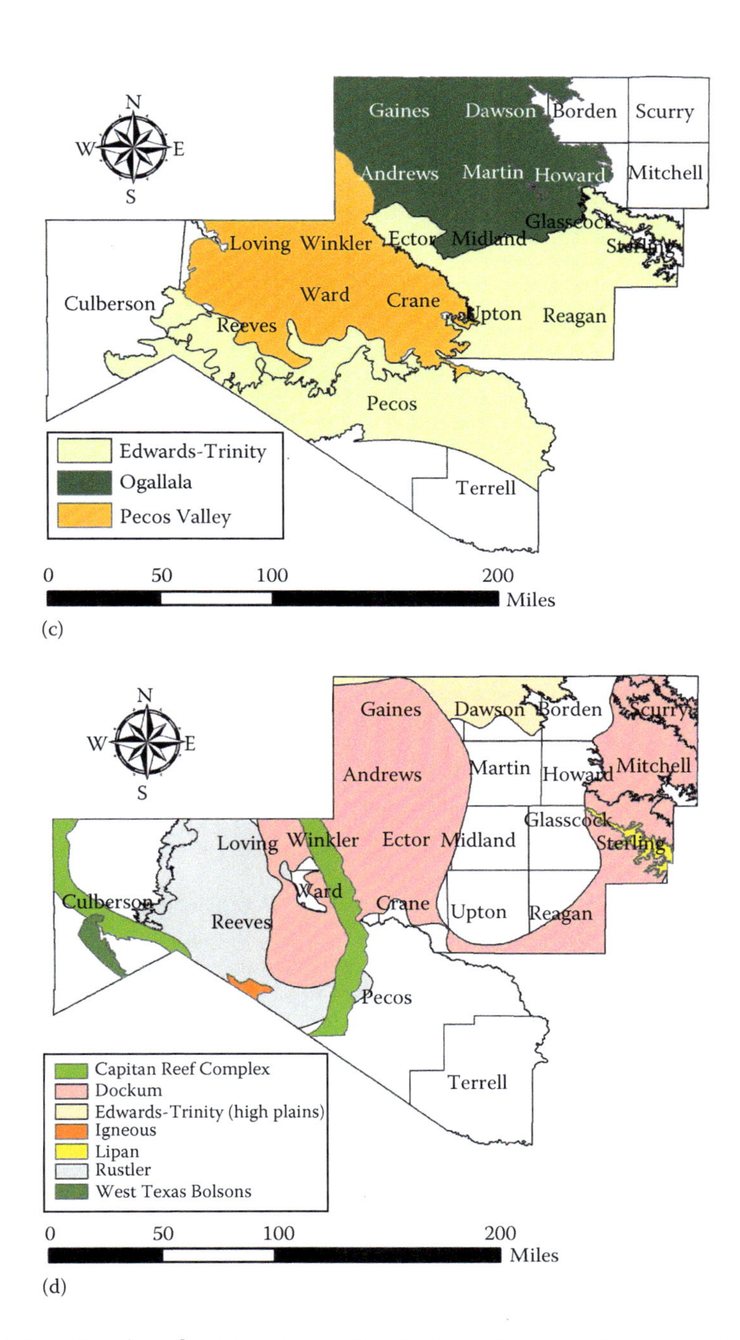

FIGURE 8.2 (*Continued*) (c) major aquifers in the region, and (d) minor aquifers in the region.

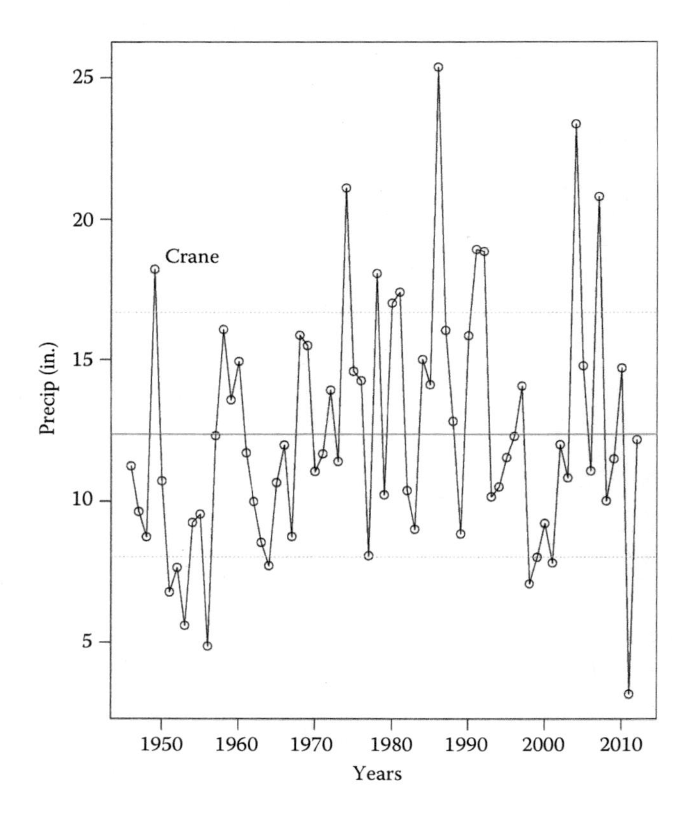

FIGURE 8.3 Historical precipitation pattern in Martin County, TX. (Data from the PRISM Database; 1 in. = 2.54 cm.)

of the Brazos and Colorado rivers. Several small reservoirs have been constructed on these rivers primarily to supply water to regional municipalities. There are several known springs (which tend to be ephemeral) in the southwestern sections of the study area. Surface depressional features (typically $<1 \text{ km}^2$), referred to as playas, can be found in the northern sections (Dawson, Martin, Midland, and Ector counties). The storage in playas also tends to be ephemeral. Playas and springs serve useful ecosystem functions and provide water for wildlife and vegetation. While these features are incapable of sustaining large municipal, agricultural, and industrial (mining) uses, they can be affected by large-scale production in their vicinity. According to Nicot et al. (2012), surface water is currently not being used for hydraulic fracturing operations in West Texas.

Groundwater is the most reliable of all water sources in the region. As shown in Figure 8.2c and d, the study area has three major and seven minor aquifers as defined by the Texas Water Development Board (TWDB). Of these, the Pecos Valley Alluvium and Ogallala are unconfined formations. The Edwards-Trinity (Plateau) occupies the largest area and occurs under both unconfined and confined conditions. The minor aquifers are unconfined in their outcrop areas and confined elsewhere.

Of the minor aquifers, the Dockum aquifer and the Rustler aquifer are present over a large portion of the study area. The subsurface area in Martin, Howard, Upton, and parts of Regan and Borden counties (shown as white area in Figure 8.2d) also contain the Dockum group sediments. These sediments are, however, not designated as an aquifer by the TWDB as they contain poor quality water typically not suited for municipal and agricultural uses (Bradley and Kalaswad, 2003).

The work of George et al. (2011) presents an overview of aquifers in West Texas, while more detailed geologic descriptions can be found in the work of Mace et al. (2001). Information pertaining to aquifer stratigraphy and hydrogeologic properties and summaries of previous local hydrogeological studies can be found in groundwater availability modeling (GAM) reports that have been developed as part of the regional water planning process (e.g., Blandford et al., 2003; Anaya and Jones, 2009; Oliver and Hutchinson, 2010). Aquifers in West Texas are generally comprised of clastic and carbonate rocks. The Ogallala aquifer and Pecos Valley Alluvium aquifers have a high percentage of sand and gravel content. The Edwards-Trinity aquifers, both Plateau and High Plains, have a high degree of crushed limestone. The Rustler aquifer is the youngest of the Permian age sediments and contains large amounts of dolomite and gypsum (Boghici and Van Broekhoven, 2001). The Dockum aquifer is comprised of mudstones, siltstones, and sandstones (Ewing et al., 2008). The Capitan Reef Complex formation is made up of large cavernous limestone and dolomite formations (Uliana, 2001). As the name suggests, the igneous aquifer formation is a hard rock aquifer made up of volcanic rocks and pyroclastic sediments. Most water in this formation is found in basalt fractures (George et al., 2011).

The water availability in West Texas aquifers exhibits considerable spatial variability. For example, Bradley and Kalaswad (2003) reported that well yields in the Dockum aquifer range from less than 1 gpm (gallons per minute) to over 4000 gpm. Similar variability has been noted in other aquifers as well. Reports of pumping tests are limited within the study area and in some cases date back to the 1960s when the saturated thickness of the Ogallala aquifer was considerably higher. Most hydrogeological information is obtained using single well tests or estimated from specific discharge measurements and, as such, aquifer transmissivity is better characterized than aquifer storage coefficient (i.e., storativity and specific yield). Because the aquifers in West Texas have very limited recharge, water levels have declined considerably in most aquifers within the study area. In particular, water levels in the Ogallala and Edwards-Trinity aquifers have dropped by more than 50 ft within the 22-county study area compared to predevelopment (prior to the 1950s) levels.

8.4 OVERALL GROUNDWATER USE WITHIN THE STUDY AREA

The historical annual water-use estimates of various counties in the study area were analyzed to assess how the resource is split among different water use groups (WUGs). The data used for this analysis were obtained from the TWDB. TWDB collects this information as part of the state water planning process using self-reported data and other standard estimation methods (TWDB, 2012a). A comparison was carried out for the years 2004 and 2011, as they represent, respectively, a relatively wet and extremely dry year in the region (see Figure 8.3). The data presented in

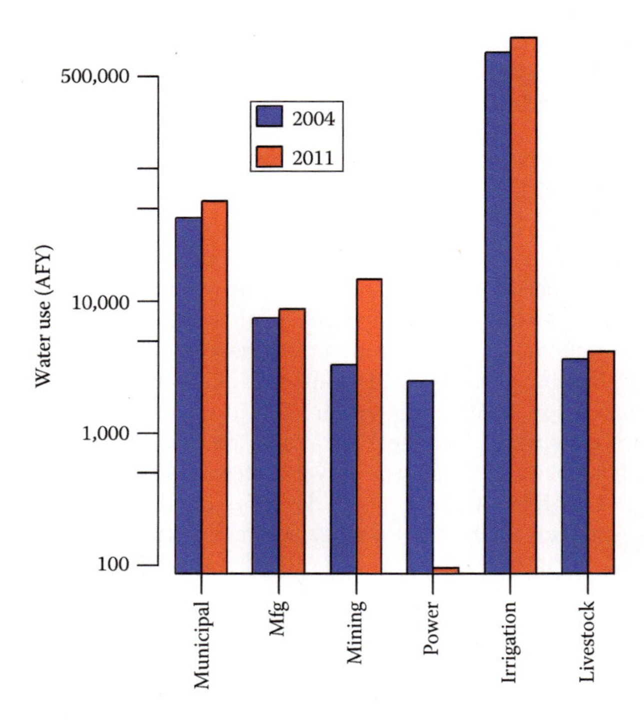

FIGURE 8.4 Aggregated groundwater use in the 22 county study area across different WUGs.

Figure 8.4 show that agriculture is the largest user of groundwater within the study area although the water use for mining (i.e., oil and gas production) has increased considerably as well.

The spatial distribution of agricultural (irrigation) and mining water use is depicted in Figure 8.5. The irrigation water use is considerably higher (an order-of-magnitude or more) than mining, except in Crane, Loving, and Ector counties. The changes in water use for irrigation and mining between the years 2004 and 2011 are shown in Figure 8.5. Mining use has increased everywhere except in Gaines and Winkler counties and now accounts for 30%–80% of non-irrigation use in these counties, highlighting the increased oil and gas activity within the region. Agricultural use has decreased in Gaines, Andrews, Ector, and Ward Counties where the Ogallala aquifer is the predominant source. This decrease can be largely attributed to declining saturated thickness of the Ogallala aquifer more so than to the transfer of water to oil and gas operations. Pecos County has seen a large increase in irrigation-related groundwater use; this could be due to limited surface water availability in 2011, the worst single-year drought in the state of Texas (Nielsen-Gammon, 2012). In general, however, both agricultural and mining use have increased in most counties within the study area. This result indicates that at least on a county scale there appears to be minimal or zero competition between agricultural water use and oil and gas water demands, although competition at more local scales cannot be ruled out.

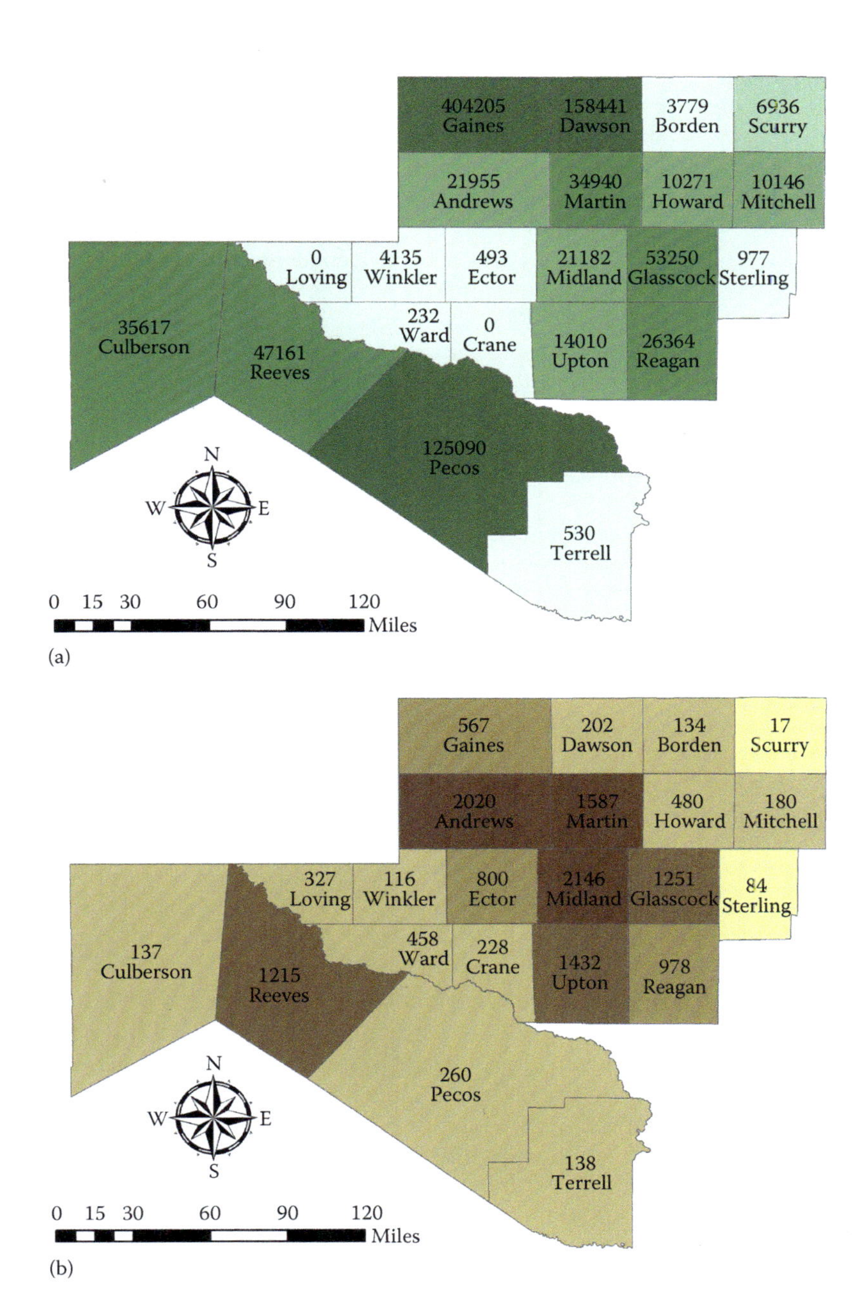

FIGURE 8.5 (a) Agriculture groundwater use in the year 2011 and (b) mining groundwater use in the year 2011 (all numbers are in acre-feet; 1 ac-ft = 0.12 ha-m).

8.5 HOW MUCH WATER IS USED FOR UNCONVENTIONAL OIL AND GAS PRODUCTION IN WEST TEXAS?

The mining-water-use data presented in Figures 8.4 and 8.5 include freshwater used by both unconventional (hydraulic fracturing) and conventional (water flooding and enhanced oil recovery) oil and gas operations. A very small fraction could be attributed to quarrying operations in the region. Freshwater use for water flooding operations has gone down considerably in recent times. According to the Texas Railroad Commission (RRC), a large portion of secondary and tertiary oil recovery is now carried out using produced water (i.e., water extracted from conventional oil and gas reservoirs). There are 35,000 active injection and disposal wells in Texas, of which 27,500 wells are used for injection in secondary or enhanced oil recovery (TRC, 2013).

Unconventional oil and gas operations rely on hydraulic fracturing to recover product trapped in tight formations. The basic idea behind hydraulic fracturing is to use water mixed with proppants (e.g., sand) and other chemicals to break open and increase the permeability of tight reservoirs. This process requires water of a certain quality in order to avoid mineral precipitation, avoid biological activity, and enhance flow of oil and gas. Recently, Nicot et al. (2012) presented data pertaining to Texas water use for hydraulic fracturing. This dataset (the Enerdeq-IHS database) is obtained from a survey of industries in the region. Based on the data presented in Figure 8.6, a total of 12,877 ac-ft of water was used in 2011 for hydraulic fracturing activities within the study area. On the other hand, the total mining use reported by TWDB was equal to 14,757 ac-ft based on the data presented in Figure 8.5. Assuming these estimates to be compatible, hydraulic fracturing accounts for nearly 90% of all the mining use. The data from Nicot et al. (2012), however, indicate that the water use for hydraulic fracturing was greater than the total estimated mining groundwater use in Culberson, Ward, Glasscock Howard, Martin, Reagan, and Upton counties. This discrepancy highlights the difficulty in obtaining reliable estimates of how much water is actually being used. On the other hand, the water use reported for hydraulic fracturing in Gaines, Borden, Mitchell, and Terrell counties is less than 25% of the total mining use reported by TWDB.

Figure 8.7 also depicts the locations of hydraulic fracturing wells drilled in 2011 as reported in the FracFocus.org database. It is important to remember that reporting to this registry was not required by the state of Texas in the year 2011 and, as such, was completely voluntary. Therefore, the database represents only a small percentage of all wells fractured within the study area. Nonetheless, the sampled well density generally correlates with the hydraulic fracturing water use estimates in these counties.

The reported water use in the FracFocus.org database was also analyzed to evaluate water use per well. The results presented in Figure 8.8 indicate that the median water use is approximately 3785 m^3 (1 million gal per well) but can vary considerably. The median of 1 million gal per well is considerably lower than what has been reported for other shale plays nationally. This lowered water use is because several wells in the area are fractured only vertically and do not extend horizontally. The typical well depth for vertical completions has increased to over 2500 ft in recent

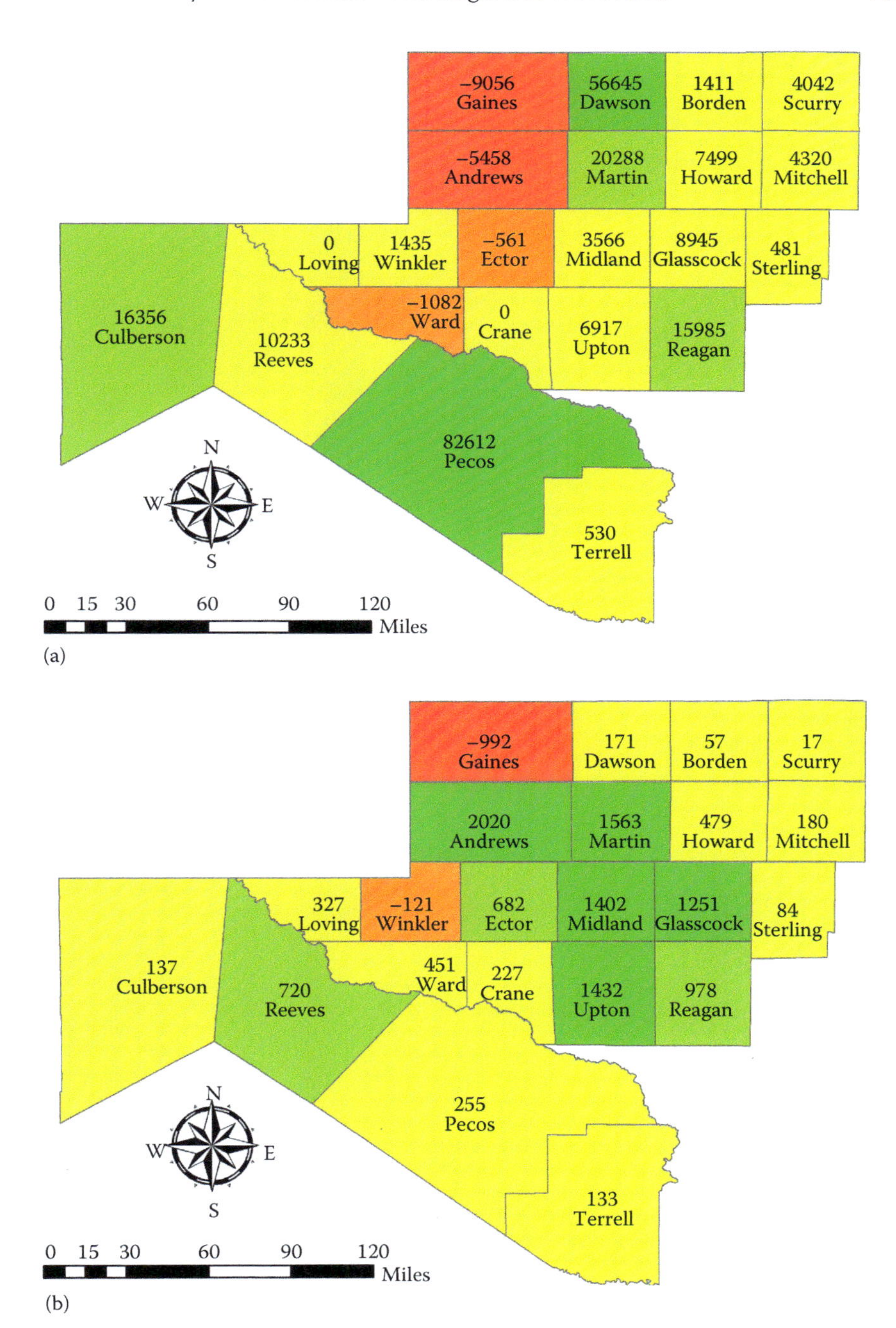

FIGURE 8.6 (a) Changes in irrigation groundwater use and (b) changes in mining groundwater use between the years 2004 and 2011. (All values in acre-feet; positive values indicate an increase in 2011 compared to 2004; 1 ac-ft = 0.12 ha-m.)

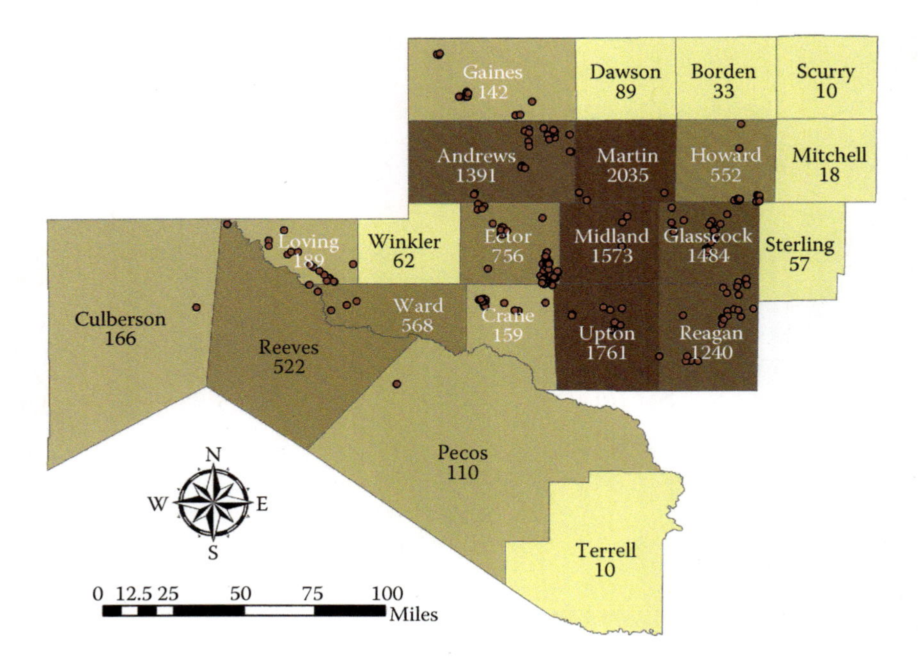

FIGURE 8.7 Water use for hydraulic fracturing operations in the year 2011 along with locations of selected wells in the FracFocus.org database. (Data from Nicot, J.P. et al., Oil and gas water use in Texas: Update to the 2011 mining water use report, Austin, TX, Texas Oil & Gas Association, 2012 and FracFocus.org; all values in acre-feet; 1 ac-ft = 0.12 ha-m.)

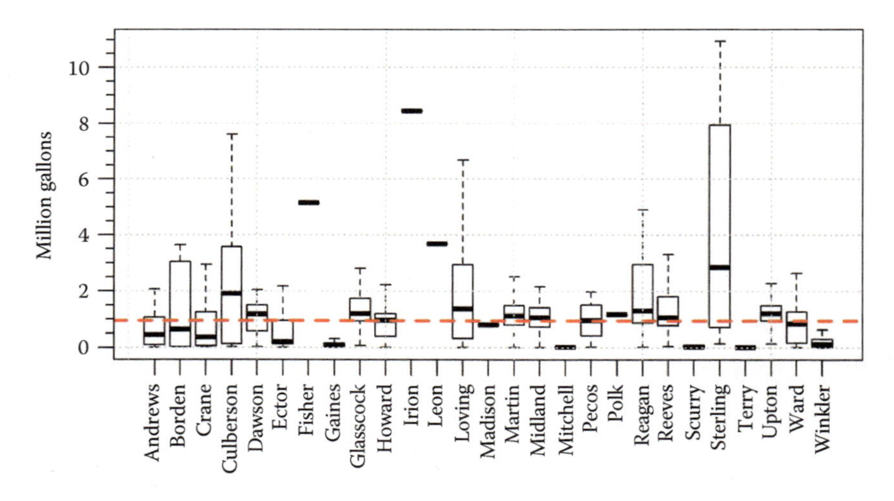

FIGURE 8.8 Reported water use per well across different counties in the study area. (Data from FracFocus.org obtained from frack.skytruth.org.)

years, resulting in a water-use density of about 400 gal/ft. The typical water use for a horizontally fractured well is estimated to be around 3–5 million gal per well with a density of 800–900 gal/lateral foot (Nicot et al., 2012).

In addition to fracturing, well-drilling operations require approximately 375 m^3 (100,000 gal of water, and this requirement can be as high as 1700 m^3 (450,000 gal) for horizontal wells.

8.6 FACTORS INFLUENCING WATER USE FOR UNCONVENTIONAL OIL AND GAS PRODUCTION AND ASSOCIATED WATER USE

The per-well water requirement for hydraulic fracturing is fairly low, particularly in comparison to other uses. Water use is directly tied to the nature (vertical versus horizontal) and extent (well density) of fracturing operations. On a county scale, the concern is often whether many such wells will be drilled in the near future, over a period of time short enough to cause significant cumulative water use. The water-use requirements for hydraulic fracturing are high at the beginning of the operations and a well that is hydraulically fractured can produce for a long period of time (King, 2012). As such, oil and gas industry groups contend that water used for hydraulic fracturing can essentially be viewed as a one-time requirement. Nicot and Scanlon (2012), however, note a strong correlation between cumulative oil and gas production and water use in Texas indicating that oil and gas production must be sustained by drilling and fracturing new wells. Comparing the extent of fracturing that has already occurred to the estimates of oil reserves in the region, fracturing of new wells (particularly horizontal wells) and associated water use is likely to continue in the near future.

The number of available drilling rigs is viewed as one of the major constraints currently limiting the proliferation of hydraulic fracturing operations. The number of drilling rigs in the United States has stayed steady over the last several years; currently, there are 1873 rigs operating in the United States, of which 412 are currently operating within the study area (Hughes, 2014). Roughly half of these rigs are used to drill vertical wells, and the remaining ones are used to drill horizontal or directional wells. It is important, however, to recognize that while rig count is a good indicator of the rate at which new wells are being drilled, it is not a good indicator of how much oil and/or gas can be produced due to advancements in extraction technologies (EIA, 2014). Therefore, higher production is possible without well drilling and associated water consumption.

Existing oil and gas transportation infrastructure is also another important factor controlling the rate at which hydraulic fracturing operations are likely to proliferate in the Permian Basin region. Oil and gas exploration in the water-limited Permian Basin, Eagle Ford and Bakken shale plays have increased significantly over the last few years, since they have a relatively lower environmental impact on sensitive surface water bodies and regional air quality compared to shale plays in urban areas (Uddameri et al., 2014). National and global socio-economic policies affect the worldwide supplies of oil and natural gas and play a crucial role in defining the extent of hydraulic fracturing in the Permian Basin region.

8.7 DOES INCREASED HYDRAULIC FRACTURING IMPLY GREATER FRESHWATER PRODUCTION IN THE PERMIAN BASIN?

The quality of water plays an important role in hydraulic fracturing operations. Anions such as phosphate and orthosilicate can bind metallic cross-linkers, and bicarbonate can bind polymers and render them ineffective. Cations such as iron, calcium, and magnesium can interfere with buffers in cross-link systems and precipitate friction reducers. While oil and grease by themselves may not be problematic, they can bring in other contaminants, particularly bacteria and microorganisms that cause clogging (Jester et al., 2013). Trace elements such as barium and strontium can form precipitates, particularly in the presence of high sulfate concentrations (Kondash et al., 2013).

The typical water quality characteristics in aquifers of West Texas are summarized in Table 8.1, along with acceptable limits for use as a fracturing fluid. The acceptable values exhibit a range since different companies use different processes (e.g., gel-based versus slickwater) as well as different chemicals for formulating their fracturing fluid. The water quality in the West Texas aquifers is highly variable as evidenced by relatively high coefficient of variation values that are presented parenthetically in Table 8.1. Generally speaking, total hardness (calcium, magnesium), sulfate, bicarbonate, and silica appear to pose problems, particularly when lower-end estimates of acceptability are to be met. Blending, buffering through additions of acids and bases, and other water treatment methods are used by the industry when formulating fracturing fluids (Joe Lee, 2013, personal communication) but add to the overall cost of procuring water. Hydraulic fracturing fluids can tolerate a high degree of total hardness (1000–2000 mg/L as $CaCO_3$) and high total dissolved solids (TDS) in the 30,000–50,000 mg/L range. Thus, there is a growing interest within the industry to use brackish water otherwise unfit for human consumption. The Permian Basin area has large amounts of brackish groundwater resources in major and minor aquifers, which could be suitable for unconventional oil and gas operations (LBG-Guyton and Associates, 2003; Meyer et al., 2012). Currently, the brackish groundwater use in the region for hydraulic fracturing appears to be 10%–30% (Nicot et al., 2012).

"Produced water" is the water in the underground oil and gas reservoir brought to the surface as part of the oil and gas production. It is estimated that three to seven barrels of water are produced for every barrel of oil produced in conventional reservoirs (Khatib and Veerbik, 2002; Lee et al., 2002). Viel et al. (2004) indicated that in conventional reservoirs, the volume of produced water is a small fraction during the early stages of production but increases significantly over time. In mature oil fields, water comprises nearly 98% of all the material brought to the surface. In hydraulic fracturing operations, fluid backflow is the activity of recovering injected fluid to initiate oil and gas production (King, 2012). Flowback water is defined as the portion of the injected fracturing fluid that is returned back to the surface once the pressure is released to initiate oil flow. The first fluid to flow back from the formation is typically the last fluid that is injected. As such, the composition of the flowback water will be very similar to that of the fracturing fluid. Boschee (2014) estimates that 20%–40% of the hydraulic fracturing fluid is returned as flowback in the Permian Basin. In most cases, the flowback water is used directly or blended with freshwater for reuse.

TABLE 8.1
Typical Values of Chemical Constituents in Various Aquifers of West Texas

Acceptable Levels[a]	Constituent	Capitan-Reef	Dockum	Edwards-Trinity	Ogallala	Pecos Valley Alluvium	Rustler
6–8	pH	7.12 (0.06)	7.54 (0.05)	7.35 (0.04)	7.43 (0.04)	7.31 (0.03)	7.1 (0.07)
30,000–50,000	TDS	3401.88 (1.09)	2185.38 (2.19)	1286.56 (0.95)	1333.95 (0.75)	1695.62 (0.93)	8278.69 (2.11)
500–2,000	Total hardness[b]	1568.69 (0.77)	547.01 (1.76)	642.48 (0.78)	654.48 (0.7)	784.56 (0.84)	2236.64 (0.76)
200	Calcium	417.14 (0.79)	138.09 (2.37)	162.9 (0.73)	123.34 (0.69)	207.65 (0.82)	562.07 (0.61)
50–1,200	Magnesium	127 (0.79)	48.91 (1.18)	56.82 (1)	82.13 (0.87)	64.34 (0.92)	198.45 (1.08)
36,000	Sodium	514.04 (1.92)	561.88 (2.93)	183.63 (1.28)	196.89 (0.98)	267.54 (1.13)	2060.6 (2.93)
	Potassium	25.51 (2.31)	7.02 (1.33)	8.18 (0.99)	13.56 (1.13)	9.2 (0.75)	29.25 (2.22)
300–500	Bicarbonate	292.81 (0.41)	295.59 (0.38)	253.9 (0.25)	287.95 (0.38)	246.98 (0.25)	159.05 (0.49)
200	Carbonate	0.53 (1.63)	1.16 (7.9)	0.02 (0.26)	0.18 (1.11)	0.02 (0.14)	0.16 (0.75)
50–400	Sulfate	1244.85 (0.79)	628.66 (1.25)	460.41 (1.05)	347.53 (1.03)	607.05 (0.93)	1895.78 (0.63)
10,000–45,000	Chloride	897.1 (1.88)	621.27 (4.05)	244.66 (1.74)	329.39 (1.07)	410.93 (1.45)	3751.13 (2.7)
10	Boron	NA	0.88 (0.47)	0.53 (0.04)	0.3 (0.03)	0.5 (0.07)	NA
10	Barium	NA	0.03 (0.03)	0.04 (0.04)	0.07 (0.07)	0.02 (0.02)	NA
10	Strontium	5.65 (0.87)	2.71 (0.92)	4.24 (1.69)	7.9 (5.06)	2.94 (0.98)	7.54 (0.6)
1	Silica	18.13 (0.48)	23.57 (0.53)	19.74 (0.54)	56.78 (0.31)	25.17 (0.35)	19.01 (0.41)
10–20	Iron	NA	0.47 (0.47)	0.04 (0.04)	0.03 (0.03)	0.07 (0.07)	NA

Based on data from TWDB, Groundwater Database. All values in mg/L unless otherwise noted; numbers in parentheses are the coefficient of variation.

[a] Bascd on values reported by Kidder (2011), Chesapeake Energy (2013), Jester et al. (2013), Cudd Energy Services (2013), and West Virginia University Industry Contact Group (2012).

[b] Expressed as mg/Lof $CaCO_3$.

Most of the flowback occurs within the first 2 weeks of production and tapers off within a month or so. Produced water in hydraulic fracturing generally refers to the connate water within the shale play that is brought to the surface along with the oil. The chemical composition of the produced water reflects the mineral composition of the shale play. The transition between flowback and produced water is not distinct and cannot be ascertained from water quality alone. The rate of produced water production (<10 barrels/day) is significantly slower than that of the flowback (approximately 50 barrels/day) and can be a useful indicator of the transition (Gregory et al., 2011). Unlike conventional production, the amount of water produced from shale plays drops over the course of production. The volume of produced water is higher in the Permian Basin compared to other active shale plays in the United States (Boschee, 2014), and it is estimated that at least 100% of the injected water volume is extracted back over the lifetime of the well (Nicot et al., 2012). A significant portion of the produced water obtained from both conventional and unconventional reservoirs is injected underground: (1) to maintain pressure in subsurface formations, (2) is used for enhanced oil recovery, or (3) is disposed in nonproductive zones.

Flowback and produced water are increasingly being viewed as valuable water resources, and there is a growing interest in using them in fracturing operations to reduce freshwater withdrawals in the Permian Basin (Jester et al., 2013). Produced water concentrations presented in Figures 8.9 and 8.10 were all based on conventional reservoirs within the Permian Basin, and there is considerable variability in the produced water quality. Generally speaking, waters produced at later times tend to have higher concentrations of dissolved solids and naturally occurring radioactive

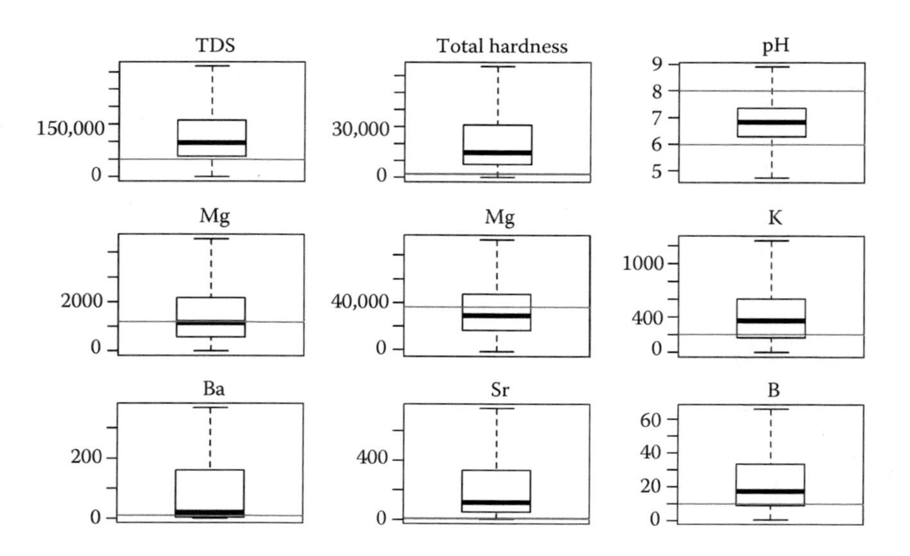

FIGURE 8.9 Typical concentration ranges of chemical constituents (bulk measures and cations) in produced waters in the Permian Basin. (All units are in mg/L and total hardness is expressed in units of mg/L as $CaCO_3$; horizontal (red) line indicates typical fracturing water requirement; data from USGS Produced Water Database.)

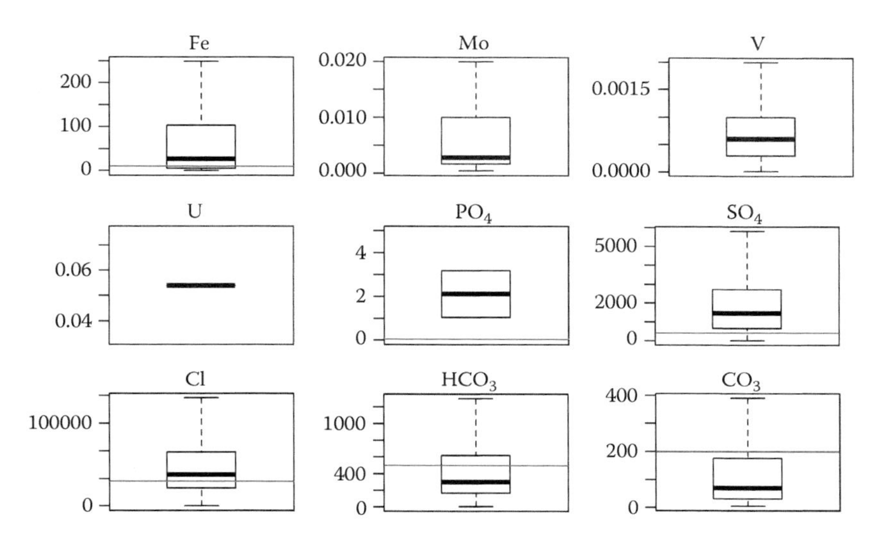

FIGURE 8.10 Concentration of chemical constituents (trace elements and anions) in produced water in the Permian Basin. (All units are in mg/L; horizontal (red) line indicates typical fracturing water requirement; data from USGS Produced Water Database.)

materials. The produced waters can be highly acidic or strongly basic, as seen from Figure 8.10. Trace elements such as iron, barium, strontium, and boron that are often present in very small quantities (<1 mg/L) in aquifers are significantly enriched in produced waters (compare data in Table 8.1 with those presented in Figure 8.10). In a similar vein, carbonate, sulfate, and phosphate concentrations are relatively higher in produced waters compared to typical ranges experienced in aquifers.

In addition to inorganic constituents, produced water also contains large amounts (at or near solubilities) of hydrocarbons (Haluszczak et al., 2013). The composition of produced water from shale formations can be different from those in conventional reservoirs. Unfortunately, however, water quality data from the Permian Basin shale plays have not been publicly documented.

The Permian Basin does have a significant amount of flowback water from conventional production that could be diverted for use in shale formations. However, direct use of produced waters for fracturing shale formations is not likely feasible based on the concentration ranges presented in Figure 8.10. Mineral precipitation, particularly of barite (barium sulfate), and the formation of $(BaSr)SO_4$ complexes are noted to be particularly problematic (Todd and Yuan, 1992; Chapman et al., 2012). Development of new treatment technologies for tackling high concentrate wastewaters and specification of process trains is an active area of interest. The reader is referred to Chapter 9 (Williams et al., 2015) for a discussion on treatment technologies. The industry has recognized that reuse of produced water is an environmentally beneficial and cost-effective option for reducing water stresses associated with hydraulic fracturing (Jester et al., 2013). Additional research is necessary to make treatment technologies cost-effective, particularly given the fact that deep well disposal of produced water is convenient and cost-effective at this point in time.

8.8 GROUNDWATER PRODUCTION FOR FRACTURING—LOCAL-SCALE EFFECTS

On a county-wide scale, water use for hydraulic fracturing remains small but problems can arise on a more local scale. The biggest challenge associated with hydraulic fracturing is the need for relatively large quantities of water (1–5 million gal) over a short period of time (3–5 days). Water for fracturing can either be procured onsite (less common) or purchased and hauled in from an outside source. The cost for raw freshwater alone is around \$0.30–\$0.50 per barrel (42 gal) and with transportation and other costs factored in, the total cost to the oil and gas producer can be as high as \$2–\$3 per barrel (Boschee, 2014). The production of water from a well causes a drop in the water level at the well and in the surrounding areas (referred to as drawdown), resulting in a cone of depression. The extent of this drawdown depends upon the rate of production (flowrate), the time over which the water is produced, and aquifer hydrogeological characteristics. Hydraulic conductivity is a measure of how easily water flows through the aquifer and, when multiplied with the aquifer thickness over which the water is produced, is referred to as aquifer transmissivity. Storage characteristics of the aquifer indicate how much water can be extracted over a cross-sectional area for a unit drop in hydraulic head. The extent to which the drawdown effects are felt depends upon whether the aquifer is unconfined (water flows under the influence of gravity) or is confined (water is stored under pressure). Recovery denotes how quickly the water level recovers at the well (i.e., comes back to its original state) upon cessation of pumping.

Following Carslaw and Jeager (1959), the practical limit of detection of a periodic wave in a diffusive system is given by its wavelength and can be expressed as

$$\lambda = \sqrt{\left(\frac{8\pi^2\alpha}{w}\right)} \tag{8.1}$$

where

- λ is the wavelength (ft)
- w is the frequency of the disturbance (1/day)
- α is the aquifer diffusivity defined as the ratio of aquifer transmissivity to storage coefficient (ft^2/day)

The wavelength calculations for select West Texas aquifers are summarized in Table 8.2 based on transmissivity and storage coefficients summarized in the work of Christian and Wuerch (2012). The results depict that the effects of production can be felt over several miles in the confined portions of Edwards-Trinity and Dockum aquifers (as they are confined, the disturbance passes virtually unattenuated). On the other hand, the disturbance (drawdown) is localized in the unconfined Ogallala aquifer.

While the wavelength analysis provides a first-cut approximation on how far the effects of pumping are likely to be felt, calculating potential drawdown and recovery can be useful for obtaining the following: (1) a more precise estimate of the drawdown effects; (2) whether the cone of depression caused by groundwater production will extend beyond the property boundaries or interfere with production

TABLE 8.2
Typical Distances over which a Cyclical Production Lasting 5 Days Can Be Felt in Different Aquifers of West Texas

Aquifer	Wavelength (ft)
Ogallala	1,073–1,858
Edwards-Trinity (confined)	4,350–7,500
Dockum (confined)	9,000–28,579

in nearby wells; and (3) how long before the water levels recover practically back to their original levels.

The idealized distance–drawdown profile shown in Figure 8.11 was calculated using the Theis solution and corroborates the earlier analysis of how far the effects of production can be felt at the end of a 5-day period over which 11,350 m^3 (3 million gal) of water are pumped. The effects are most pronounced in the Dockum aquifer, which has extremely high aquifer diffusivity. Drawdowns in excess of 1.5 m (5 ft) can be felt almost a mile from the production well. The drawdown effects

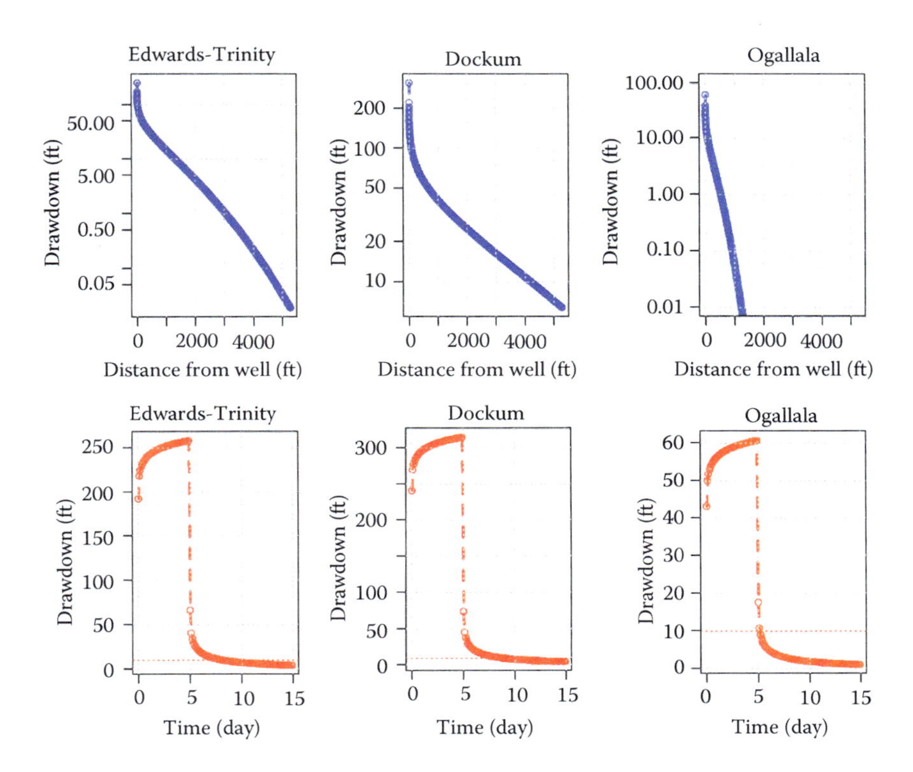

FIGURE 8.11 Distance–drawdown and drawdown and recovery at the production well in different aquifers of West Texas.

are contained within a relatively short distance in the Ogallala aquifer, which is an unconfined formation with high storage characteristics. The drawdown in both Edwards-Trinity and Dockum aquifers can be in excess of 76–91.44 m (250–300 ft) and over 18 m (60 ft) in the Ogallala aquifer in the vicinity of the well. The analysis does not include entrance and friction losses within the well, which can increase the drawdown by another 50% (Todd and Mays, 2005). The drawdowns can be 50%–80% of the respective aquifer thicknesses, and indicate that such intense production cannot be sustained over long periods since submersible pumps must be placed at least 7.5–9.0 m (25–30 ft) below the maximum possible drawdowns to avoid damage and mechanical failure. The recovery curve presented in Figure 8.11 depicts how the water level in the well recovers after the cessation of pumping. Recovery proceeds in two stages where water from larger pores in the vicinity of the well is drawn into the well quickly and then slows down as water from smaller pores of the aquifer and at larger distances flow into the well. Again, recovery is relatively quick in the Ogallala aquifer due to its unconfined characteristics and slowest in the Dockum aquifer, where it may take more than a month for water levels to come back to within 1.5 m (5 ft) of preproduction levels. While water flows back into the well upon cessation of pumping in response to a hydraulic gradient, water is removed from the aquifer during pumping, which will cause well yields to diminish over time due to removal of water and lack of sufficient recharge.

Well yield refers to the groundwater production rate that can be sustained over a long period of time. Typically, groundwater production is kept below the well yields to avoid excessive drawdowns that can damage the pumps. The measured well yields for aquifers in West Texas are summarized in Table 8.3 and exhibit considerable spatial heterogeneity. Based on the reported well yields, it can be seen that sustainable extraction of groundwater required for fracturing (say 13,250 m^3 or 3.5 million gal) will take anywhere from over a year (5.45 m^3/day or 1 gpm) to about 5 days at 2,726 m^3 or 500 gpm. These calculations explain the need for storage

TABLE 8.3
Reported Well Yields for Various Aquifers in the Study Area

Aquifer Formation	Typical Well Yield (gpm)	Remarks
Edwards-Trinity Plateau	50–200	From USGS (2014)
Dockum	5–100	Bradley and Kalaswad (2003); higher values noted in some places
Ogallala	250–500	Based on current saturated thickness and projections from Bell and Morrison (1982)
Pecos Valley Alluvium	300–1000	Interpretation of moderate to high classification by TWDB (2012b)
Rustler	100–300	Higher values up to 4000 gpm reported in some places (USGS, 2014)
Edwards-Trinity High Plains	50–100	Higher values reported north of the study area (USGS, 2014)

of water and transporting it over relatively large distances, and highlight the role of geologic heterogeneity in our ability to extract groundwater.

8.9 GROUNDWATER PRODUCTION FOR FRACTURING AND COMPETITION WITH AGRICULTURE

Irrigated agriculture is the largest user of water within the study area. Most irrigation in the region is carried out using center-pivot irrigation systems (Warrick et al., 2002), although more efficient technologies such as the mid-elevation spray application (MESA), low-elevation spray application (LESA), low energy precision application (LEPA), and surface and subsurface drip irrigation methods have been proven to reduce water requirements while maintaining crop yields in West Texas (Colaizzi et al., 2004). The typical irrigation efficiencies for various irrigation systems are summarized in Table 8.4.

The estimated water savings in Table 8.4 suggests that at least on a macro (county-wide) scale, upgrading around 400 ha (1000 ac) of cotton farmland from center-pivot to a more efficient system can provide all the water necessary for hydraulic fracturing in many counties of West Texas. According to the recent agricultural census, nearly 267,000 ac of land were irrigated for cotton production alone within the study area (USDA, 2013). Therefore, increasing agricultural efficiencies can bring in significant water savings in the region.

Low profit margins and risks associated with irrigation failure are major deterrents for adoption of more efficient irrigation methods. The real options analysis carried out by Seo et al. (2008) indicated that farmers in the Texas high plains are likely to stay with their existing irrigation technologies due to high entry costs of adoption. Therefore, the creation of efficient market mechanisms, where oil and gas companies bear the cost of implementation of these technologies in return for rights to the water saved, could lead to win–win situations. For example, the typical costs of drip irrigation are estimated to be around \$1300 per acre (Stalcup, 2007). Converting 1000 ac from center-pivot to drip would save roughly 290 ac-ft of water at a cost of \$1.3 million, resulting in a water cost of about \$0.60/barrel. This is competitive with current

TABLE 8.4
Typical Efficiencies of Irrigation Systems

Irrigation System	Efficiency (%)	Water Savings per Upgrade (in)[a]
Center-pivot	75	
LESA	85	3.5
LEPA	90	4.9
Drip	85	3.5
Subsurface Drip	95	6.2

Source: Efficiency data from Porter, D., Irrigation for small farms, Texas AgriLife Research, Austin, TX, 2012.

[a] Assumes a cotton crop requiring 22 in of water over a growing season.

market rates of water in the area (Boschee, 2014) and suggests that any such investments can be recovered within a short time-period. Conversion to new technologies also requires shifts in farming practices and willingness of farmers to adapt. This may require additional investments for demonstration of water conservation methods and dissemination of this knowledge.

While the benefits of saving water by switching to better irrigation methods are evident on a macroscale, the transfer of this saved water from agriculture to oil and gas operations must carefully consider the spatio-temporal disconnects. Geologic heterogeneity plays a critical role in determining whether water saved at one location (farm) can be extracted at another location (drilling pad). The drawdown impacts could be different at these two locations due to variations in local geology. The timing of irrigation is a critical parameter for increasing crop yields (Steger et al., 1998) and, thus, water transfers must also take into account production by agriculture and oil gas operations to reduce drawdown effects.

A water balance model was developed to estimate the crop water requirements and timing of irrigation events at a cotton field in Gaines County, Texas. Simulations were performed using meteorological data for the years 2011 and 2004 to simulate dry and wet years, respectively. The antecedent soil moisture was assumed to be 0.05 (close to wilting point) for both years, an effective root depth of 0.6 m (2 ft) was used in these simulations, and the planting date was assumed to be May 15 for both years. The dry year had 1.27 cm (0.5 in.) of rainfall, while the wet year had 8.89 cm (3.5 in.) of rainfall over the growing season. The crop water requirements were estimated from cumulative growing degree days (GDD) using relationships suggested by Porter (2012). The wet year (2004) had a slightly longer growing season compared to 2011 due to cooler temperatures. Irrigation scheduling was based on the idea of maintaining soil moistures in the root zone between 50% and 100% of the field capacity (taken as 0.2 for the sandy loam soil), which is a recommended practice in the region. The results presented in Figure 8.12 indicate the need for a preplanting irrigation (around May 15) and four to five irrigation events around first square, first bloom, and maximum bloom stages of plant growth. The theoretical irrigation requirement (not accounting for technology inefficiencies) during the dry year was estimated to be over 40 cm (16 in.), while the wet year required 38 cm (15 in.) of irrigation. This result indicates that irrigation requirements are somewhat independent of rainfall volumes due to the storm arrival patterns during the growing season. The demand for water during the growing season unfortunately coincides with times when the demand for oil and gas is also high. Therefore, water extraction for oil and gas production could locally compete with irrigation needs during dry summer months. However, switching to more efficient irrigation technologies will reduce the overall water footprint and still allow for both irrigation and hydraulic fracturing demands to be met. In addition to changing irrigation infrastructure, deficit irrigation and smart irrigation scheduling technologies that activate irrigation activities by remotely sensing soil moisture deficits will help reduce the overall water use in the region and ameliorate any competition for water between food and energy sectors. Therefore, water management and availability are best viewed in a holistic manner and cooperative efforts between oil and gas and agricultural sectors can lead to long-term sustainability of the region.

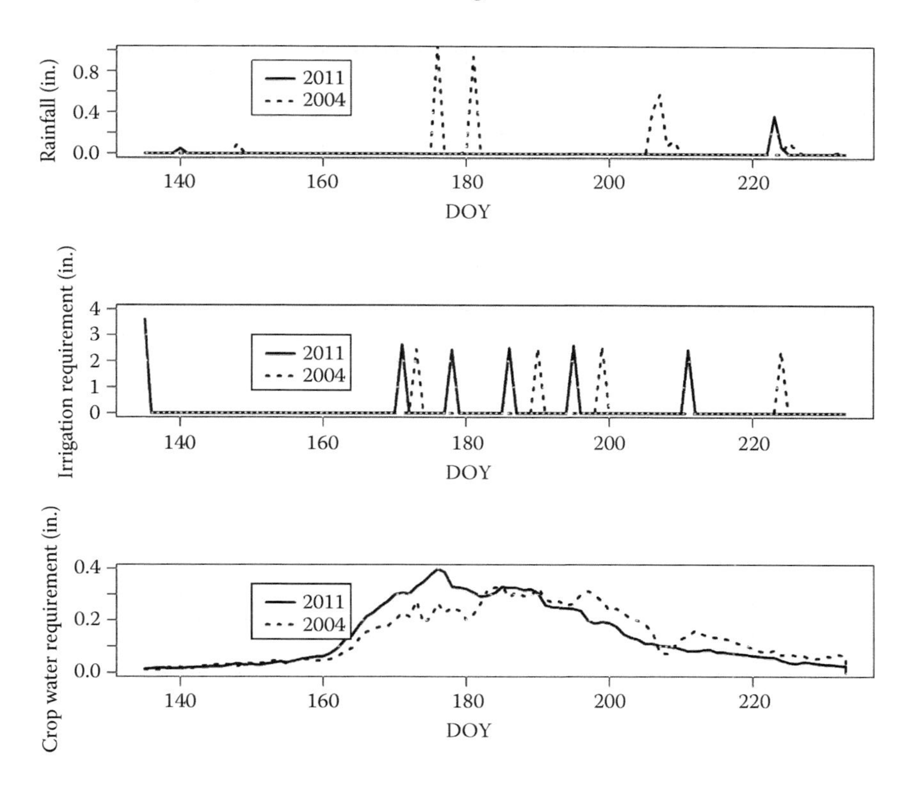

FIGURE 8.12 Rainfall, irrigation, and crop water requirements at a cotton field near Gaines County, Texas.

8.10 SUMMARY AND CLOSING REMARKS

The overall goal of this study was to conduct a broad assessment of water use within a 22-county study area in the Permian Basin region of West Texas. Oil and gas operations have increased significantly in recent years due to hydraulic fracturing activities, which, in turn, have increased the requirements of water in this agriculturally intensive, water-parched region that is currently experiencing a prolonged drought. With a focus on sustainable water resources management, the study sought to address the following major questions: (1) How much water was being used for hydraulic fracturing operations? (2) What are potential sources of water? (3) How to assess competition for this water between other users? (4) How to identify innovative strategies which reduce the overall water footprint in the region? Theoretical distance–drawdown and time–drawdown relationships were used to gain insights into potential local-scale effects of water withdrawals for hydraulic fracturing. Hydrogeological and geochemical data compilation effort was also undertaken to establish baseline values, characterize variability, and identify critical data gaps. There is a paucity of pumping test data for estimating hydrogeological parameters and understanding aquifer response. Data are also lacking on flowback and produced water chemistry from unconventional reservoirs. The analysis presented herein emphasizes the need for holistic water management and cooperation among different WUGs in the region. As part of this study three broad strategies were identified to sustain water availability for hydraulic fracturing operations. These strategies include the following: (1) explore and utilize brackish groundwater resources available in minor aquifers of West Texas, (2) utilize produced water from both conventional and unconventional oil and gas production in the region, and (3) reduce the overall water footprint of the region by increasing water use efficiencies in the agricultural sector, which is the largest water user in the region.

ACKNOWLEDGMENT

Financial support from Apache Corporation through a gift to Texas Tech Whitacre College of Engineering is gratefully acknowledged.

NOTE ON UNITS

Water resources data in the United States are often presented in customary units. While every effort was made to also present the information in both SI and U.S. customary units in the text, certain figures were developed using data in U.S. customary units to be readily accessible to policy makers in the region.

REFERENCES

Anaya, R. and I. C. Jones (2009). Groundwater availability model for the Edwards-Trinity (Plateau) and Pecos Valley Alluvium aquifer systems. Austin, TX: Texas Water Development Board, p. 103.

Bell, A. E. and S. Morrison (1982). Analytical study of the Ogallala aquifer in Potter and Oldham counties, Texas. Austin, TX: Texas Department of Water Resources, Texas Water Development Board, Texas Water Commission.

Blackmon, D. (2013). Shale, fracking are not the main cause of Texas water shortages. Forbes. Available online: http://www.forbes.com/fdc/welcome_mjx.shtml, accessed August 15, 2014.

Blandford, T. N., D. J. Blazer, K. C. Calhoun, A. R. Dutton, T. Naing, R. C. Reedy, and B. R. Scanlon (2003). Groundwater availability of the southern Ogallala Aquifer in Texas and New Mexico—Numerical simulation through 2050. Albuquerque, NM: Daniel B. Stephens and Associates, Inc.

Boghici, R. and N. G. Van Broekhoven (2001). Hydrogeology of the rustler aquifer, Trans—Pecos Texas, R. E. Mace, W. F. I. Mullican and E. S. Angle (eds.). Austin, TX: Texas Water Development Board, pp. 207–225.

Boschee, P. (February 2014). Produced and flowback water recycling and reuse—Economics, limitations and technology. *Oil and Gas Facilities*, pp. 17–22.

Bradley, R. G. and S. Kalaswad (2003). Groundwater resources of the Dockum aquifer in Texas. Austin, TX: Texas Water Development Board.

Carslaw, H. and J. Jaeger (1959). *Heat in Solids*. Oxford, U.K.: Clarendon Press.

Chapman, E. C., R. C. Capo, B. W. Stewart, C. S. Kirby, R. W. Hammack, K. T. Schroeder, and H. M. Edenborn (2012). Geochemical and strontium isotope characterization of produced waters from Marcellus Shale natural gas extraction. *Environmental Science and Technology* **46**(6): 3545–3553.

Christian, B. and D. R. Wuerch (2012). Compilation of results of aquifer tests in Texas. Austin, TX: Texas Water Development Board.

Colaizzi, P., A. Schneider, S. Evett, and T. Howell (2004). Comparison of SDI, LEPA, and spray irrigation performance for grain sorghum. *Transactions of the ASAE* **47**(5): 1477–1492.

EIA (2014). Annual energy outlook 2014. U.S. Energy Information Administration, Washington, DC.

Ewing, B. T., M. C. Watson, T. McInturff, and D. Liang (2014). Current and future economic impacts of the Texas oil and gas pipeline industry. Tech University and Texas Pipeline Association, Lubbock, TX.

Ewing, J. E., T. L. Jones, T. Yan, A. M. Vreugdenhil, D. G. Fryar, J. F. Pickens, K. Gordon et al. (2008). Groundwater availability model for the Dockum aquifer. Austin, TX: Texas Water Development Board, p. 510.

George, P. G., R. E. Mace, and R. Petrossian (2011). Aquifers of Texas, S. Doenges. Austin, TX: Texas Water Development Board (TWDB).

Goldenberg, S. (2013). A Texan tragedy; ample oil, no water—Barnhart, Texas. *The Guardian*, London, U.K. Available online: http://www.theguardian.com/environment/2013/aug/11/texas-tragedy-ample-oil-no-water, accessed August 15, 2014.

Gregory, K. B., R. D. Vidic, and D. A. Dzombak (2011). Water management challenges associated with the production of shale gas by hydraulic fracturing. *Elements* **7**(3): 181–186.

Haluszczak, L. O., A. W. Rose, and L. R. Kump (2013). Geochemical evaluation of flow-back brine from Marcellus gas wells in Pennsylvania, USA. *Applied Geochemistry* **28**: 55–61.

Hughes, B. (2014). Baker Hughes: US Drilling Count Jumps to 1873. *Editorial, Oil and Gas Journal* June Issue, Houston, TX. Available online: http://www.ogj.com/articles/2014/06/baker-hughes-us-drilling-rig-count-jumps-to-1-873.html, accessed on August 15, 2014.

Jester, S., R. Sharma, and A. Shields (2013). Cost effective reuse of produced water in the Permian Basin. *Oil & Gas Environmental Conference*, Conoco Phillips, Dallas, TX.

Khatib, Z. and P. Verbeek (2002). Water to value-produced water management for sustainable field development of mature and green fields. *SPE International Conference on Health Safety and Environment in Oil and Gas Exploration and Production*, Society of Petroleum Engineers, Richardson, TX.

King, G. E. (2012). Hydraulic fracturing 101: What every representative environmentalist regulator reporter investor university researcher neighbor and engineer should know about estimating frac risk and improving frac performance in unconventional gas and oil wells. *SPE Hydraulic Fracturing Technology Conference*, Society of Petroleum Engineers, Richardson, TX.

Kondash, A. J., N. R. Warner, O. Lahav, and A. Vengosh (2013). Radium and barium removal through blending hydraulic fracturing fluids with acid mine drainage. *Environmental Science and Technology* **48**(2): 1334–1342.

LBG-Guyton and Associates (2003). Brackish groundwater manual for Texas regional water planning groups. Austin, TX: Texas Water Development Board.

Lee, R., R. Seright, M. Hightower, A. Sattler, M. Cather, B. McPherson, L. Wrotenbery, D. Martin, and M. Whitworth (2002). Strategies for produced water handling in New Mexico. *Ground Water Protection Council Produced Water Conference*, Colorado Springs, CO.

Mace, R. E., W. F. I. Mullican, and E. S. Angle (2001). Aquifers of West Texas. Austin, TX: Texas Water Development Board.

Meyer, J. E., M. R. Wise, and S. Kalaswad (2012). Pecos valley aquifer, West Texas: Structure and brackish groundwater. Austin, TX: Texas Water Development Board.

Nicot, J. P., R. C. Reedy, R. A. Costley, and Y. Huang (2012). Oil and gas water use in Texas: Update to the 2011 mining water use report. Austin, TX: Texas Oil & Gas Association.

Nicot, J.-P. and B. R. Scanlon (2012). Water use for shale-gas production in Texas, US. *Environmental Science and Technology* **46**(6): 3580–3586.

Nielsen-Gammon, J. W. (2012). The 2011 Texas drought. *Texas Water Journal* **3**(1): 59–95.

Oliver, W. A. and W. R. Hutchinson (2010). Modification and recalibration of the groundwater availability model of the Dockum aquifer. Austin, TX: Texas Water Development Board.

Porter, D. (2012). Irrigation for small farms. Austin, TX: Texas AgriLife Research.

Ruppel, S. C. (2009). Integrated synthesis of the Permian basin: Data and models for recovering existing and undiscovered oil resources from the largest oil-bearing basin in the US. Austin, TX: Bureau of Economic Geology.

Seo, S., E. Segarra, P. D. Mitchell, and D. J. Leatham (2008). Irrigation technology adoption and its implication for water conservation in the Texas High Plains: A real options approach. *Agricultural Economics* **38**(1): 47–55.

Stalcup, L. (2007). Drip pays dividends. Retrieved October 7, 2014. Available online: http://hayandforage.com/mag/farming_drip_pays_dividends, accessed August 15, 2014.

Steger, A. J., J. C. Silvertooth, and P. W. Brown (1998). Upland cotton growth and yield response to timing the initial postplant irrigation. *Agronomy Journal* **90**(4): 455–461.

Todd, A. and M. Yuan (1992). Barium and strontium sulfate solid-solution scale formation at elevated temperatures. *SPE Production Engineering* **7**(1): 85–92.

Todd, D. K. and L. W. Mays (2005). *Groundwater Hydrology*, 3rd edn. Hoboken, NJ: Wiley.

TRC (2013). Saltwater disposal wells. Retrieved January 8, 2014. Available online: http://www.rrc.state.tx.us/about-us/resource-center/faqs/oil-gas-faqs/faq-saltwater-disposal-wells/.

TWDB (2012a). Historical water use estimates. Austin, TX: Texas Water Development Board.

TWDB (2012b). Water for Texas. *State Water Plan*. Austin, TX: Texas Water Development Board.

Uddameri, V., A. Morse, and D. Reible (2014). Unconventional oil and natural gas resources development and their potential environmental impacts. *Environmental Manager*, **July Issue**: 18–25.

Uliana, M. M. (2001). The geology and hydrogeology of the Capitan Aquifer—A brief overview. Aquifers of West Texas, R. E. Mace, W. F. I. Mullican and E. S. Angle (eds.). Austin, TX: Texas Water Development Board, pp. 207–225.

USGS (2014). United States Geological Survey. Retrieved January 8, 2014. Available online: http://www.usgs.gov/.

Veil, J. A., M. G. Puder, D. Elcock, and R. J. Redweik Jr. (2004). A white paper describing produced water from production of crude oil, natural gas, and coal bed methane. Argonne National Laboratory for the US Department of Energy, National Energy Technology Laboratory, Lemont, IL, January. Available online: http://www.ead.anl.gov/pub/dsp_detail.cfm, accessed August 15, 2014.

Warrick, B. E., C. Sansone, and J. Johnson (2002). Grain sorghum production in west central Texas. Texas Coop. Ext., San Angelo, TX. Available online: http://sanangelo.tamu.edu/agronomy/sorghum/gsprod.htm, accessed August 15, 2014.

Williams, J. H., D. Reible, R. Darvari, T. Vercillino, and A. Morse, eds. (2015). Reuse and recycling flowback and produced waters. *Fracturing Impacts and Technology*. London, U.K.: Taylor & Francis.

9 Reuse and Recycling of Flowback and Produced Waters

John H. Williams, Danny Reible, Roxana Darvari, Tony Vercellino, and Audra Morse

CONTENTS

9.1 Overview 159
9.2 Introduction 159
9.3 Exploring Water Consumption in Hydraulic Fracturing 160
9.4 Water Storage and Delivery for Hydraulic Fracturing Activity 163
9.5 Water Produced in Hydraulic Fracturing Activity 163
9.6 Making the Water "Fracable" 165
9.7 Opportunities for Recycling and Reuse 166
9.8 Water Treatment as a Solution 169
9.9 Chemical Requirements for Reuse of Flowback and Produced Waters 170
9.10 Conclusion 173
References 173

9.1 OVERVIEW

Water treatment technologies and solutions to recycle and reuse produced waters are discussed in this chapter.

9.2 INTRODUCTION

Hydraulic fracturing is a petroleum mining technique used to bring oil and natural gas trapped in underground shale and rock formations to the surface for capture. As the name implies, water typically carries a proppant—such as sand—down into a wellbore to fracture shale or rock formations containing oil and gas. The proppant is utilized to keep the fractures open, which allows the trapped oil and gas to flow back to the surface with the water. The fracturing process uses, on average, 230,000 barrels (36,563 m^3/9,660,000 gal) of water per well. The average pad drilling setup typically contains two to six wells (Figure 9.1), which could result in water consumption ranging between 460,000 barrels (73,126 m^3/19.3 million gal) and 1,380,000 barrels (219,378 m^3/57.9 million gal) of water.

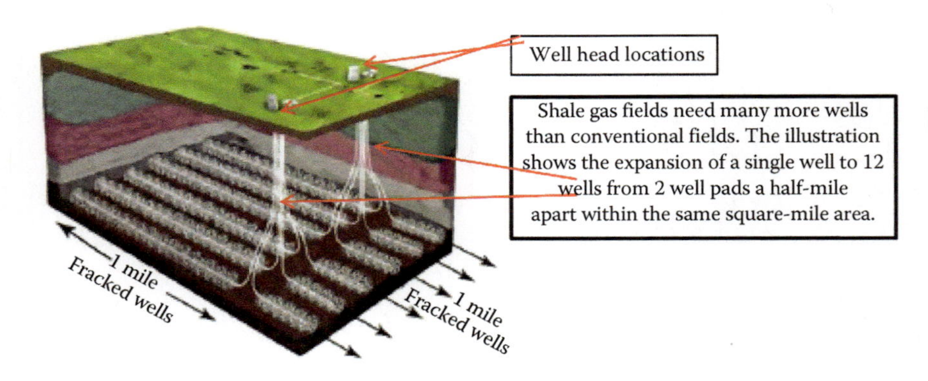

FIGURE 9.1 Typical frac well-field configuration. (Images courtesy of frackingboom.com.)

Typically, a medium-to-large energy producer will have more than 100 wells completed in a given year. Using the water consumption values given earlier, an energy producer with 100 wells would need over 23,000,000 barrels (3.66 million m^3/966 million gal) of water each year. If 15 producers are all producing wells within the same area, 345,000,000 barrels (54.8 million m^3/14.5 billion gal) of water would be needed to conduct the hydraulic fracturing activity in the area. Most areas where oil and gas wells are being completed do not have the water resources necessary to support the needs of current development, let alone support the hydraulic fracturing growth that is projected over the next several decades (EIA, 2014). Thus, the purpose of this chapter is to outline water usage during hydraulic fracturing operations, to identify whether water produced during hydraulic fracturing activities could be reused and recycled, and to identify the water quality constraints and technology needs necessary for recycling and reuse of hydraulic fracturing water.

9.3 EXPLORING WATER CONSUMPTION IN HYDRAULIC FRACTURING

As of May 30, 2014, RigData listed 1865 land drill rigs within the United States (Figure 9.2) working on a 4-week average; of those drill rigs, 890 were located in the state of Texas (Table 9.1). These data imply that nearly 50% of the entire active drilling fleet is working in Texas, and that hydraulic fracturing activity in Texas will use more water than any other state to support the completion of oil and gas wells. If an average well takes 14 days to complete, then the 890 drilling rigs could complete 23,203 wells in 1 year. If only half of these wells use the average water consumption listed earlier, then 2.7 billion barrels 424 million m^3 (112 billion gal of water) of water would be needed to support hydraulic fracturing activity during 1 year in the state of Texas. Once this water is used in the hydraulic fracturing process, the wastewater can no longer be used without treatment and would typically be disposed.

Fresh water is in high demand due to droughts, population growth in active hydraulic fracturing areas, and separate water withdrawals from local agriculture. As such, the water available for hydraulic fracturing is quite scarce in some areas.

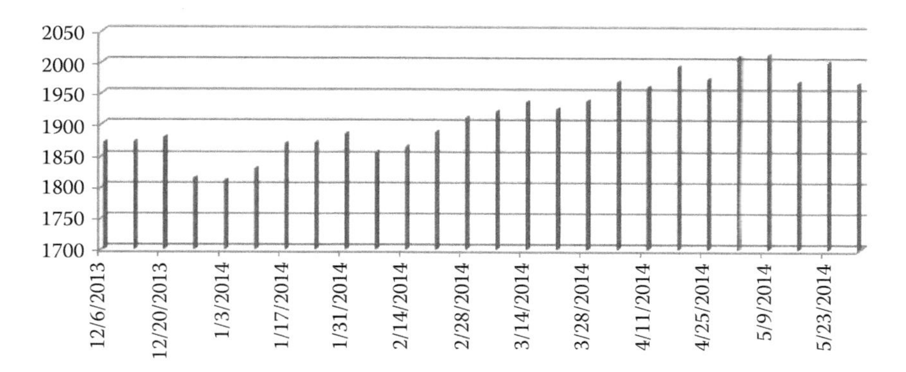

FIGURE 9.2 U.S. land working rig count between December 2013 and May 2014.

Water availability issues in today's society requires that the oil and gas industry recycle and reuse the water that under normal operations would be disposed of in saltwater disposal (SWD) wells. The conventional method of using water in the development of oil and gas wells must change as a consequence of the large volume of water needed by the hydraulic fracturing process.

The oil and gas industry has always used water in the development and completions of wells. Until the advent of directional drilling, the volume of water used was only a fraction of what is needed by today's oil and gas exploration endeavors. The business of oil and gas wells has evolved over time into separate units, replacing the older model of one company overseeing the life of a well. These business units are normally subdivided into several groups, including well drilling, well completion, and well production. Each of these units has water demands needed to achieve their individual unit goals, and all units have to dispose of the water after use within their processes. The process which consumes the largest amount of fresh water (non-saltwater chlorides usually below 1500–2000 ppm) is the well hydraulic fracturing completion process, commonly known as "fracing."

The reuse and recycling of fracing water will reduce the total disposal volume of water, and will thereby offset the additional fresh water needed to fracture a well. Every gallon not disposed offsets another gallon of fresh water needed in the completions process. To coin a new term, the water needs to be made "fracable." By examining the water used in the hydraulic fracturing process, we can provide a better understanding of "fracable water."

The rule of thumb about whether a fracing operation can move forward is based on logistics: how much water and sand are needed to fracture and open the shale formation. Or more simply, "no sand, no water, no frac." Planning and supply chain logistics ensure the delivery of proppants (sand) and water so that the hydraulic fracturing process occurs. While both of these mediums are important, the more important of the two is water. The process of fracturing a shale formation occurs when water mixed with proppant is pumped and pressured up to extremely high pressure rates which can approach 9000 PSI. The pressurized fluid will travel deep into the surface of the earth, sometimes up to three miles. If the mixing of the proppant stops

TABLE 9.1
Rig Count Data for Texas as of May 30, 2014

	Baker–Hughes Rig Count				RigData Rig Count				
	4 Week Average				4 Week Average				
District	**2013**	**2014**	**Last Week**	**This Week**	**2013**	**2014**	**Last Week**	**This Week**	**Waiting To Spud**
Texas RRC District 1	137	118	116	114	136	123	125	123	14
Texas RRC District 2	87	90	91	90	81	93	91	89	4
Texas RRC District 3	48	55	53	58	50	64	62	64	3
Texas RRC District 4	34	35	33	35	31	29	29	25	5
Texas RRC District 5	13	9	9	9	12	11	12	13	1
Texas RRC District 6	26	33	34	31	24	33	36	35	1
Texas RRC District 7B	13	9	10	9	18	18	16	20	7
Texas RRC District 7C	81	97	97	95	74	102	101	100	7
Texas RRC District 8	278	326	325	334	265	320	318	316	11
Texas RRC District 8A	36	38	41	38	34	41	42	37	3
Texas RRC District 9	21	16	16	16	31	28	30	28	2
Texas RRC District 10	60	64	62	63	64	67	66	65	0
Texas total	*834*	*890*	*887*	*892*	*820*	*929*	*928*	*915*	*58*

and/or is cut while fracing, the fracing process can still continue. The pump rates and pressures may still be maintained, because the water is still flowing through the pumping process, keeping the pumps primed and maintaining the pressure on the formation. If, however, the water source is suddenly lost, pressure would drop drastically and the flow rates would stop. This could result in a catastrophic failure and loss of the well.

To avoid operational failures, planning and logistics revolve around the supply and storage of water on site. The frac design requires a minimum amount of water to be on the location at all times. The amount of water volume needed must take into account a safety factor allowing for flushing of the pumping equipment and for capping of the well to maintain hydraulic pressure on the formation. This amount of water can fluctuate depending on many variables, but at least four available 500 barrels (21,000 gal) tanks totaling 2000 barrels (84,000 gal) of water storage are on site at all times.

9.4 WATER STORAGE AND DELIVERY FOR HYDRAULIC FRACTURING ACTIVITY

In a perfect "frac completions world," all of the water needed to complete an oil and gas well would already be on location and within 91 m (100 yards) of the well head. To approach the reality of this concept, companies usually dig a large frac pit and line the pit with a single liner, or they will build an above-ground tank farm to hold the necessary water volume. The above-ground tank farm can be made up of multiple frac tanks that hold up to 79.48 m^3 (21,000 gal) of water each, or a single field tank that can hold more than 5678 m^3 (1,500,000 gal) of water.

The water must be supplied to the hydraulic fracturing location by pumping or by trucking. If the water is pumped, the time to procure the water will be limited by water withdrawal permits, distances, and elevation changes from the withdrawal points. Additionally, the points of water withdrawal could pull water from multiple sources at the same time, compounding the aforementioned factors. These factors will play a role in determining the length of time water will be stored on location. Due to the length of time needed to procure the necessary water, water may be stored on these locations for weeks and/or months at a time prior to fracing (Figure 9.3).

If the water is trucked and not pumped, the water can be delivered with two basic types of trucks: a straight truck has an average load capacity of 100 barrels (15.9 m^3/4200 gal) of water, whereas a tanker truck has an average load capacity of 130 barrels (20.7 m^3/5460 gal). If the water is trucked and only tanker trucks were used, it would take over 1753 separate truck loads at 70 loads per day, which would take 25 days to supply the water needed to complete the well-fracturing process.

9.5 WATER PRODUCED IN HYDRAULIC FRACTURING ACTIVITY

Over the life of a fracing well, some of the water pumped in will eventually return to the surface. There are some formations that consume water and yield no more than 10%–15% return of the water. On the other end of the spectrum, due to formational waters, some wells will yield 25%–50% return of water for the life

FIGURE 9.3 Water storage methods for hydraulic fracturing.

of the well. Formational water is water present in the formation at the time the formation was formed. However, in the middle of the spectrum, the majority of fracing wells will bring back almost all of the water pumped into the well by the end of the hydraulic fracturing process. Therefore, in most cases, the 230,000 barrels (36,563 m^3/9,660,000 gal) of water pumped through the well will be trapped in the sand and fracture of the formation. The well will then be opened to allow water "flowback." As this water flows back up to the surface, it will carry oil and gas for capture. The ratio of water to oil and gas is initially high, with water as the largest percentage of the fluid that the well yields. For up to several weeks, the ratio varies anywhere from 10% to 40% of the total water pumped downhole returned with the completion of the "fracturing process." Of the 230,000 barrels (36,563 m^3/9,660,000 gal) approximately 23,000 barrels (36,563 m^3/966,000 gal) of water to 92,000 barrels (14,625 m^3/3,864,000 gal) of water will be considered flowback waters. The injected water was fresh water with chloride levels of less than 2000 ppm (parts per million) and total dissolved solids (TDS) less than 3000 ppm. The flowback water will now typically have a chloride level between 10,000 and 50,000 ppm and a TDS level between 15,000 and 60,000 ppm. The change in chloride and TDS levels in the flowback water is due to the water sitting in the formation for a period of time and to the water establishing a chemical equilibrium with the minerals in the surrounding rock formation. In many cases, salinity or TDS limits reusing and recycling options.

Once the ratio of water to oil and gas changes and the water volume decreases until it does not exceed the minimum capacity of the production equipment, the flowback water will move through oil–gas–water separation devices, such as gun-barrels, heater-treaters, and separators. The well will now be put in production and no longer be considered "flowback" water. The water flowing through the production equipment is separated from the oil and gas, and is now considered produced/production water. The produced water will have a chloride level in the range of 50,000 ppm to >300,000 ppm and a TDS range of 50,000 ppm to >300,000 ppm. The water is extremely saline, and just like the flowback waters, will require treatment before reuse. Typically, rather than reuse, these fluids are disposed by saltwater well injection where the water will not affect or impact the surface environment.

9.6 MAKING THE WATER "FRACABLE"

In hydraulic fracturing, water's second most important role is to carry the sand (proppant) downhole and into the formation, keeping the fractures open so that the trapped oil and gas can flow out and up the well to the surface. The amount of sand and the types of sand added to a well during the fracturing process varies, and an average well can use up to 136,076 kg (300,000 lb) of sand during a hydraulic fracturing event. The goal of the hydraulic fracturing process is to carry the sand down into the wellhole, out into the formation, and to leave the sand in the formation. To do this, chemicals are used to keep the sand particles suspended in the water stream when being injected into the formation. If the sand comes out of solution during transport into the formation, the fracturing process will stop due to elevated pressures. This process is referred to as a "screen out" or "sanding off the well." The most common chemicals used to keep the sand suspended are friction reducers, gels, and cross-linkers. The properties of the water must not inhibit the hydration of these proppant carriers, and each of these proppants has different threshold limit values that can potentially restrict the reuse of untreated and processed flowback/produced waters.

Making water "fracable" for reuse and recycling requires some well-defined water quality and water recycling process goals and must meet the following criteria:

- Not affect the potential proppant carriers of gels, friction reducers, and cross-linkers
- Not change the pH of the water
- Not allow biological fouling of formation and distribution system following treatment
- Not change the scaling index or chemically foul the formation and distribution system
- Not cost more than disposal, and cost less than the cost of fresh water

Scaling caused by cations and anions, corrosion (due to pH and presence of dissolved gases), and biological growth (anaerobic and aerobic) are all issues in the fracturing process, affecting fresh waters, and reused and recycled waters. Thus, water quality is an important challenge in reuse and recycling of flowback and produced water.

Another big challenge is that the hydraulic fracturing activity never consumes or produces the same amount of water at each activity location. Some would say "It is easy, just take that water to the next frac and put it away downhole." The problem with this is that the hydraulic fracturing location may move only a few thousand feet on the surface, but it could move 100 mi in any area at any given time within the formation. The variable amount of water volume necessary to complete a frac is in a state of constant flux between sites, adding to the complexity of solving the reuse and recycle problem.

Once a well is completed, time is required to install the production equipment, take the water from the well, and collect the oil and gas. The rate of water produced and the rate of water consumption are not synchronized, necessitating storage. The storage method must be decided in the frac design process, and may be above or below ground, just as with the water stored prior to the start of the fracing activity. Storage tanks or a bladder expansion system introduce costs of moving the portable storage system and increase the challenge of making hydraulic fracturing water reuse and recycling a cost-effective strategy. However, from the temporary storage tanks, stored water could be pumped to the next location via temporary water lines. State regulatory agencies have regulations and special permitting requirements, which add additional planning, and costs to pumping frac water from point to point. If flowback or produced water leaks onto the ground, the state environmental agency must be notified, and the spill must be maintained and cleaned up, increasing the cost and risk associated with any recycle and reuse process.

9.7 OPPORTUNITIES FOR RECYCLING AND REUSE

To solve problems and overcome obstacles related to the recycling and reuse of water in the fracing process, it is necessary to find and identify specific opportunities in the hydraulic fracturing process (Figure 9.4).

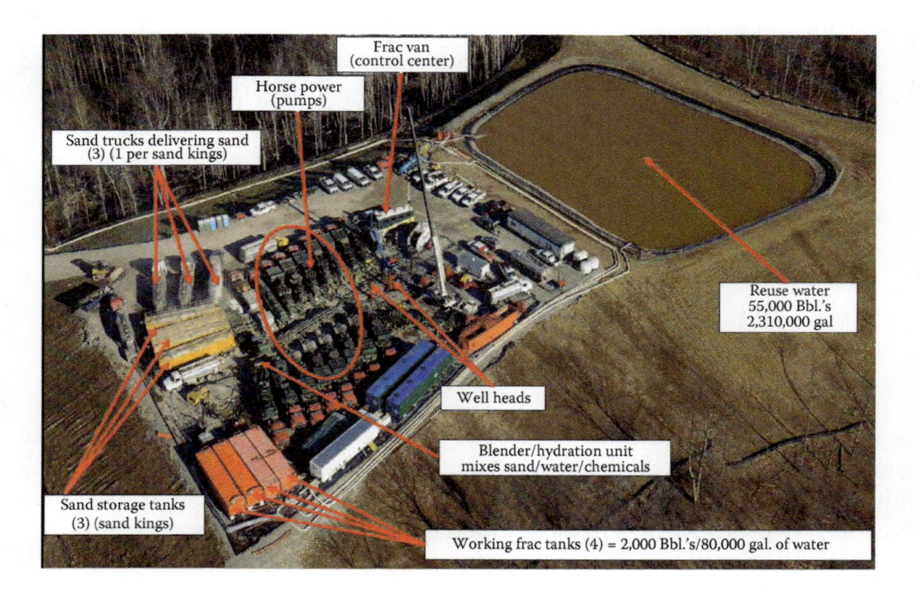

FIGURE 9.4 The water and sand distribution process at a typical hydraulic fracturing site.

All water flowing into the well has to pass through the working tanks, and every frac location has some number and configuration of working frac tanks. The water pumped to the hydraulic fracturing process will pass through a water distribution manifold and will reside in the working frac tanks that number from 4 to 8 on most hydraulic fracturing jobs (Figure 9.4). The distribution process was built to deliver water and not measure, change, or influence the water properties as the water enters the fracturing process.

The working tanks are an ideal location to monitor water quality. Currently, the common practice in the oil and gas industry is to only measure the water quality in storage tanks weeks prior to the hydraulic fracturing event. This is done in order to evaluate water compatibility between sites, the scaling index, and the biological activity of the stored water. The water quality (namely, ion concentrations and biological activity) can change within hours of water quality testing. As such, the fracing industry needs to sample the water in the storage tanks more frequently to assess water quality changes and to better understand reuse and recycled water quality.

The fracing industry should include compatibility studies to determine if the raw flowback and produced waters would negatively impact proppant-carrying chemistry. From an author's fieldwork experience, early test results showed that the raw flowback and product waters were not exchangeable with freshwater, as the hydraulic fracturing fluid could not hydrate the gels or cross-linkers. In fact, the raw flowback and product waters caused flocs to form in the tanks and bind up the proppant carriers. A solution was developed based on dilution with some very big misunderstanding of the chemistry of the flowback and produced waters, which affected operators' understanding on how to recycle the flowback and produced waters. The dilution model went forward and is currently used even though the chemistry is not fully understood.

One operational solution to allow for reuse of water via dilution (Figure 9.5) is to add an additional frac tank and isolate the stored flowback and produced waters. If the hydraulic fracturing pumping process started to develop pressure losses and/or lose flow rates while pumping flowback or produced waters, the tank with untreated recycled waters could be quickly shut down to prevent a screen out. The idea behind the operational configuration is that the untreated recycled waters would be diluted by up to 80%, which would keep the properties of the untreated water from negatively changing the properties of the bulk fresh water. This, in turn, would minimize the impact on the proppant placement. The specific challenges of this operational solution are as follows:

- The pump company must connect the frac tanks at the same point every time.
- To balance water flow from the frac tanks, an isolation valve will require adjustment.
- The water level in the frac tanks will vary with time, even with the valve position static.
- The water blending amount would have to be adjusted to maximize the amount of recycled water.
- Trucked or piped deliveries of water would be limited by storage and the actual frac operations.

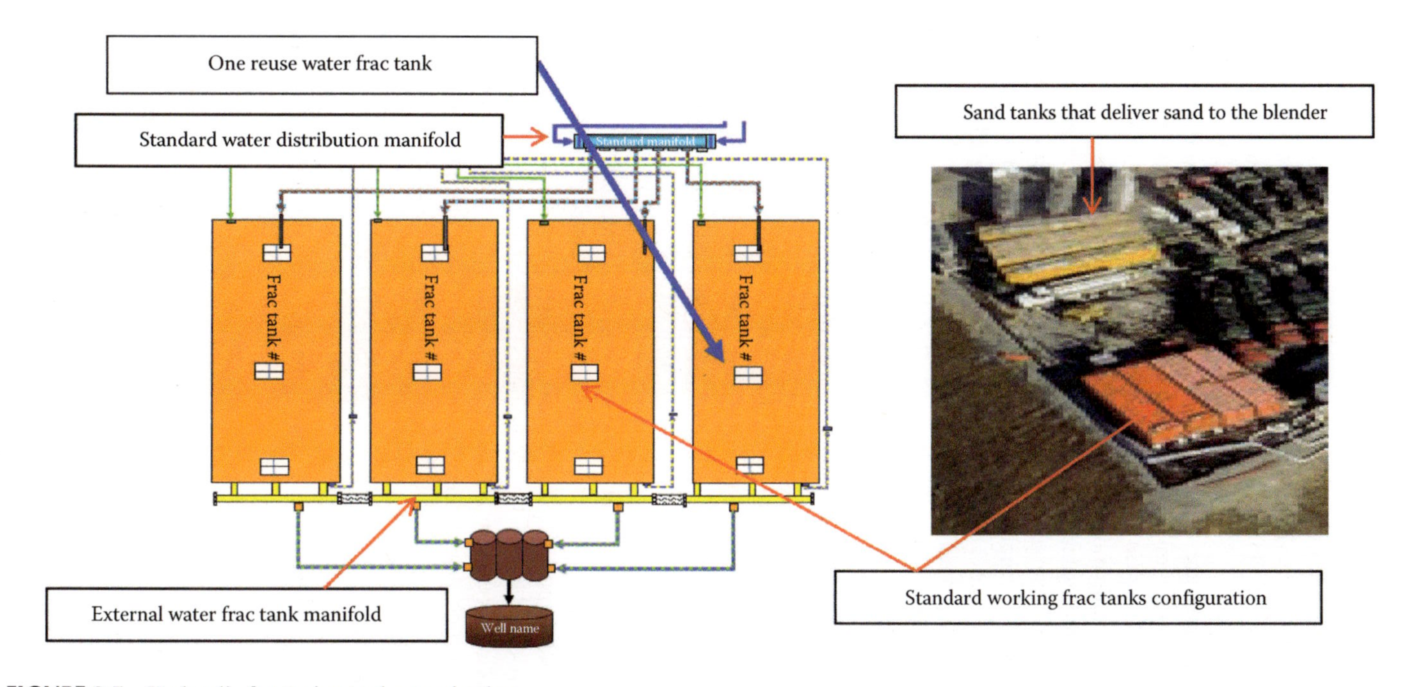

FIGURE 9.5 Hydraulic fracturing tank organization.

FIGURE 9.6 Reusing and recycling flowback water on site through dilution of fresh water.

Figures 9.5 and 9.6 show how this type of dilution process would work and the addition of the additional frac tanks.

9.8 WATER TREATMENT AS A SOLUTION

To minimize the impacts of flowback and produced water quality on reuse and recycling potential, the oil and gas industry has explored centralized treatment facilities. With centralized treatment, the flowback and produced water is transported to a central treatment facility (Figure 9.7), and the facility treats the frac water from many different wells to resemble fresh water. The concept is similar to centralized treatment of wastewater in cities. This concept has been explored within the industry, but the cost for this type of solution is very expensive with capital, personnel, and transportation costs. The transportation of the water is also of concern since the water has to be collected and delivered to the treatment facility, and once treated it must be delivered to the next frac job. Another option is on-site treatment for recycling and reuse. On-site treatment reduces the potential of environmental impacts during transportation to a SWD well, and would also mitigate the issues surrounding transport of the water. On-site wastewater treatment relies on portable wastewater treatment systems that move from well to well during hydraulic fracturing operations.

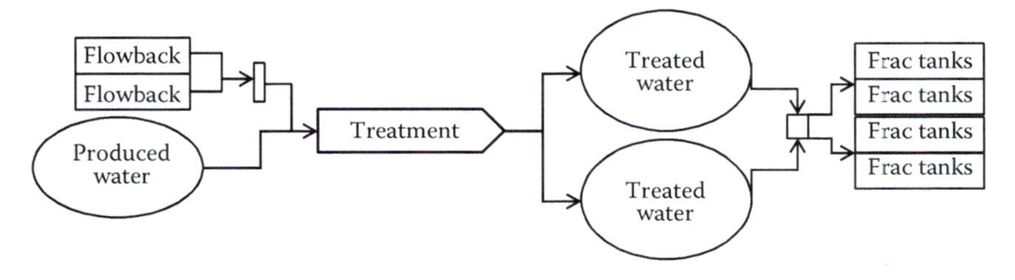

FIGURE 9.7 On-site treatment of flowback water.

Most research into flowback or produced water treatment has focused on the reduction of TDS by evaluating the performance of individual processes such as membrane treatment (Hutchings et al., 2010; Jiang et al., 2013; Miller et al., 2013) and thermal systems such as evaporation, distillation, or crystallization (Gregory et al., 2011). Beneficially, many of these processes remove bacteria and other bio-fouling agents using extreme environments (UV, high temperature, high salinity) or size exclusion (membrane filtration). The NOMAD 200 MVR Evaporator system has been employed to process highly variable oilfield wastewater, resulting in a water recovery rate of 85% (Horner et al., 2011). Other options use conventional coagulation and flocculation processes to treat water produced during hydraulic fracturing operations. The Aqua-Pure ROVER Mobile Clarifier removes organics and suspended solids using coagulation and clarification technologies mounted onto portable trailers (Horner et al., 2011). Halliburton developed the CleanWave® system to treat flowback and produced water using electrocoagulation (Halliburton, 2014). Evaporation, distillation, and crystallization technologies can be energy intensive but may prove beneficial if coupled with existing renewable energy sources such as solar and wind, or to simply use the fracturing products themselves.

9.9 CHEMICAL REQUIREMENTS FOR REUSE OF FLOWBACK AND PRODUCED WATERS

As described previously, flowback and produced waters associated with shale gas development in the United States are managed via injection into disposal wells, reuse in other fracturing operations, or treatment for specific reuse purposes. The treatment option provides considerable opportunity to minimize new water use, which can be particularly important in water-stressed regions. Identifying the status of produced water treatment technology, water chemistry, and recycling in different U.S. shale plays is key in that it controls the oil and gas industry's ability to reuse the water.

The primary limitation to recycling is the cost of produced water treatment relative to disposal (Accenture, 2012). The goals of produced water treatment include

- Reducing TDS for discharge
- Reducing the volume of produced water for disposal
- Reducing scaling and bio-fouling for reuse or injecting into underground injection control wells (ALL consulting, 2011) and
- Minimizing the amount of makeup or new water required to develop shale gas resources.

Water concerns are particularly acute in the western United States, which has faced a severe drought in recent years. The overall water needs for power generation by hydraulic fracturing for shale gas are quite small compared to other fuel sources (Meldrum et al., 2013) but that does not mean that local dislocations in water supply versus demand may not be driven by technology (Nicot and Scanlon, 2012). The rapid expansion of shale gas development has led to increasing concerns for the

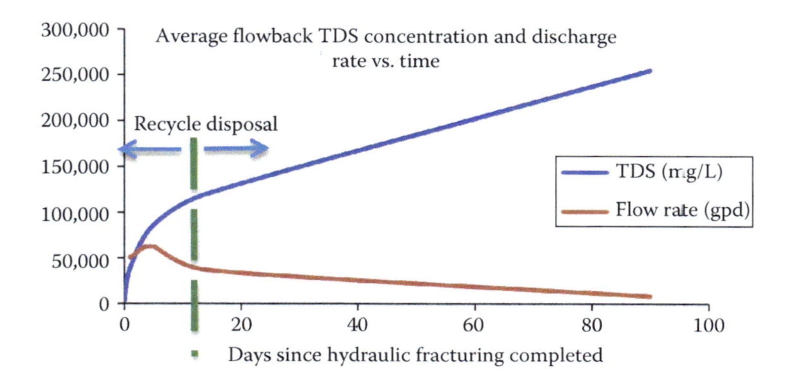

FIGURE 9.8 TDS and water flow in flowback and produced waters (Yoxtheirmer 2012) with hypothetical dividing line between efficient treatment and reuse versus disposal.

water required to drive that development and treatment, and reuse may be a critical tool to minimize the impact of the technology.

The optimal approaches for treatment are a strong function of the chemical composition of the produced and flowback water. This is highly variable in different shale plays, and is a function of time after a well is fractured. Initial flowback waters are very similar to those used to fracture a well, while over time the flowback and waters produced after a well is put into production take on more of the characteristics of the formation waters as shown in Figure 9.8. The volume of water is also a strong function of time and formation characteristics. All of these characteristics complicate the issues surrounding water treatment in fracing operations. In general, it is expected that waters generated soon after fracturing a well can be cost-effectively treated or reused directly, while later produced waters require extensive treatment and are likely to be disposed of via deep-well injection, if available.

The quality of water required for hydraulic fracturing has been studied by Ziemkiewicz et al. (2012) and the conclusions are summarized in Table 9.2. This shows that relatively poor quality waters can be used for hydraulic fracturing activity, and that many initial flowback waters would meet these requirements. Late flowback and produced waters will typically require some treatment to meet these standards. Optimal treatment strategies would involve minimal, low-cost treatment schemes designed to achieve these standards, rather than attempt to achieve near potable waters for hydraulic fracturing make-up water.

Given the nature of the challenge for recycling and reusing flowback and produced water, the expectations for recycling and reuse in selected major shale plays are summarized here:

- *Barnett Shale (Texas)*: Limited reuse has been done in the Barnett Shale (Duncan, 2013) due to the availability of low-cost SWD wells. Reuse has made up approximately 6% of total water needed to hydraulically fracture a new well (Mathis, 2011). Mathis (2011) also noted that in comparison

TABLE 9.2
Industry Contact Group Chemical Criteria for Frac Water Makeup

Chemical Parameter	Maximum Value (mg/L)[a]
TDS	50,000
Hardness	26,000
Bicarbonate (HCO_3)	300
Sulfate (SO_4)	50
Chloride (Cl)	45,000
Sodium (Na)	36,000
Calcium (Ca)	8,000
Magnesium (Mg)	1,200
Potassium (K)	1,000
Iron (Fe)	10
Barium (Ba)	10
Strontium (Sr)	10
Manganese (Mn)	10

[a] Requirements of industry contact group members.

with injection wells, the main limiting factors in preventing higher levels of reuse are logistics and economics. Recycling and reuse are not currently considered cost-effective methods for flowback and produced water management in this area.

- *Haynesville Shale (Louisiana)*: Haynesville Shale exhibits a very low quality (high salinity) flowback and produced water. It also has a relatively low volume of flowback water (Mathis, 2011). Due to the limited volume of the returned water and its high salinity, recycling and reuse of flowback/produced water has not been practical in this shale play (Nicot and Scanlon, 2012). Recently, operators in the Louisiana area have been trucking flowback and produced water to injection wells in east Texas as a result of new laws limiting the maximum injection pressure in Class II disposal wells (Bruyninckx, 2010).
- *Fayetteville Shale (Arkansas)*: The flowback and produced water of the Fayetteville Shale play has low TDS and chloride as compared to other plays (TDS from 10,000 to 20,000 ppm) (Mathis, 2011). Therefore, reuse is a suitable method for flowback and produced water management. As a result, approximately 80% of initial flowback and produced water (approximately 946 m^3/well or 250,000 gal/well) has been reused. Fayetteville uses 100% surface water in its operations (36% from their own constructed impoundments, 26% from private ponds, 21% from reuse of flowback, and 17% direct withdrawals from streams) (SWN, Undated).
- *Marcellus Shale (Pennsylvania, West Virginia, New York)*: Injection wells in the Marcellus Shale region have limited injection capacity, and drilling

> new disposal wells is not cost effective due to the geology of the area. Use of Class II injection wells in Ohio is decreasing with approximately 20% of the wastewaters disposed in this manner in 2011 and 10% in 2012 (Yoxtheimer, 2012). Flowback and produced water is either partly treated and reused, trucked out of state to saline water disposal wells, or sent to licensed saline waste water treatment plants.

Mathis (2011) reported that Chesapeake is currently reusing nearly 100% of initial flowback and produced water in the Marcellus Shale via improved filtering processes. Reuse makes up about 10% of the total water requirements for fracturing a well (Rassenfoss, 2011). Rassenfoss (2011) notes that the cost ($6/barrel) of trucking produced water to the nearest cluster of disposal wells in Ohio can be compared with the cost of desalination treatment ($7.5/barrel). Cagle (2012) also states that currently in the Marcellus Shale region, 100% of flowback water and the majority of produced water are reused. The water needed for hydraulic fracturing in the Marcellus Shale region is a mixture of 80% of fresh water, ponds, and rivers with 20% of flowback and produced water. According to the PaDEP analysis for 2012, approximately 87% of flowback and produced water were reused and 13% were disposed in Class II injection wells. Eighty-five percent of recycling water was treated in the field and 15% treated at centralized treatment facilities (Yoxtheimer, 2013).

9.10 CONCLUSION

Overall, as the reliance of fossil fuels in this country continues to rise, the oil and gas industry will have to meet these needs and will continue to pursue methods to extract oil and gas reserves from the earth. Due to this, the water resources of the areas surrounding these extraction activities will continue to be stressed as they attempt to meet the water demands of the oilfield industry. To mitigate the issues surrounding the scarcity of water in some areas, the oilfield industry must move toward embracing water reuse and recycling techniques to preserve the regional water resources. Simple techniques such as utilizing portable on-site treatment systems to achieve the necessary water quality, using well-established technologies, or constructing centralized treatment systems in areas of high drilling activity would greatly help to preserve water resources. Further research into the area of fracing water reuse must focus on finding a cost-effective solution to treat the produced and flowback waters, and to establish a minimum set of water quality criteria that result in a reusable water source beneficial to the fracing industry.

REFERENCES

ALL Consulting. 2011. Treatment of shale gas produced water for discharge. http://www.nxtbook.com/nxtbooks/naylor/LOGT0211/index.php?startid=13#/12. Accessed on February 20, 2015.

Bruyninckx, J. (2010). Director of North Louisiana LOGA, Haynesville Shale, Louisiana Pipeliner Association, May 2010. https://www.louisianapipeliners.org/uploads/5.17.10 PipelinersPresentationsmall.pdf. Accessed February 20, 2015.

Cagle, C. 2012. Sustainable solutions for flowback/produced water challenges in oil and gas fields. Presented at *AIChE & AWMA Joint Workshop: Shale Oil and Gas E & P*, Pittsburg, PA, November 1–2, 2012.

Duncan, I.J. 2013. Evaluating the sustainability of water usage for hydraulic fracturing of shale gas reservoirs: A life cycle approach, in *Effective and Sustainable Hydraulic Fracturing*, eds., A.P. Bunger, J. McLennan, and R. Jeffery. InTech.

EIA. 2014. AEO 2014 early release overview. http://www.eia.gov/forecasts/aeo/er/early_production.cfm. Accessed October 15, 2014.

Gregory, K.B., D.V. Radisav, and D. Dzombak. 2011. Water management Challenges Associated with the production of shale gals by hydraulic fracturing. *Elements* 7: 181–186.

Halliburton. 2014. CleanWave® frac flowback and produced water treatment. http://www.halliburton.com/en-US/ps/multi-chem/production-challenges/cleanwave.page. Accessed February 20, 2015.

Horner, P., B. Halldorson, and J. Slutz. 2011. Shale gas water treatment value chain—A review of technologies, including case studies. *Proceedings of the 2011 SPE Annual Technical Conference and Exhibition*, Denver, CO, October 30–November 2, 2011.

Hutchings, N.R., E.W. Appleton, and R.A. McGinnis. 2010. Making high quality frac water out of oilfield waste. *The International Society of Petroleum Engineers 2010 Annual Technical Conference and Exhibition*, Florence, Italy, September 19–22, 2010.

Jiang, Q., J. Rentschler, R. Perrone, and K. Liu. 2013. Application of ceramic membrane and ion-exchange for the treatment of the flowback water from Marcellus shale gas production. *Journal of Membrane Science* 43: 55–61.

Mathis, M. 2011 Regulatory affairs of Chesapeake Energy, Shale Natural Gas-Water Use Management. Presented at *ICWP Annual Meeting*, St. Louis, MO, October 11–14, 2011. http://www.icwp.org/cms/conferences/Mathis14Oct2011.pdf. Accessed February 20, 2015.

Meldrum, J., S. Nettles-Anderson, G. Heath, and J. Macknick. 2013. Life cycle water use for electricity generation: A review and harmonization of literature estimates. *Environmental Research Letters* 8(1): 015031.

Miller, D.J., X. Huang, H. Li, H. et al. 2013. Fouling-resistance membranes for the treatment of flowback water from hydraulic shale fracturing: A pilot study. *Journal of Membrane Science* 437: 265–275.

Nicot, J.P. and B.R. Scanlon. 2012. Water use for shale-gas production in Texas, U.S. *Environmental Science and Technology* 46(6): 3580–3586. http://www.beg.utexas.edu/staffinfo/Scanlon_pdf/Nicot+Scanlon_ES&T_12_SI.pdf. Accessed March 2012.

PG Accenture. 2012. Water and shale gas development: Leveraging the US experience in new shale developments. http://www.accenture.com/SiteCollectionDocuments/PDF/Accenture-Water-And-Shale-Gas-Development.pdf. Accessed February 20, 2015.

Rassenfoss, S. 2011. From flowback to fracturing: Water recycling grows in the Marcellus Shale, JPT. http://www.spe.org/jpt/print/archives/2011/07/12Marcellus.pdf. Accessed July 2011.

Yoxtheimer, D. 2012. Water resources management for shale energy development. Presented at *AIChE & AWMA Joint Workshop: Shale Oil and Gas E & P*, Pittsburg, PA, November 1–2, 2012.

Ziemkiewicz, P., J. Hause, B. Gutta et al. 2013. Final Report, Water quality literature review and field monitoring of active shale gas wells, Phase I; for "Assessing Environmental Impacts of Horizontal Gas Well Drilling Operations." Prepared for the West Virginia Department of Environmental Protection Division of Air Quality by the West Virginia Water Research Institute, West Virginia University. February 15, 2013.

10 Impact of Hydraulic Fracturing on Transportation Infrastructure

Sanjaya Senadheera

CONTENTS

10.1 Overview 175
10.2 Introduction 175
10.3 Transportation Demand from a Fracturing Well 177
10.4 Pavement Distresses Caused by Heavy Vehicles 180
10.5 Traffic Flow and Safety Impacts 184
10.6 Summary and Concluding Remarks 185
References 185

10.1 OVERVIEW

The transportation infrastructure in the rural United States is not designed to handle high volume loads associated with hydraulic fracturing. Proactive management is necessary to minimize further deterioration to rural roadways and reduce the risks of accidents.

10.2 INTRODUCTION

Oil and gas production has dramatically increased over the last few years due to widespread use of horizontal drilling and hydraulic fracturing in tight reservoirs that are not amenable to production using conventional drilling and pumping techniques. Over the past decade, the amount of natural gas produced from shale plays has grown from 2% to over 37% of domestic energy supply, and it has been predicted by some to grow to over 75% within the next 20 years (US EIA, 2014). Due to the rapid development of shale gas exploration, the steady coexistence between the transportation infrastructure and the fossil fuel industry has been disturbed over the past few years, particularly in areas where hydraulic fracturing is employed. This stress is particularly felt in Texas, Pennsylvania, West Virginia, and Kansas where hydraulic fracturing activities have been in full swing in the last few years.

Hydraulic fracturing is a material-intensive operation. A typical hydraulic fracturing pad has a large site footprint with areas for transportation vehicle movement and storage, as well as storage for water, chemicals, mixers, pumps, and other appurtenances needed for fracturing. Water is a critical resource in the hydraulic fracturing process. A typical hydraulic fracturing operation requires anywhere between 11,356 and 18,927 m^3 (3–5 million gal) of water. Water may be brought to the site from several sources including surface water, groundwater, municipal water supply, treated wastewater, power plant cooling water, and recycled (flowback) water. Water is usually the first material brought to the site, mostly using tanker trucks. While a variety of factors such as water needs of local communities and local policies and regulations are considered during sourcing of water supplies, ease of transportation is a major factor as it controls the costs of water. In the Permian Basin of Texas, raw water costs are currently around $0.25–$0.50 per barrel of water. However, water sometimes has to be transported over tens of kilometers, which increases the costs to over $2.00 per barrel.

Sand (or other proppants) is used to keep the hydraulically fractured shale pores open to let the oil and gas flow out. Greater proppant volumes lead to larger openings and greater flow of oil and gas. Nearly 4,990–1,134 kg (11,000–25,000 lb) of proppants are used during hydraulic fracturing operations (Powell et al., 1999). The best frac-sands can be found in mid-western states of Minnesota, Wisconsin, and Illinois, and, as such, frac-sands are often transported over large distances to their fracturing destinations (Masalowski, 2012). In addition to proppants, hydraulic fracturing needs several other chemicals, which are brought in trucks and mixed on site. In addition, hydraulic fracturing is a labor-intensive activity and employs 10–50 on the site. These people typically stay in nearby towns and cities and often drive tens of kilometers to get to the fracturing pad.

Once the fracturing process is complete, a portion of the water injected into the formation during fracturing flows back to the surface. The amount of flowback and produced water ranges from 5% to 40% of the total volume that is injected and can be several million gallons (4,000–20,000 m^3). This water needs to be transported to either a wastewater disposal or a treatment facility. In addition, hydraulic fracturing creates solid wastes that are often disposed in landfills (Gilmore et al., 2014). Typically, hydraulic fracturing operations last for 5–15 days at a well. Drilling rigs and fracturing operations move from one fracturing pad to the next within a shale play. A conceptual model of the transportation flow into and out of the hydraulic fracturing operations is schematically depicted in Figure 10.1. Finally, upon completion of their useful operational life, oil and gas wells need to be plugged and abandoned properly. Cement slurry and other additives are often used to plug these wells, which need to be transported onto the well site and thus add to the overall transportation demand generated at the fracturing well.

These activities have led to a significant increase in the vehicular traffic in areas overlying shale plays currently being used to produce oil and gas through unconventional drilling. What impact does this enhanced oil and gas production have on the transportation infrastructure? This chapter seeks to provide a basic understanding of the pavement engineering concepts and uses it to discuss the effects of fracturing operations on existing transportation infrastructure. The chapter concludes

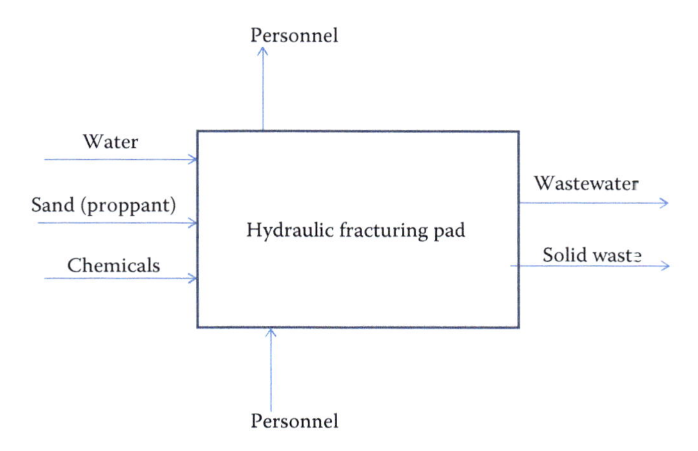

FIGURE 10.1 Transportation flows in hydraulic fracturing.

by providing certain recommendations for improving traffic safety and reducing energy-related transportation impacts.

10.3 TRANSPORTATION DEMAND FROM A FRACTURING WELL

By considering all aspects of a fracturing well including drilling and operations, the Texas Department of Transportation (TxDOT) estimated the following traffic generation numbers (TxDOT, 2012a–c):

- Average number of truck trips to bring a natural gas well into production is 1184.
- Average number of truck trips per year for fracturing operation is up to 353.
- Average number of truck trips to refracture a well every 5 years is 997.

At first glance, these numbers may not seem like very much, particularly considering the millions of vehicles that travel on our nation's highways. However, the numbers do add up to create significant strain on the highway transportation infrastructure, especially those designed for low traffic volumes. For example, there were 2764 new wells fractured in the Permian Basin region of Texas (Hernandez et al., 2015) during January 2012–May 2013, which translates to roughly 4.5 million truck trips in the area that would have otherwise not occurred if there was no oil and gas activity. As will be discussed later in the paper, even a few hundred vehicles each year can make a significant difference on pavement performance.

Federal and State laws limit the magnitude of both gross vehicle weight and axle load of vehicles that can operate on the US State highway system. For example, the limit in Texas for single axles is 9,072 kg (20,000 lb) and for tandem axles, it is upwards of 15,422 kg (34,000 lb), depending on the spacing between the two single axles in tandem. Similar limits also exist for tridem and quad axle groups. These limits are enforced by the States to ensure that load intensity due to axle loads is maintained within reasonable levels for which the highway pavements are designed. The limit on gross vehicle weight is generally enforced. Unless grandfathered provisions exist, a permit needs to be obtained from the State highway agency to transport a "super-heavy" weight that makes the gross vehicle weight exceed the State regulated limit. The permit seeks to ensure that the gross vehicle weight is distributed in such a way that the stipulated axle load limits are not exceeded.

In addition to vehicle and axle weight limits, vehicle size limits are also enforced by transportation agencies at the local, state, and federal levels. Setting of vehicle height limits is an obvious step in ensuring that vehicles can clear the overhead bridges along the travel path. Limits on vehicle width are important to ensure that the vehicle and its cargo can fit within the vehicle right-of-way available. Even though most major roads in Texas have lane widths of 12 ft, there are many roads that have lower lane widths from 9 to 11 ft. This narrower width may be due to several reasons, including (a) there has not been a need to require a wider roadway lane, and (b) a lack of funds in situations where wider lanes are desired. In addition to lane width, use of a shoulder and its width are also very important considerations with regard to highway geometry, particularly in rural roads that are not designed to carry wider and heavier loads during their design life. A narrower roadway lane or the lack of a shoulder of sufficient width may cause the edge of the roadway to be damaged due to the wander of the outer axle tires toward the edge of the road. This is referred to as pavement edge-damage and occurs frequently when wider vehicles travel on rural roads.

Figure 10.2 shows three commonly used pavement layer configurations for flexible pavements that have asphalt-based material as their surface (wearing) layer. The first pavement configuration with only a surface treatment as the wearing course (Figure 10.2a) is only used in very low traffic roadways for which pavements can be designed and built at a much lower cost than other more robust designs depicted in Figure 10.2b and c. In the first pavement layer configuration (Figure 10.2a), the surface course on which the vehicles ride is a surface treatment that consists of a thin sprayed asphalt film over which one layer of small coarse aggregate is applied before rolling to seat the spread aggregate in the asphalt at its most stable position. This surface layer has no structural, or load carrying, capacity and the vehicle axle loads have to be carried by the base layer below. Since the base layer is limited in its strength, this type of pavement is only suitable for a very low traffic volume, with little or no heavy truck traffic. If heavy trucks do travel on such roadways, they will fail much earlier than the anticipated design life.

Pavement damage observations from road tests as well as general roadway performance have revealed that when it comes to structural damage to the pavement, a heavier truck such as a large eighteen-thousand-pound tractor trailer (18-wheeler) will cause five- to six-thousand times more damage than a four-thousand-pound passenger car. TxDOT has estimated that the damage from truck traffic generated by one fracturing well is equivalent to damage caused by over 3.5 million car trips. A highway designed to carry only a small number of passenger vehicles in a day could easily be damaged very quickly with the introduction of even a small amount of heavy truck traffic.

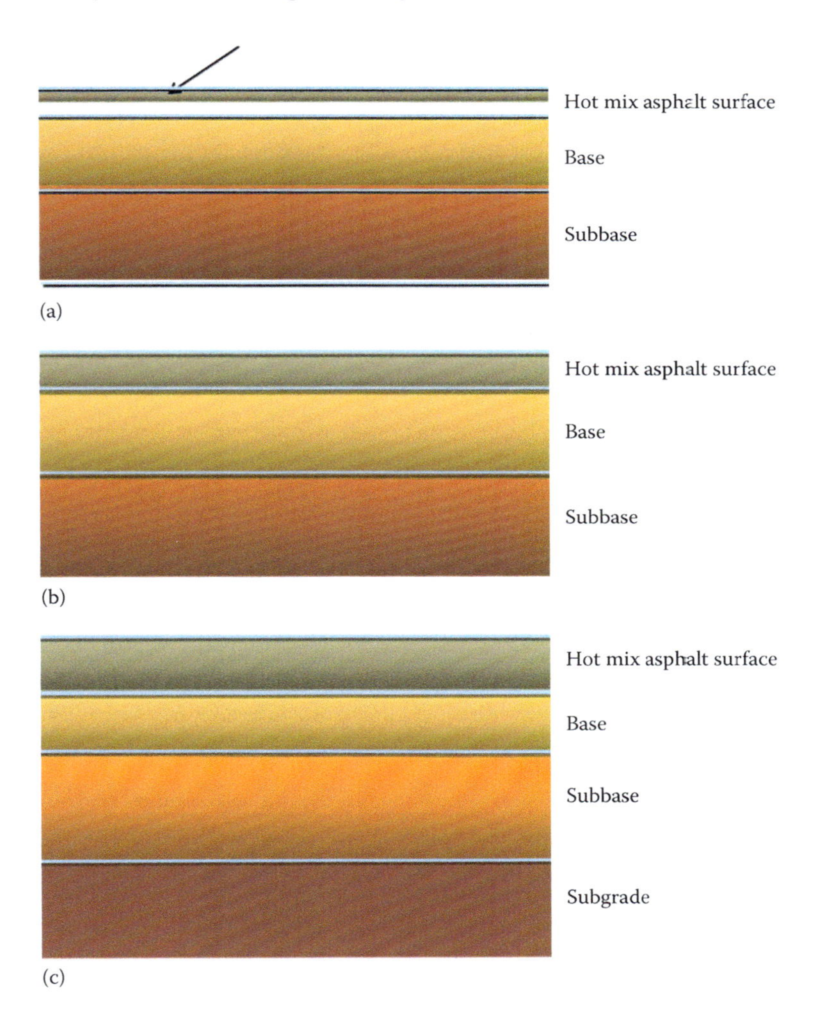

FIGURE 10.2 Typical layer configurations for flexible pavements, with (a) low traffic, (b) medium traffic, and (c) heavy traffic.

Unfortunately, most of the rural highways now affected by high levels of fracturing truck traffic fall into this category of rural roads. The introduction of heavy truck traffic will undoubtedly accelerate the deterioration of such roads. Significant deterioration of roadways can have many negative consequences including but not limited to

1. Slowing traffic down and increasing the user cost for the traveling public
2. Increasing the fuel consumption and maintenance costs of the vehicles
3. Increasing the costs of roadway repair
4. Compromising the safety of the traveling public

If significant truck traffic increases are anticipated, a prudent approach would be to strengthen the low-volume roadways that will be affected by the activity. Strengthening the roadways will allow the highway agency and the taxpayer to save significant amounts of money as well as improve the safety of the traveling public.

10.4 PAVEMENT DISTRESSES CAUSED BY HEAVY VEHICLES

Roadway pavements always deteriorate with time. However, engineers have the responsibility to design and build a road that will withstand traffic loads and the climatic elements. Problems associated with pavement performance are referred to as distresses, and there are many different distresses resulting from a variety of mechanisms. A few of the more common distresses in areas experiencing high traffic volumes due to unconventional oil and gas production are presented here.

Figure 10.3 shows pavement edge damage due to wider wheel-base trucks used at fracturing sites. The damage is due to both heavier vehicles and to roads whose lanes are not wide enough to safeguard the roadway. This distress can be prevented by widening of the lane and/or introduction of a roadway shoulder following guidelines available in the AASHTO's *A Policy on Geometric Design of Highways and Streets* (AASHTO, 2011).

Figure 10.4 shows more severe forms of structural distresses caused by unanticipated truck traffic. These pictures show that the distresses such as alligator cracking, rutting, and patches are fairly widespread in areas experiencing high-intensity oil and gas activity. These are structural distresses and are more expensive to repair and time-consuming to execute the repairs. Figure 10.5a shows the asphalt flushing distress on a highway that has very high truck traffic levels. Figure 10.5b shows a pavement that is badly deteriorated due to heavy truck traffic from fracturing sites.

In addition to identifying pavement distresses due to truck traffic in low volume roads, it is also useful to determine an overall quality of service by the roadway pavement. The "Pavement Condition Index" is a metric that allows engineers during the design stage to predict pavement performance over time. Peterson et al. (1976) were the first to recognize that the longer a distress is left untreated, the more costly it will be to fix distresses and improve their ride quality to a more acceptable level.

(a)

(b)

FIGURE 10.3 Pavement edge damage due to fracking well truck traffic. (Courtesy of Texas Department of Transportation, Education Series, Energy Sector, 2013.)

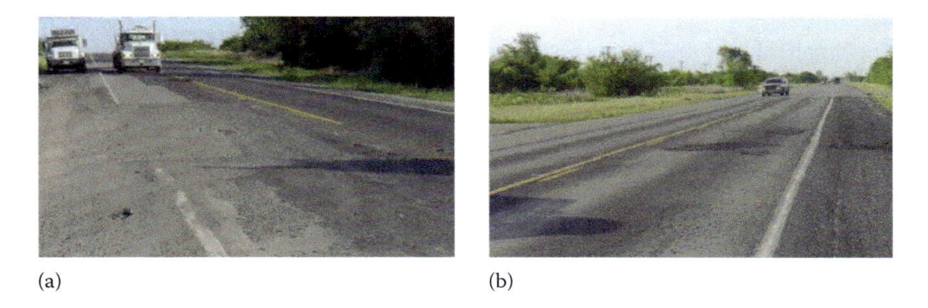

(a) (b)

FIGURE 10.4 Structural damage to asphalt (flexible) pavement caused by gas field traffic. (Courtesy of Texas Department of Transportation, Education Series, Energy Sector, 2013.)

(a) (b)

FIGURE 10.5 (a) Pavement flushed with asphalt due to excessive track traffic volume and (b) significant surface and base layer damage. (Courtesy of Texas Department of Transportation, Education Series, Energy Sector, 2013.)

This idea is conceptually depicted in Figure 10.6. Furthermore, the pavement performance curve shows two segments, a slow gradual drop in condition over the first few years followed by a steeper drop as years go by. Peterson et al. (1976) theorized that the pavement condition can be maintained at a desirable level by using more cost-effective maintenance treatments before the pavement condition reaches the steeper part of the performance curve.

When a pavement structure is subjected to a higher level of traffic, particularly heavy truck traffic, than it was designed for the pavement condition, it will deteriorate sooner than was projected at the design stage. Figure 10.7 illustrates this scenario with the curve A (in green) being the performance curve predicted during the design phase and curve B (in red) being the actual performance curve with higher than anticipated traffic levels.

The previous discussion on pavement deterioration shows how heavy vehicles generated from hydraulic fracturing wells impact the highway infrastructure. Available literature indicates studies that have estimated such impacts. Abramson et al. (2014) estimated the cost of additional heavy truck traffic on Pennsylvania state-maintained roadways due to Marcellus Shale natural gas development at about

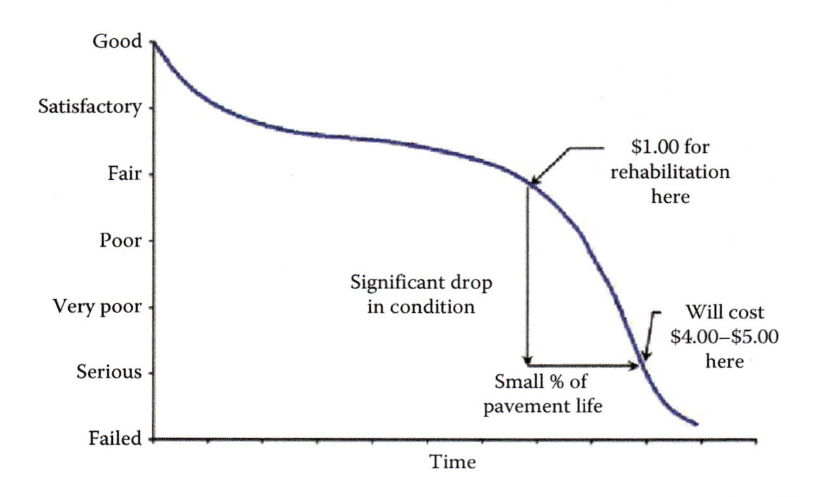

FIGURE 10.6 Typical pavement performance curve showing the two stages of condition deterioration. (FAA, Airport Obligations: Pavement Maintenance, Federal Aviation Administration, published December 11, 2014, http://www.faa.gov/airports/central/airport_compliance/pavement_maintenance/, accessed January 20, 2015.)

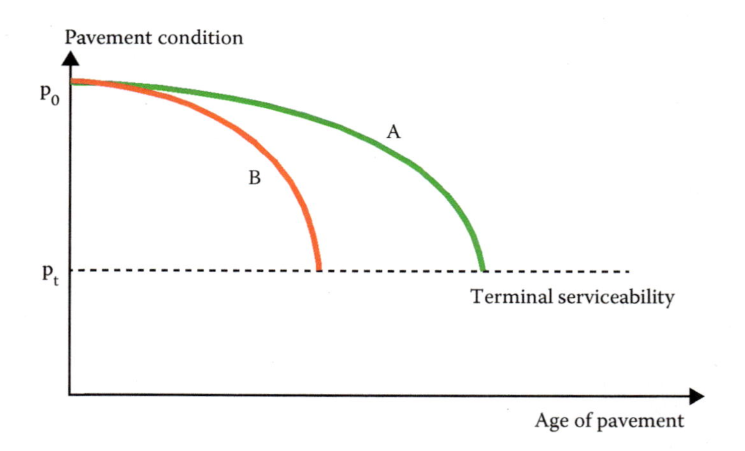

FIGURE 10.7 Illustrative pavement performance curves for two traffic scenarios. (Red—high loads; green—design loads).

$13,000–$23,000 per well for all state-maintained roads, and $5,000–$10,000 per well if state roads with the lowest traffic volumes are excluded. TxDOT (2013) reports from its own studies on specific roads that the pavement life expectancy is reduced during construction of a hydraulic fracturing well by anywhere from 4% to 53%, the movement of the rig causes a reduction from 1% to 16%, and the disposal of the saltwater causes a reduction from 1% to 34%. TxDOT's estimate of the average overall impact on pavement life expectancy is 30% per unconventional (hydraulic fracturing) well. In comparison, a conventional oil well is estimated to cause a reduction on life expectancy of 16%.

A highway agency entrusted with the task of managing a highway that is going to experience significantly increased traffic levels compared to what was assumed during its design has two options. One option is to conduct reactive maintenance by fixing the road as problems appear. The other is to take a proactive approach where the agency will retrofit the highway before the increased traffic hits the road. Estimated cost data from TxDOT simulations for an actual 20.3 km (12.6 mi) long rural highway, as shown in Figure 10.8, indicate that the proactive approach will be five to nine times more cost-effective for the highway agency and the taxpayer. The proactive approach, however, requires a coordinated effort to plan and execute the retrofitting of highways that are going to experience increased heavy vehicle volumes. The challenges for the highway agency in such situations are (1) to accurately forecast how the hydraulic fracturing operations will spread within a shale play and (2) to secure funding for such retrofitting work in a timely manner so that the work can precede energy development activities. The oil and gas production is directly tied to market prices which in turn are controlled by a variety of factors ranging from climate to geopolitical upheaval and as such are hard to predict a priori. In a similar vein, securing funding for proactive transportation infrastructure is also challenging, given other competing priorities.

In addition to traffic levels that are not considered at the design stage, another factor that contributes to roadway damage is overweight vehicles. Strict enforcement of legal limit rules can be adopted to alleviate this problem. However, traditional

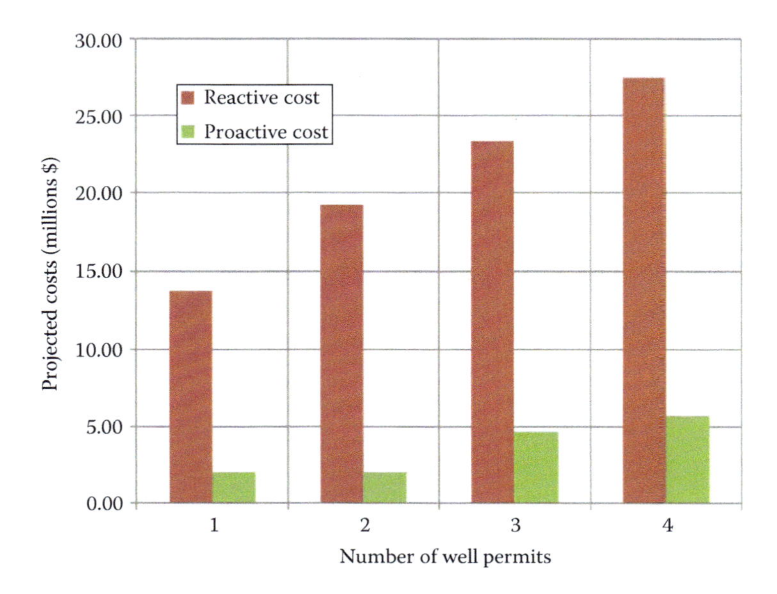

FIGURE 10.8 Reactive and proactive maintenance costs estimated for a 20-year period for a 12.6-mile rural farm-to-market road. (Based on Texas Department of Transportation, Education Series, Energy Sector, 2013.)

vehicle inspection methods use fixed facilities that are designed for major highway corridors and not readily available on rural roads that service one or more fracturing wells. Portable weight measurement pads are increasingly being used innovatively to check for and enforce weight limits (TxDOT, 2013).

10.5 TRAFFIC FLOW AND SAFETY IMPACTS

In addition to distress on transportation infrastructure, high volume truck activity causes disruptions to traffic flow and has the potential to create safety concerns. Some of the flow and safety concerns include (1) slow acceleration which causes traffic jams, (2) slow merging and diverging actions, (3) larger turning radius which can lead to damage of pavements as well as property, (4) creation of slow moving platoons on highways, (5) kicking up of gravel and other debris on the road that can damage vehicles by becoming projectiles, and (6) spilling of hazardous materials due to negligence. While the energy sector has created a large number of jobs, the influx of people and heavy vehicles in rural areas affected by the shale boom has also increased the number of accidents and fatalities.

Quiroga et al. (2012) conducted a comprehensive review of transportation impacts connected to energy developments in Texas including hydraulic fracturing wells. They analyzed crash data from the Texas Crash Records Information System (CRIS) database for years 2003–2009 in the three TxDOT regional districts Abilene, Lubbock, and Fort Worth that are significantly affected by the energy sector activities involving oil, natural gas, and wind. They focused on rural roads that were designed for low levels of heavy vehicles, and combined rural major and minor collectors in the analysis, but analyzed rural minor arterials at the Fort Worth District separately. Their data showed increasing crash frequencies in these three districts. TxDOT reports 3450 traffic crashes and 238 traffic fatalities in 2013 in the Eagle Ford shale play area, representing an increase of 7% over the previous year. In a similar vein, the Permian Basin region experienced a 15% increase in the roadway fatalities between 2012 and 2013, resulting in a total of 4411 fatal and serious injury crashes and 365 traffic fatalities (TxDOT, 2014). Even though not all of these accidents and fatalities are caused by energy-related traffic, the unexpected driving situations arising from heavy traffic due to fracturing activities significantly adds to the risks associated with the driving experience in such areas. Such risks are applicable not only to the state highway system, but also to the county and city transportation infrastructure.

Another important highway-related issue connected to fracturing wells is the provision of access to and from the well site. Quiroga et al. (2012) indicated that there is an urgent need to provide guidance to engineers who evaluate access permits to fracturing sites. They suggested that such improvements should address not only new permits, but also consider situations where access to existing nonenergy development property can be made to accommodate energy development activity. The significant impacts from heavy vehicles also call for revisiting intersections between state highways and county roads in terms of culverts and roadway width requirements. In light of this situation, it is important for both the highway agencies and the well operators to ensure that both parties are aware of the anticipated vehicle types and traffic volumes that will be generated by the well site, so the transportation facilities do not suffer any short-term or long-term negative impacts that will be counterproductive to the general public and the well operator.

10.6 SUMMARY AND CONCLUDING REMARKS

The oil and gas boom in the United States fostered by unconventional drilling has ushered a new era of economic prosperity and reduced energy dependence in the United States. However, this growth has disturbed the steady coexistence between the transportation infrastructure and the fossil fuel industry. Unconventional oil and gas production using horizontal drilling and hydraulic fracturing is a material-intensive activity requiring large volumes of water, sand, and personnel. Most of the oil and gas production is taking place in rural areas where roads are designed only to handle low volume and lighter load traffic. The introduction of large-scale heavy duty truck traffic to transport materials into and out of hydraulic fracturing sites has created distress to transportation infrastructure and increased risks of accidents and fatalities. A proactive approach wherein suitable enhancements are made to existing transportation infrastructure to meet increased traffic demands is perhaps the best course of action.

In a similar manner, finding suitable alternatives to reduce truck traffic such as constructing temporary pipelines to transport water and wastewater will ameliorate the deterioration of transportation infrastructure. Additionally, developing driveway standards that minimize the amount of debris that is brought onto the road and provide a safe way for trucks to enter and exit the highway would help increase safety. It is also important to recognize that increased truck traffic is not limited to hydraulic fracturing activities alone. Wind energy operations, for example, are also material intensive and could lead to increased heavy load traffic on rural roads. Hydraulic fracturing operators must emphasize safety and must hire drivers who are committed to the safety and well-being of the public. In a similar manner, the public must exercise caution in areas undergoing intense hydraulic fracturing activities and must take proper precautions and plan for exigencies to ensure their own safety and that of other drivers on the highway.

REFERENCES

(December 2011). Kansas Geological Survey Public Information Circular 32, Revised May 2012.

Abramzon, S., Samaras, C., Curtright, A., Litovitz, A., and Burger, N. (2014). Estimating the consumptive use costs of shale natural gas extraction on Pennsylvania roadways. *J. Infrastruct. Syst.*, 20(3), 06014001.

American Petroleum Institute. (April 2014). *Hydraulic Fracturing: Unlocking America's Natural Gas Resources.*

FAA. (2015). "Airport Obligations: Pavement Maintenance," Federal Aviation Administration, published December 11, 2014, http://www.faa.gov/airports/central/airport_compliance/pavement_maintenance (accessed January 20, 2015).

National Gas Subcommittee of the Secretary of Energy Advisory Board: Safety of Shale Gas Development. (June 2010). *Water Management Associated with Hydraulic Fracturing. American Petroleum Institute Guidance Document HF2*, 1st edn. Washington, DC: API Publishing Services, pp. 9–20, 20–23. Retrieved at: http://www.shalegas.energy.gov/resources/HF2_e1.pdf.

National Petroleum Council. (September 15, 2011). Prepared by the Technology Subgroup of the Operations and Environment Task Group. Paper #2–25: Plugging and Abandonment of Oil and Gas Wells. Retrieved at: http://www.npc.org/Prudent_Development-Topic_Papers/2-25_Well_Plugging_and_Abandonment_Paper.pdf.

NTC Consultants. (September 2009). Impacts on community character of horizontal drilling and high volume hydraulic fracturing in the Marcellus Shale and other low-permeability gas reservoirs.

Peterson, D.E. (1974). Utah's pavement design and evaluation system, Transportation Research Record #512. Washington, DC: Transportation Research Board.

Penn Environment Research and Policy Center. (2012). pp. 25.

Shlachter, B. (July 3, 2012). Drilling trucks have caused an estimated $2 billion in damage to Texas roads. Fort Worth Star-Telegram.

Texas Department of Transportation. (June 26, 2012a). Impact of Energy Development Activities on the Texas Transportation Infrastructure: Testimony before the House Committee on Energy Resources. By Phil Wilson.

Texas Department of Transportation. (December 13, 2012c). Task Force on Texas' Energy Sector Roadway Needs: Report to the Texas Transportation Commission.

Texas Department of Transportation. (2013). Education Series, Energy Sector.

Texas Department of Transportation: Research and Technology Implementation Office, and Texas Transportation Institute: The Texas A&M University System. (March 2012b). Energy developments and the transportation infrastructure in Texas: Impacts and strategies. By Cesar Quiroga, Emmanuel Fernando, and Jeongho Oh. FHWA/TX-12/0-6498-1.

U.S. Department of Energy, Office of Fossil Energy, and National Energy Technology Laboratory. (April 2009). Modern Shale gas development in the United States: A primer. Prepared by Ground Water Protection Council and ALL Consulting, DE-FG26-04NT15455, pp. 62, 64. Retrieved at: http://energy.gov/sites/prod/files/2013/03/f0/ShaleGasPrimer_Online_4-2009.pdf.

U.S. Department of Transportation. (August 2000). Comprehensive truck size and weight study, FHWA-PL-00-029, Vol. 2, Chapter 3. Retrieved from: https://www.fhwa.dot.gov/reports/tswstudy/Vol2-Chapter3.pdf.

U.S. Energy Information Agency (US EIA). (2014). Retrieved from: http://www.eia.gov/dnav/ng/NG_SUM_LSUM_A_EPG0_XDG_COUNT_A.htm (accessed on March 30, 2015).

11 GIS-Based Assessment of Wastewater Disposal Impacts in Permian Basin, Texas

Elma Annette Hernandez, Sreeram Singaraju, Abdullah Karim, Jorge Ruiz de Viñaspre, and Venkatesh Uddameri

CONTENTS

11.1 Overview ... 187
11.2 Introduction ... 187
11.3 Study Area ... 190
11.4 Methodology—Spatial Analysis ... 190
11.5 Results and Discussion ... 193
  11.5.1 Closest Facility Analysis ... 193
  11.5.2 Transportation Routes for Wastewater Disposal ... 199
  11.5.3 Air Quality Impacts Due to Wastewater Disposal ... 200
11.6 Summary and Closing Remarks ... 202
References ... 203

11.1 OVERVIEW

Disposal of produced water via deep well injection is a commonly used approach in Texas. A majority of the transportation associated with wastewater disposal happens on rural roads not designed for high-volume traffic. Air quality is also impacted by wastewater transportation and disposal.

11.2 INTRODUCTION

A typical hydraulic fracturing operation at a well requires anywhere between 11,356 and 18,927 m^3 (3–5 million gal) of water or nearly 12.4 m^3/m (1000 gal per linear foot) of well (Gregory et al., 2011). During hydraulic fracturing of a well, water is mixed with proppant (sand) and other chemicals (e.g., cross-linking gels, friction reducers, biocides) and injected under pressure to break open shale plays. Once the fractures are established, the pressure is released, which causes the injected water to flowback to the surface. Nearly all the injected water and perhaps some of the connate water is likely to come back to the surface over the entire production life (several years) of a fractured well (Nicot and Scanlon, 2012). However, shale plays are typically undersaturated with respect to water, so about 5%–40% of the injected water is recovered over the first month (King, 2012). The flow of water from the shale play is seen to significantly taper off after approximately the first 2 weeks. The water flowing out of the shale play during hydraulic fracturing is referred to as flowback water, and the water flowing out after the fracturing is referred to as produced water. To avoid notational difficulties, flowback and produced water are jointly referred to as recovered water in this study following Hernandez and Uddameri (in press).

The recovered water is in contact with the shale play, which causes its chemistry to be remarkably different from the freshwater used for hydraulic fracturing operations (Chapman et al., 2012). These waters often contain elevated levels of barium, strontium, iron, and sulfates that can precipitate and interfere with hydraulic fracturing operations (Kondash et al., 2013). In addition, the recovered water can also contain large quantities of organic chemicals such as benzene, toluene, ethylbenzene, and xylene (BTEX), which are classified as either known or potential carcinogens. Furthermore, recovered water can also contain naturally occurring radioactive materials (NORMs). Colborn et al. (2011) have compiled a list of over 600 chemicals associated with oil and gas operations. Therefore, recovered water must be properly handled to ensure the safety of personnel working at the hydraulic fracturing pad (site). Improper storage and disposal of recovered water can have deleterious effects on the environment and cause groundwater and surface water contamination, as well as increase air quality emissions.

Reuse of recovered water has both economic and environmental benefits and is, therefore, actively encouraged by oil and gas industry as well as environmental groups. Nonetheless, there are several barriers associated with reuse. The recovered water has very high total dissolved solids (TDS) often exceeding 50,000 mg/L. The theoretical pressure required for desalinating these high concentration brines is higher than what available membrane modules can handle (Kim et al., 2007). The recovery efficiency also drops markedly at these high concentrations significantly increasing the risk of membrane fouling (Greenlee et al., 2009). Considerable pretreatment may be necessary, which adds to treatment costs. While other innovative desalinating technologies such as forward osmosis and mechanical vapor compression are promising, they are not considered fully mature and consequently not readily adopted by the industry. Generally, though desalination of recovered water may technically be feasible, the decision to treat and reuse water comes down to economics (Shaffer et al., 2013).

Disposal of recovered water in deep wells (i.e., deep well injection) is a common strategy for managing wastewater generated during oil and gas operations. These deep well injections take three basic forms. (1) Wastewater is injected back into the formation for enhanced oil recovery operations, called water flooding. These wells are referred to as injection/recovery wells. (2) Wastewater may also be injected back into the depleted formation (i.e., after secondary and tertiary oil recovery) to maintain formation pressures, which is referred to as disposal into productive zones. (3) Finally, oil and gas wastewater are also disposed into commercial deep disposal wells that do not tap into the oil and gas formation. This type of disposal is referred to as disposal into a nonproductive zone. It is estimated that a barrel of oil can produce anywhere between 3 and 10 barrels of water during conventional oil and gas production (Veil et al., 2004). As the quality of water is not a major concern during conventional production, strategies 1 and 2 (injection into production zones for secondary oil recovery and pressure maintenance) are used extensively as disposal options. The reuse of wastewater is constrained by water quality requirements in unconventional production and as such disposal post hydraulic fracturing is typically accomplished using strategy 3 (i.e., disposal into nonconventional zone). According to the Texas Railroad Commission, as of 2013, there were 35,000 active injection wells in Texas of which 7,500 wells were classified as disposal wells (RRC, 2014). Deep well injection wells used for oil and gas wastewater disposal are classified as UIC class II injection wells and are referred to as saltwater disposal wells (SWDs) in Texas.

The disposal of wastewater into SWDs is currently around \$0.45–\$0.60 per barrel (1 barrel = 42 US gal) excluding transportation costs in Texas (City of Fort Worth, 2014), and the cost for a barrel of freshwater is estimated to be about \$0.25–\$0.75 per barrel (excluding transportation costs). The cost of reusing flowback and produced (recovered) water is estimated to be upward of \$2.75 per barrel (Boschee, 2012) and also requires operators to assume environmental liabilities for its storage and eventual disposal of unused water, which adds to the total costs. Therefore, disposal of recovered water into SWDs is often the most economically efficient strategy and is widely used in Texas. It is estimated that the amount of recovered reuse is around 10% in Texas (Nicot and Scanlon, 2012), which is significantly lower than other shale plays like the Marcellus shale where deep disposal wells are not readily available. This situation is likely to persist at least in the near term as the ratio of disposal wells to unconventional oil and gas production wells remains favorable.

In addition, there are limited regulations on freshwater production for oil and gas operations in Texas. Groundwater extractions for oil and gas exploration activities (including hydraulic fracturing operations) are exempt from any permitting process when they are carried out on the same oil and gas lease. Therefore, while most active shale plays in Texas (e.g., Permian Basin and Eagle Ford) lie in water-parched arid and semi-arid regions, recent droughts have highlighted the need for water conservation across all sectors, and there is a growing recognition of the need to reuse recovered water (Jester et al., 2013). The current administrative and economic environment in Texas offers little incentive for reuse of recovered water in oil and gas operations.

Transportation of water for hydraulic fracturing and moving recovered water to SWDs for disposal not only add costs to water management but also impact the regional transportation infrastructure and air quality. Life cycle assessment (LCA) is a comprehensive cradle-to-grave approach to characterize the environmental consequences of a product, process, or service (Hendrickson et al., 2006). LCA seeks to quantify the multimedia environmental impact beginning with the extraction of raw materials from the earth and tracking it through its entire production, operation, and eventual disposal back into the earth (USEPA, 2006). LCA is increasingly being used to quantify the environmental impacts associated with hydraulic fracturing operations (Jiang et al., 2011).

An important component of the hydraulic fracturing waste management life cycle is the transportation sector (Gilmore et al., 2014). This includes transportation of both water and wastes, which contributes to global greenhouse gas emissions due to the burning of fossil fuels (Howarth et al., 2011). Geographic information systems (GIS) can play a vital role in evaluating optimal water management and waste transportation strategies (Uddameri and Honnungar, 2010; Gilmore et al., 2014; Uddameri et al., 2014). The present study therefore uses GIS to provide a preliminary assessment of transportation-related impacts of hydraulic fracturing wastewater disposal in the Permian Basin region of Texas.

11.3 STUDY AREA

Contained in its 14 different shale plays, the Permian Basin is projected to have the largest volume of oil reserves in the world (Ruppell, 2009). The present study focuses on a 21-county area within the Permian Basin where oil and gas production has increased significantly since 2009 and hydraulic fracturing is being used extensively for enhancing vertical and horizontal fractures. The geographic characteristics of the study area are depicted in Figure 11.1. The terrain is hilly and undulating in the southwestern sections and tends to be more uniform in the northern portions of the study area (high plains). The land cover in the region is predominantly desert-like shrub land in the southwestern portions and cropland and herbaceous rangelands in the north. Midland and Odessa are the two major cities in this otherwise predominantly rural area. Figure 11.1c depicts the location of SWDs in the area. There are over 1900 such facilities reported by the Texas Railroad Commission. While the disposal wells are rather uniformly spread out across the study area, there is a greater concentration of them in the east-central portion of the study area. Figure 11.1d depicts locations of the hydraulic fracturing wells completed within the study area during January 2012–May 2013.

The study area has a relatively dense network of roadways spanning over 107,432 km (66,755 mi). The major roads in the area include the Interstate Highway 10 (IH 10) and several other state highways and farm-to-market (FM) roads shown in Figure 11.1c and d totaling 26,211 km (16,287 mi) in length. In addition, there are several smaller city and rural roads that connect the major roads to interior areas and facilities not depicted in the maps in the interest of clarity.

11.4 METHODOLOGY—SPATIAL ANALYSIS

Two spatial network analysis techniques were employed in this study. The first technique, referred to as the closest facility analysis, seeks to identify the closest disposal wells (facilities) for each hydraulic fracturing well in the study area. Transportation costs of wastewater are directly proportional to the distance traveled for disposal. Tanker rental charges for wastewater disposal are approximately \$80–\$100 per hour

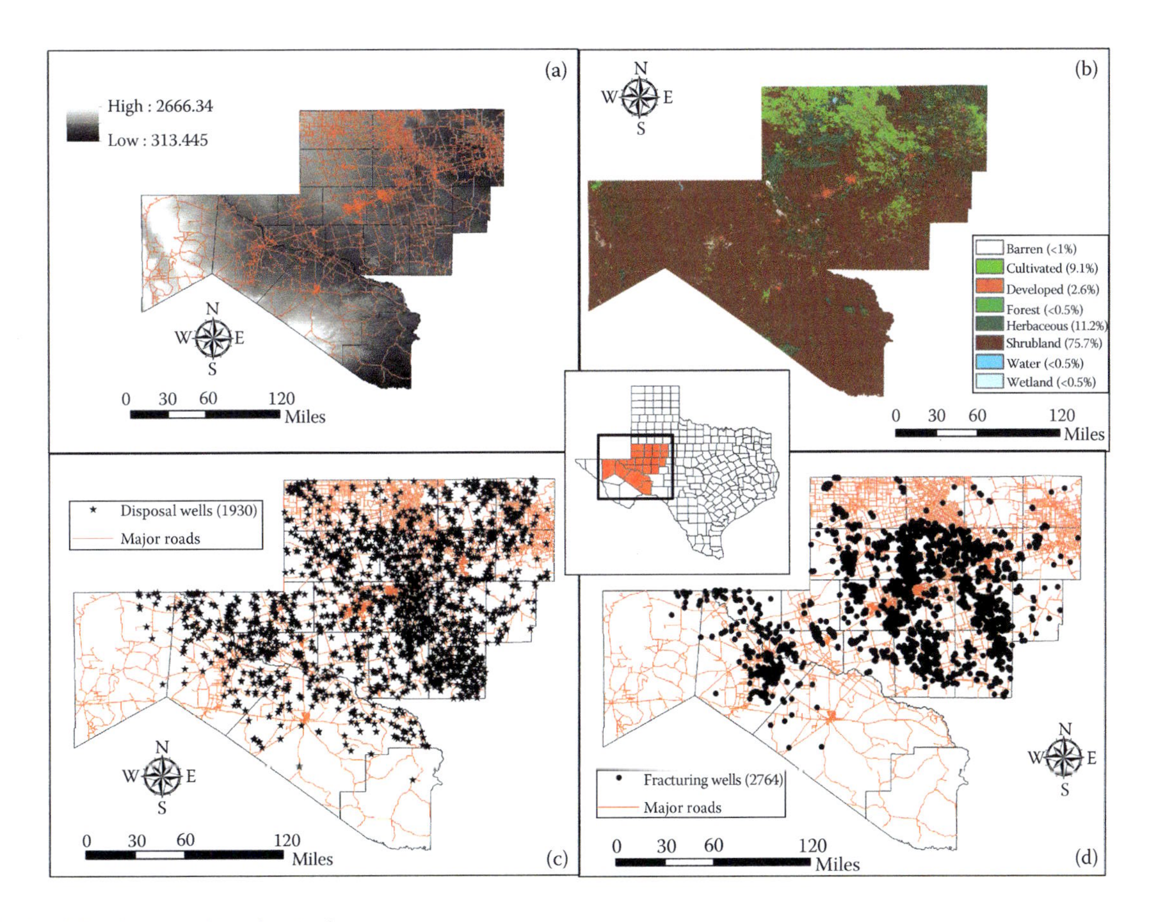

FIGURE 11.1 (a–d) Study area—Permian Basin.

for a 30.3 m^3 (8000 gal) truck (GE, 2014). Essentially, there is an economic incentive to dispose the wastewater at the closest possible facility. In addition, transportation of wastewater leads to emissions of harmful air quality pollutants such as oxides of nitrogen (NO_x) and sulfur (SO_x) and other volatile organic compounds (VOCs) that can lead to elevated ground level ozone (Kim et al., 2011). Clearly, the shorter the travel distance, the lower the emissions. Disposal of wastewater at closest facilities also has environmental benefits. Furthermore, shorter-travel distances decrease the potential for road damage and reduce the risk of accidents and, therefore, are beneficial to the society at large.

The closest facility analysis was implemented within ArcGIS v 10.2 (ESRI Inc., Redlands, CA) spatial analysis software using the network extension add-in. The closest facility analysis uses the Diijkstra algorithm that can be viewed as a successive implementation of dynamic programming. The algorithm uses concepts from graph theory to efficiently calculate the shortest possible route connecting an origination point (hydraulic fracturing well) to a facility (disposal well). The analysis can be carried out for all facilities to determine the one closest to the origination point of interest. The closest facility analysis requires (1) spatial locations of hydraulic fracturing wells of interest, (2) spatial locations of saltwater disposal facilities, and (3) a road network connecting the hydraulic fracturing wells and disposal facilities. Table 11.1 lists the sources of data sets used in this analysis. In addition to documented major and minor roads, there is some travel that takes place on unpaved or ad hoc roads, for example, travel from a hydraulic fracturing pad to the nearest road at a drilling site and travel from the gate of the facility to the disposal well. A tolerance of 500 m (1640 ft) was used to account for these undocumented roads within the facilities as well as to overcome the limitations of cartographic errors associated with available spatial datasets. The acceptable tolerance used in this study is much smaller than those reported in previous similar studies (e.g., Gilmore et al., 2014). The closest facility analysis was carried out separately for each of the 21 counties within the study area. However, as the closest disposal facility can be outside of the county, all the disposal facilities within the study area were included in the analysis. Sensitivity analysis was also performed by restricting the disposal to occur in the same county where the fracturing well was located.

TABLE 11.1
Datasets Used in the Analysis

Data	Source	Link	Comments
Fracking wells	Frac Focus Sky Truth	http://frack.skytruth.org/fracking-chemical-database/frack-chemical-data-download	January 2013–May 2014
Disposal wells	Texas Railroad Commission	http://webapps2.rrc.state.tx.us/EWA/uicQueryAction.do	Nonproductive zone
Roads	TNRIS	http://www.tnris.org/get-data	Strat Map Transportation

11.5 RESULTS AND DISCUSSION

11.5.1 Closest Facility Analysis

The empirical cumulative distribution function for all the closest facilities for wells within the study area is shown in Figure 11.2. The results indicate that the median distance between the fracturing well and the disposal well is less than 4.0 km (2.5 mi) and 90% of the fracturing wells are within 8 km (5 mi) of a disposal well.

The close proximity of the fracturing well to a disposal well is perhaps the primary reason why this mode of wastewater management is preferred in Texas. Summary statistics for the closest facility from disposal wells in each county is presented in Table 11.2.

The results presented in Table 11.2 indicate that the median distance to the disposal well varies among counties and can range from less than 2.4 km (1.5 mi) to as high as 8 km (5 mi). In addition, the intra-county variability of closest distances is also high particularly in large, predominantly rural Pecos and Sterling counties. However, the maximum distance to a closest disposal well is greater than 16 km (10 mi) in only 7 of the 21 counties and greater than 24 km (15 mi) in Reeves county alone. About 44 hydraulic fracturing wells appear to be co-located or neighbor disposal wells and have distances less than 0.16 km (one-tenth of a mile).

Travel routes to the closest facility for hydraulic fracturing wells in nine representative counties are depicted in Figure 11.3. These representative counties were chosen to capture the geographic, fracturing, and transportation diversity within the study area. As can be seen from Figure 11.3, Mitchell, Howards, and Gaines counties have a dense network of roads. Midland County has a denser network in its

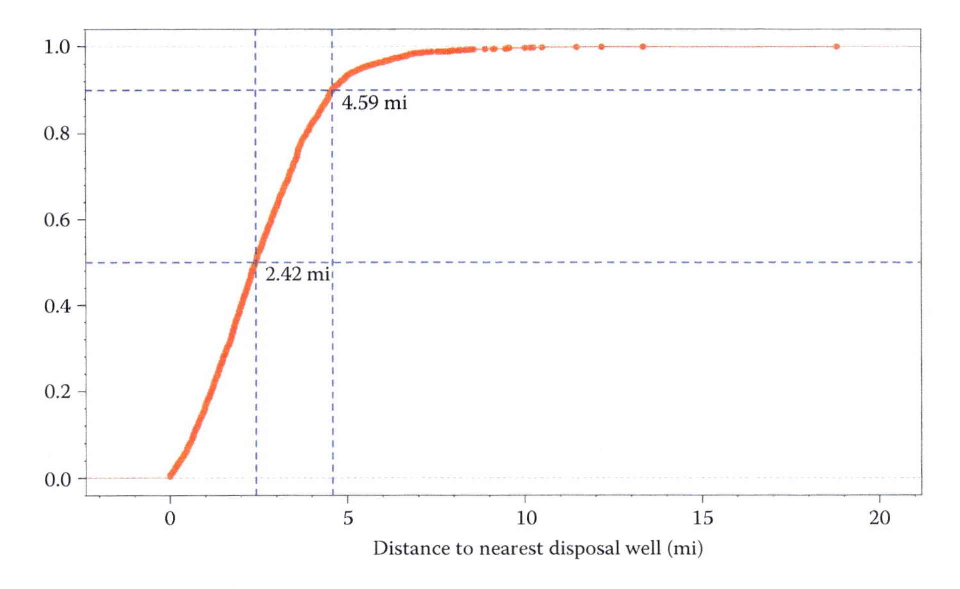

FIGURE 11.2 Empirical cumulative distribution of the closest distance for fracturing wells in the Permian Basin.

TABLE 11.2
Summary Statistics for Closest Facility in the Permian Basin (the Counties in Bold are Depicted in Figure 11.3)

		Distance Statistics—All Units in km (mi)					
County	**Total Wells**	**Total Distance**	**Mean**	**Median**	**Standard Deviation**	**Maximum**	**Minimum**
Andrews	375	1773 (1102)	4.74 (2.95)	4.47 (2.78)	2.68 (1.64)	12.9 (8.00)	<0.16 (0.10)
Borden	6	39.7 (24.6)	6.61 (4.11)	5.31 (3.30)	4.99 (3.10)	16.4 (10.2)	2.91 (1.81)
Crane	68	473 (294)	6.95 (4.32)	5.47 (3.40)	4.00 (2.49)	16.1 (10.0)	<0.16 (0.10)
Cullberson	19	101 (62.8)	5.31 (3.30)	4.42 (2.75)	3.86 (2.40)	18.5 (11.5)	1.29 (0.80)
Dawson	30	113 (70.0)	3.74 (2.33)	2.96 (1.84)	2.78 (1.73)	9.41 (5.85)	0.40 (0.25)
Ector	340	1580 (982)	4.65 (2.89)	4.79 (2.98)	2.57 (1.60)	16.9 (10.5)	<0.16 (0.10)
Gaines	75	311 (193)	4.15 (2.58)	3.97 (2.47)	2.08 (1.29)	9.16 (5.69)	0.50 (0.31)
Glasscock	238	924 (574)	3.89 (2.42)	3.93 (2.44)	2.04 (1.27)	9.90 (6.15)	<0.16 (0.10)
Howard	188	739 (459)	3.93 (2.44)	3.28 (2.04)	2.96 (1.84)	15.2 (9.47)	<0.16 (0.10)
Loving	65	224 (139)	3.49 (2.17)	2.59 (1.61)	2.86 (1.78)	11.5 (7.13)	<0.16 (0.10)
Martin	365	1342 (834)	3.68 (2.29)	3.38 (2.10)	2.20 (1.37)	9.96 (6.19)	<0.16 (0.10)
Midland	200	746 (464)	3.81 (2.37)	3.49 (2.17)	2.56 (1.59)	11.2 (6.94)	<0.16 (0.10)
Mitchell	36	123 (76.5)	3.42 (2.13)	1.98 (1.23)	4.03 (2.51)	21.5 (13.35)	0.95 (0.14)
Pecos	13	118 (73.6)	9.11 (5.66)	8.00 (4.97)	5.73 (3.56)	19.6 (12.18)	1.75 (1.09)
Reagan	156	473 (294)	3.02 (1.88)	2.80 (1.74)	1.75 (1.09)	12.3 (7.64)	<0.16 (0.10)
Reeves	141	735 (457)	5.21 (3.24)	4.76 (2.96)	3.31 (2.06)	30.3 (18.81)	<0.16 (0.10)
Scurry	11	35.2 (21.9)	3.20 (1.99)	2.91 (1.81)	1.34 (0.83)	5.37 (3.34)	1.37 (0.85)
Sterling	25	148 (92.2)	5.94 (3.69)	5.21 (3.24)	3.35 (2.08)	14.3 (8.90)	0.31 (0.19)
Upton	230	924 (574)	4.04 (2.51)	3.80 (2.36)	2.45 (1.52)	10.5 (6.53)	<0.16 (0.10)
Ward	129	366.22	4.57 (2.84)	4.46 (2.77)	2.30 (1.43)	11.3 (7.04)	<0.16 (0.10)
Winkler	54	102.66	3.06 (1.90)	2.38 (1.48)	1.77 (1.10)	7.22 (4.49)	0.95 (0.59)

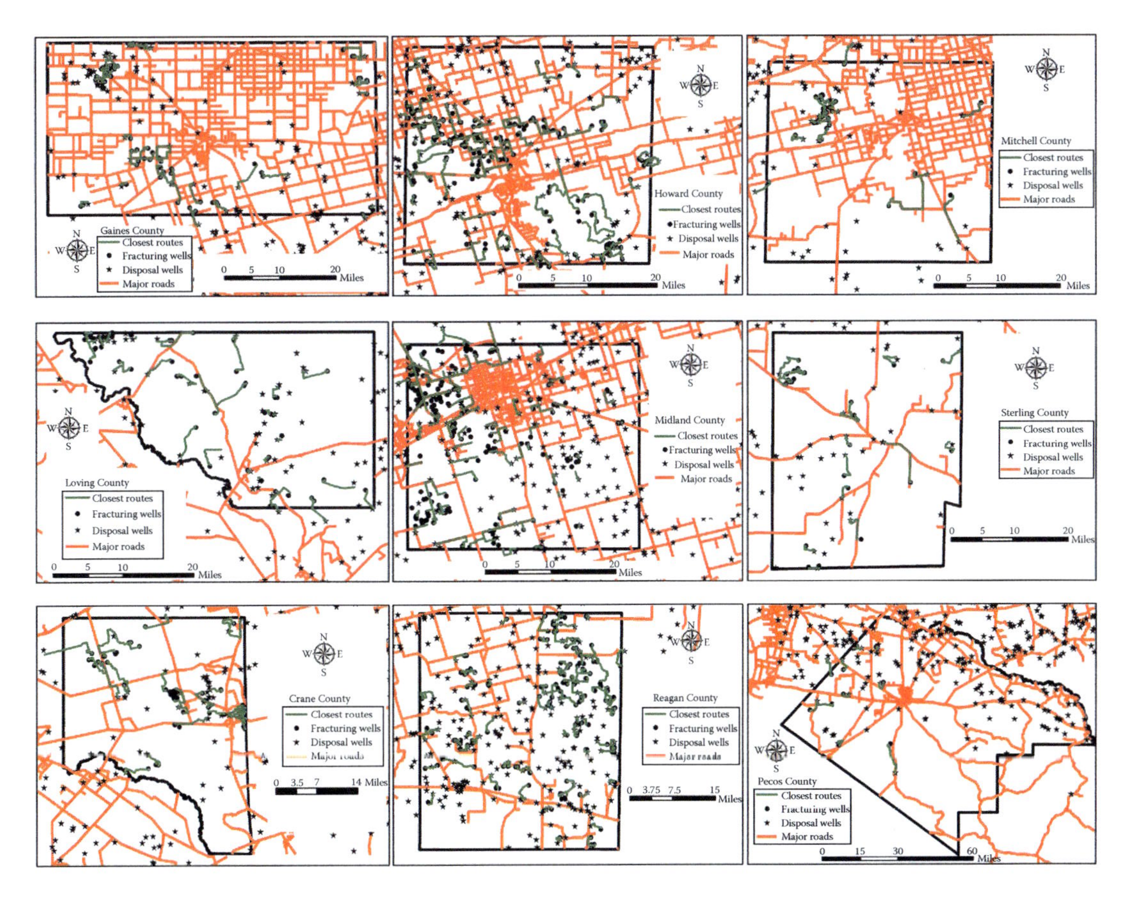

FIGURE 11.3 Travel routes for fracturing wells in nine representative counties.

population center (City of Midland) while Sterling, Pecos, and Crane counties have sparse networks. Reagan, Mitchell, and Howard counties have greater fracturing activities compared to others within the study area. Hydraulic fracturing in Sterling County is unevenly spaced and limited by the hilly terrain. Fracturing activities in Pecos County are largely concentrated in areas north of I-10.

The results of the closest distance facilities depicted in Figure 11.3 illustrate that there are many disposal facilities that are underutilized within each county. This result was seen in all other counties within the study area as well. There were a total of 674 wells that were identified as being closest out of the available 1930 wells (see Figure 11.4). By the same token, there were usually a few disposal wells that were closest to more than one fracturing well. The presence of disposal wells that are currently not optimal to existing hydraulic fracturing wells indicates that the region is well equipped to receive wastewater loads in the future as fracturing spreads out geographically. It can also be seen from Figure 11.3 that the closest facility for many hydraulic fracturing wells was in the neighboring counties. Table 11.3 summarizes the number of wells that have their closest disposal facility in a neighboring county. Typically the fraction of wells with their closest facility in a neighboring county ranges from 1% to 5%. Wells in Sterling, Scurry, and Borden counties all have facilities within the same county, while over 10% of the wells in Culberson, Ward, and Winkler counties have their closest facility in a neighboring county.

A sensitivity analysis was carried out to assess how the distance to the closest facility would change if the disposal of produced water were to occur within the same county where it was generated. The analysis, carried out for nine representative counties, is summarized in Table 11.4. Results indicate that the restriction would add 6.4 km (4 mi) on average to the total travel in each county (i.e., summed over all fracturing wells within the county). The impact of such a policy would be highly heterogeneous within the study area. Clearly, it would have no impact in Sterling, Scurry, and Borden counties where the closest facilities all lie within the same county. However, Midland county would incur the greatest impact and add more than 97 km (60 mi) of additional total travel.

It is important to recognize that the closest distance analysis carried out here uses regional-scale information. While the scale is certainly appropriate for broad-based, inter- and intra-county comparisons and for estimating distances traveled on federal, state, and county highways, the calculated distances do not fully account for travel inside the hydraulic fracturing and disposal facilities. Figure 11.5a depicts the closest distance calculation carried out for a hydraulic fracturing well (labeled A) in Sterling County, TX. Enhancing the area around the fracturing well indicates that the well is actually at some distance from the highway, which is not accounted for in the calculation (Figure 11.5b). Figure 11.5c also shows the satellite image of the fracturing well A, which depicts a private (unpaved) road that the trucks use to get to the highway. Preliminary assessment of selected sites indicates that the travel within the facilities can add up to 0.8 km (half a mile) in some cases but are generally small. The issue is more pronounced in counties with hilly terrains (e.g., Sterling) and those having a less dense network of roadways. It is therefore important that the impacts assessed based on this analysis are limited to transportation on designated roadways and not the entire transportation lifecycle.

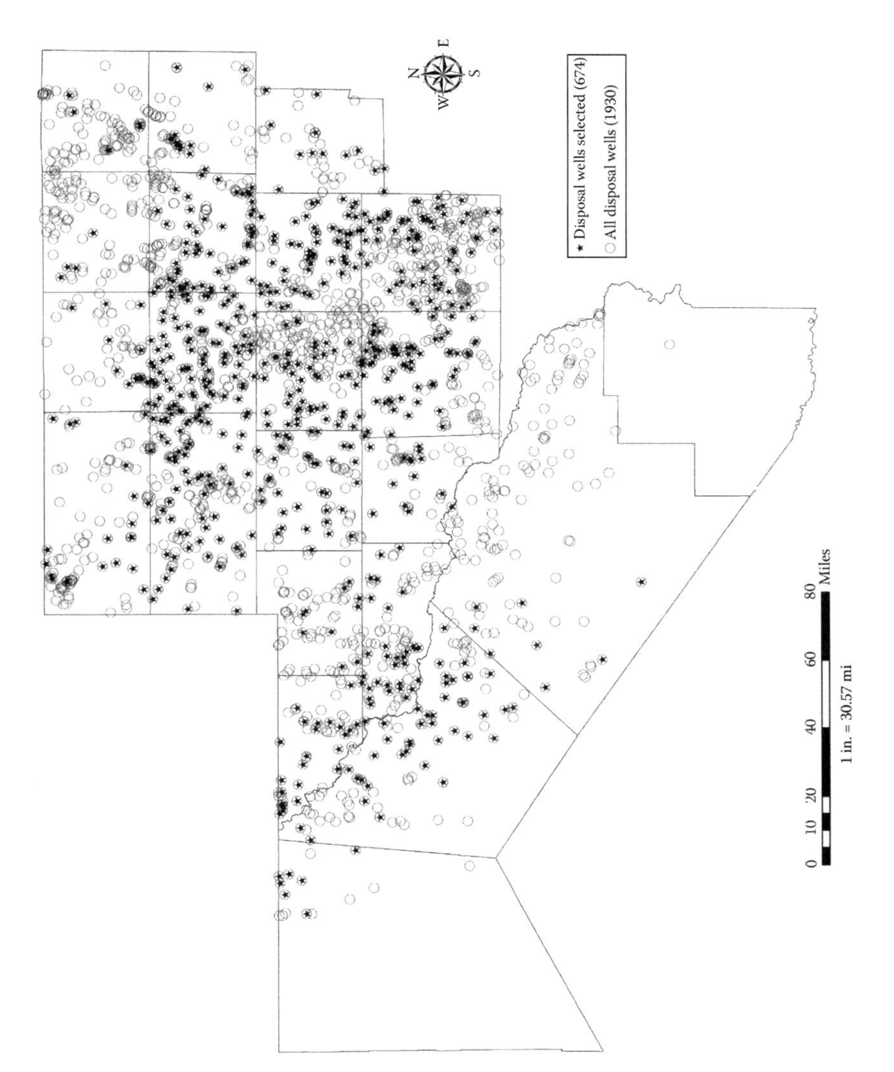

FIGURE 11.4 Disposal wells selected by the closest facility analysis.

TABLE 11.3
Number of Wells with Closest Facility in a Neighboring County

County	No. of Fracking Wells	No. of Wells with Closest Facility Outside the County
Andrews	375	21
Borden	6	0
Crane	68	3
Culberson	19	5
Dawson	30	2
Ector	340	21
Gaines	75	5
Glasscock	238	16
Howard	188	6
Loving	65	5
Martin	365	10
Midland	200	11
Mitchell	36	2
Pecos	13	1
Reagan	156	1
Reeves	141	1
Scurry	11	0
Sterling	25	0
Upton	230	16
Ward	129	18
Winkler	54	6

TABLE 11.4
Summary Statistics when Fracturing Wells Dispose in the Same County

	Distance Statistics—All Units in km (mi)					
County	Total	Mean	Median	Standard Deviation	Maximum	Minimum
Crane	481 (299)	7.06 (4.39)	5.47 (3.40)	4.15 (2.58)	16.2 (10.04)	<0.16 (0.10)
Gaines	336 (209)	4.49 (2.79)	3.97 (2.47)	2.45 (1.52)	10.9 (6.79)	0.50 (0.31)
Howard	745 (463)	3.96 (2.46)	3.41 (2.12)	2.96 (1.84)	15.2 (9.47)	<0.16 (0.10)
Loving	236 (147)	3.62 (2.25)	2.93 (1.82)	3.06 (1.90)	11.5 (7.13)	<0.16 (0.10)
Midland	848 (527)	4.25 (2.64)	3.54 (2.20)	3.83 (2.38)	23.2 (14.4)	<0.16 (0.10)
Mitchell	127 (79.2)	3.54 (2.20)	1.98 (1.23)	4.15 (2.58)	21.5 (13.4)	0.23 (0.14)
Pecos	125 (77.6)	9.61 (5.97)	8.00 (4.97)	6.32 (3.93)	19.6 (12.2)	1.75 (1.09)
Reagan	476 (296)	3.06 (1.90)	2.80 (1.74)	1.82 (1.13)	12.3 (7.64)	<0.16 (0.10)
Sterling	148 (92.2)	5.94 (3.69)	5.21 (3.24)	3.35 (2.08)	14.3 (8.90)	0.31 (0.19)

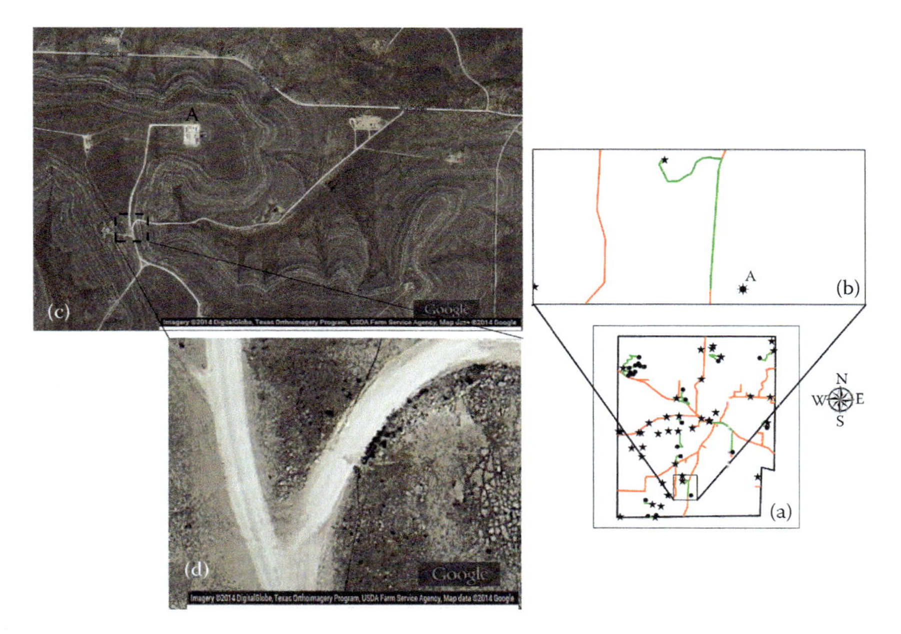

FIGURE 11.5 Illustrative example of unpaved/private roads not included in the analysis: (a) closest facility analysis for one well in the Sterling county; (b) closer look at the well location indicates its distance from the highway; (c) satellite image of the fracturing well showing the true landscape; (d) illustration of the usage of local roads with sharp turns not used in the analysis.

11.5.2 Transportation Routes for Wastewater Disposal

Major roads (federal and state highways and farm-to-market roads) are typically designed to carry heavy-duty trucks while minor roads (county roads and city streets) are not. Therefore, characterizing the amount of travel that occurs on these minor roads when wastewater is transported to the closest disposal facilities provides useful clues with regard to transportation safety as well as the structural integrity of the transportation infrastructure. A GIS analysis was therefore performed to estimate the fraction of the distance to the closest facility that was traversed on a major road. The fraction of the distance traveled on a major road was obtained by clipping each of the closest facility distances for the 2764 fracturing wells with the major routes in GIS. A bracketing analysis with tolerances ranging from 7.6 to 15.2 m (25 to 50 ft) had to be used to account for cartographic limitations of the available data. The closest distances were aggregated over each county and are presented in Table 11.5.

As can be seen from Table 11.5, the majority of the travel between the hydraulic fracturing well and the disposal facility happens on minor roads. Only Dawson, Pecos, Sterling, Ward, and Midland counties have nearly 10%–50% of the travel on major roads. The Pecos County has the highest percentage of major road travel length because most of the hydraulic fracturing activity is in the vicinity of IH 10.

TABLE 11.5
Distance Covered on Major Roads

County	Total Distance—km (mi)	% Traveled on Major Roads
Andrews	1174 (1102)	0.9
Borden	39.7 (24.7)	7.8
Crane	473 (294)	1.3–1.9
Culberson	101 (62.8)	1.6–1.8
Dawson	113 (70.02)	12.8–22.4
Ector	1581 (983)	3.1–4.5
Gaines	311 (194)	3.6–6.3
Glasscock	925 (575)	1.5–2.6
Howard	739 (459)	3.6
Loving	224 (139)	2.0
Martin	(1342 834)	4.2–6.3
Midland	747 (465)	4.5–9.4
Mitchell	123 (76.6)	3.5–3.6
Pecos	118 (73.6)	8.8–46.7
Reagan	472 (294)	4–4.2
Reeves	736 (457)	6.6–8.9
Scurry	35.2 (21.9)	6.50
Sterling	148 (92.2)	11–11.6
Upton	924 (574)	1.8–2.2
Ward	589 (366)	6.8–9.9
Winkler	165 (103)	3.3

In terms of actual kilometers traveled, Martin, Ector, Midland, and Reeves counties have over 64 km (40 mi) of travel on major highways while a total of less than 8 km (5 mi) is traversed on major highways in Winkler, Loving, Scurry, Mitchell, and Culberson counties that are on the western sections of the study area and have lower transportation network density. Large-scale travel on minor roads increases edge-damage, rutting, and pavement distress. The Texas Department of Transportation (TxDOT) estimates service life reduction of 1%–34% for major roads (Federal and state highways and farm to market roads) due to wastewater disposal operations associated with hydraulic fracturing (Wilson, 2012) and the loss of service life is even higher for minor roads, which are not designed to handle heavy traffic loads.

11.5.3 Air Quality Impacts Due to Wastewater Disposal

The closest facility distances were used in conjunction with the hydraulic fracturing water production and vehicular emission factors (Dunn et al., 2013) to obtain a preliminary estimate for the potential air quality impacts in the region due to trucking of produced water. The estimate obtained provides a lower limit for the emissions than reality because it measures the transport distances to the optimal (closest) facility and not suboptimal facilities, where at least some of the produced water is likely disposed. Based on estimates provided by Nicot and Scanlon (2012), 35% of the total water used for hydraulic fracturing in the Permian Basin was assumed to be flowback from the well and needed to be disposed. Trucking of wastewater was assumed to be in 30.3 m^3 (8000 gal) heavy-duty diesel trucks based on discussions with some operators in the area. The emission factors for seven criteria pollutants—volatile organic compounds (VOC), nitrogen oxides (NO_x), sulfur oxides (SO_x), particulate matter less than 10 μm (PM10), particulate matter less than 2.5 μm (PM2.5), methane (CH_4), and nitrous oxide (N_2O)—were obtained from Dunn et al. (2013). Following Gilmore et al. (2014), the calculation was also made for the return trip from the facility to the site. The total emissions of criteria pollutants due to disposal of produced water from 2764 wells drilled between January 2012 and May 2013 within the study area are tabulated in Table 11.6 and this information was also used to estimate average emission per well, depicted in Figure 11.6.

TABLE 11.6
Emission Factors (EF) and Emissions of Criteria Air Pollutants Due to Transportation of Produced Water

Chemical	VOC	CO	NO_x	PM10	PM2.5	CH_4	N_2O
EF (g/mi)	0.64 (0.4)	2.9 (1.8)	6.0 (3.7)	0.41 (0.26)	0.32 (0.2)	0.08 (0.05)	4.8×10^{-03} (3×10^{-03})
County	VOC	CO	NOx	PM10	PM2.5	CH^4	N^2O
Andrews	52.2	235.0	483.1	33.9	26.1	6.5	0.4
Borden	1.8	8.2	16.8	1.2	0.9	0.2	0.0
Crane	7.7	34.5	70.9	5.0	3.8	1.0	0.1
Culberson	6.6	29.5	60.7	4.3	3.3	0.8	0.0
Dawson	4.6	20.7	42.6	3.0	2.3	0.6	0.0
Ector	30.0	135.0	277.4	19.5	15.0	3.7	0.2
Gaines	2.0	8.9	18.2	1.3	1.0	0.2	0.0
Glasscock	51.4	231.3	475.4	33.4	25.7	6.4	0.4
Howard	18.4	82.7	170.0	11.9	9.2	2.3	0.1
Loving	10.6	47.8	98.2	6.9	5.3	1.3	0.1
Martin	39.2	176.3	362.3	25.5	19.6	4.9	0.3
Midland	15.6	70.3	144.5	10.2	7.8	2.0	0.1
Mitchell	2.0	9.2	18.9	1.3	1.0	0.3	0.0
Pecos	1.1	5.1	10.4	0.7	0.6	0.1	0.0
Reagan	29.2	131.2	269.7	19.0	14.6	3.6	0.2
Reeves	10.8	48.4	99.5	7.0	5.4	1.3	0.1
Scurry	0.2	0.9	1.9	0.1	0.1	0.0	0.0
Sterling	3.3	14.8	30.3	2.1	1.6	0.4	0.0
Upton	9.4	42.1	86.5	6.1	4.7	1.2	0.1
Ward	1.2	5.3	10.8	0.8	0.6	0.1	0.0
Winkler	0.1	0.3	0.5	0.0	0.0	0.0	0.0
Total	297.2	1337.3	2748.9	193.2	148.6	37.1	2.2

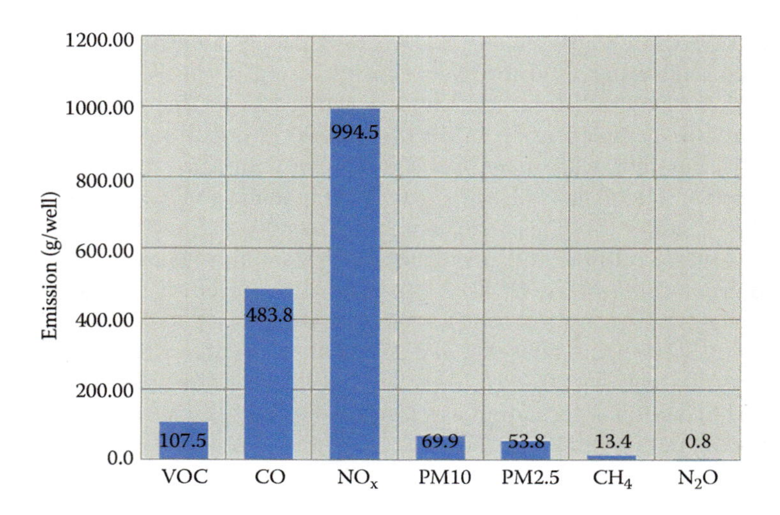

FIGURE 11.6 Average air pollution emissions per well due to the movement of produced water in the Permian Basin.

As illustrated, the transportation of produced water results in nearly 1 kg of NO_x emissions and about 500 g of carbon monoxide and 100 g of volatile organic compounds. It is important to recognize that the calculated emissions account for only the travel from the hydraulic fracturing well site to the disposal facility and back. Emissions also occur during the loading and unloading of trucks (idling) as well as other short-term stops (rest areas, refueling) that may occur during the trips but are not accounted for in this study. As discussed previously, there could be distances traveled on unpaved roads (as depicted in Figure 11.5) inside the facilities; this is not included in this analysis. Also, in some instances, the trucks may have to make sharp turns (see Figure 11.5d), which could warrant additional fuel consumption and as such increase emissions. Emission factors for idling vary widely with the following typical values—NO_x 144 g/h; VOC 36.4 g/h, PM 2.19, CO 118 g/h (Zietsman and Perkinson, 2003) and add to the overall life-cycle emissions of the wastewater disposal process. The estimates provided here should therefore be viewed as lower end (floor) estimates rather than an upper conservative value.

11.6 SUMMARY AND CLOSING REMARKS

The overall goal of this study was to evaluate the transportation impacts associated with the disposal of produced water using the deep well injection method commonly used in Texas. Transportation of produced water for disposal increases truck traffic, which not only reduces the lifespan of roads but also increases air emissions. As such, efforts must be made to dispose of the produced water at the closest disposal facility from the fracturing well (i.e., minimize the travel distance). The Permian Basin region has nearly 2000 such wells within the 21-county study area in West Texas. This study employed a GIS-based closest facility analysis to identify the optimal closest disposal well (facility) for a given hydraulic fracturing well. The analysis was carried out for 2764 hydraulic fracturing wells that were drilled during the period of January 2012–May 2013. The results of the study indicate that the median distance between a hydraulic fracturing well and the closest disposal well is approximately 4 km (2.5 mi) in the Permian Basin region and less than 8 km (5 mi) for 90% of the fracturing wells. The analysis also indicated that while a large number of disposal wells are available in the region, only a fraction of them are located in close proximity to the existing wells and represent the optimal choice for current hydraulic fracturing activities. However, the availability of other disposal wells indicates that the Permian Basin has the wastewater disposal infrastructure to meet future fracturing needs. This availability makes the disposal cheap and possibly limits the reuse and recycle of produced water on a large scale. It is however important to recognize that deep well disposal effectively removes the water from the hydrologic cycle.

The GIS analysis carried out here also indicated that only a small fraction of the produced water transportation occurs on major roads intended for heavy loads and high volume traffic. Therefore, this wastewater disposal strategy poses risks of damage to transportation infrastructure and increases the chances of fatalities and accidents. While the Permian Basin region of Texas is currently under attainment of ambient air quality standards, transportation of produced water, at a minimum, adds nearly 1 kg of NO_x, 500 g of CO, and 100 g of VOCs per well. The study focuses only on disposal of wastewater and does not consider other transportation impacts associated with hydraulic fracturing such as transportation of water, proppant (sand) for fracturing, and disposal of solid wastes in landfills all of which increase air emissions and transportation risks as well. The costs to the society associated with this produced water management practice must therefore be weighed against economic benefits accrued both to the individuals (oil and gas producers and waste management operators) and the society as a whole (e.g., counties) to make sure the strategy does not result in net negative externalities.

REFERENCES

Boschee, P. (2012). Produced and flowback water recycling and reuse. *Oil & Gas Facilities*, 16–21.

Chapman, E. C. et al. (2012). Geochemical and strontium isotope characterization of produced waters from Marcellus shale natural gas extraction. *Environmental Science and Technology* 46(6): 3545–3553.

Colborn, T. et al. (2011). Natural gas operations from a public health perspective. *Human and Ecological Risk Assessment* 17(5): 1039–1056.

Dunn, J. B. et al. (2013). *Update to Transportation Parameters in GREET*, Argonne National Laboratory, Lemont, IL.

ESRI Inc., R., Ca ArcGIS v 10.2. Redlands, CA.

GE. (2014). Economics of water reuse and recycling in hydraulic fracturing. 2014 Texas Water Summit Report. The Academy of Medicine, Engineering and Science of Texas, Austin, TX.

Gilmore, K. R. et al. (2014). Transport of hydraulic fracturing water and wastes in the Susquehanna river basin, Pennsylvania. *Journal of Environmental Engineering* 140(5): B4013002-1–B4013002-10.

Greenlee, L. F. et al. (2009). Reverse osmosis desalination: Water sources, technology, and today's challenges. *Water Research* 43(9): 2317–2348.

Gregory, K. B. et al. (2011). Water management challenges associated with the production of shale gas by hydraulic fracturing. *Elements* 7(3): 181–186.

Hendrickson, C. T. et al. (2006). *Environmental Life Cycle Assessment of Goods and Services: An Input-Output Approach*, Routledge, New York.

Hernandez, E. A. and V. Uddameri. Simulation—Optimization model for water management in hydraulic fracturing operations. *Hydrogeology* (in press).

Howarth, R. W. et al. (2011). Methane and the greenhouse-gas footprint of natural gas from shale formations. *Climatic Change* 106(4): 679–690.

Jester, S. et al. (2013). Cost effective reuse of produced water in the Permian Basin. *Oil & Gas Environmental Conference*, Conoco Phillips, Dallas, TX.

Jiang, M. et al. (2011). Life cycle greenhouse gas emissions of Marcellus shale gas. *Environmental Research Letters* 6(3): 034014.

Kim, N. et al. (2007). Effect of silane coupling agents on the performance of RO membranes. *Journal of Membrane Science* 300(1): 224–231.

Kim, S. et al. (2011). Evaluations of NO_x and highly reactive VOC emission inventories in Texas and their implications for ozone plume simulations during the Texas Air Quality Study 2006. *Atmospheric Chemistry and Physics* 11(22): 11361–11386.

King, G. E. (2012). Hydraulic fracturing 101: What every representative environmentalist regulator reporter investor university researcher neighbor and engineer should know about estimating frac risk and improving frac performance in unconventional gas and oil wells. *SPE Hydraulic Fracturing Technology Conference, Society of Petroleum Engineers*, The Woodlands, TX.

Kondash, A. J. et al. (2013). Radium and barium removal through blending hydraulic fracturing fluids with acid mine drainage. *Environmental Science and Technology* 48(2): 1334–1342.

Nicot, J.-P. and B. R. Scanlon. (2012). Water use for shale-gas production in Texas, US. *Environmental Science and Technology* 46(6): 3580–3586.

RRC. (2014). Oil and gas well records. http://www.rrc.state.tx.us/oil-gas/research-and-statistics/obtaining-commission-records/oil-and-gas-well-records/. Accessed on July 25, 2014.

Ruppel, S. C. (2009). Integrated synthesis of the Permian Basin: Data and models for recovering existing and undiscovered oil resources from the largest oil-bearing basin in the US. Bureau of Economic Geology, Austin. U.S. Department of Energy Contract No. DE-FC26-04NT15509.

Shaffer, D. L. et al. (2013). Desalination and reuse of high-salinity shale gas produced water: Drivers, technologies, and future directions. *Environmental Science and Technology* 47(17): 9569–9583.

Uddameri, V. et al. (2014). Unconventional oil and natural gas resources development and their potential environmental impacts. *EM Magazine*, July 2014.

Uddameri, V. and V. Honnungar. (2010). An optimisation model for transportation of hazardous wastes. *International Journal of Environmental Technology and Management* 13(1): 4–20.

USEPA. (2006). Life cycle assessment. http://www.epa.gov/nrmrl/std/lca/lca.html. Retrieved October 6, 2014.

Veil, J. A. et al. (January 2004). A white paper describing produced water from production of crude oil, natural gas, and coal bed methane. Argonne National Laboratory for the US Department of Energy, National Energy Technology Laboratory, Argonne, IL. http://www.ead.anl.gov/pub/dsp_detail.cfm. Accessed on August 10, 2004.

Wilson, P. (2012). Testimony before the house committee on energy resources, Texas Department of Transportation.

Worth, C. o. F. (2014). Salt water disposal terms and data. Retrieved October 8, 2014, from http://fortworthtexas.gov/uploadedFiles/Gas_Wells/SWD_questions.pdf.

Zietsman, J. and D. G. Perkinson. (2003). Technical note Heavy-Duty Diesel Vehicle (HDDV) idling activity and emissions study: Phase 1—Study design and estimation of magnitude of the problem, Texas Transportation Institute, College Station, TX.

12 Challenges and Opportunities for Increasing Guar Production in the United States to Support Unconventional Oil and Gas Production

Noureddine Abidi, Sumedha Liyanage, Dick Auld, Robert K. Imel, Lewis Norman, Kulbhushan Grover, Sangu Angadi, Sudhir Singla, and Calvin Trostle

CONTENTS

12.1 Overview....208
12.2 Introduction....208
12.3 Guar Application in the Petroleum Industry....210
12.4 Guar Breeding....212
12.5 Guar Production and Processing in the Thar Desert Region of India and Pakistan....214
12.6 Processing....215
12.7 Guar Production in the Semi-Arid U.S. Southwest....216
12.7.1 Guar Production Basics for the U.S. Southwest....217
12.7.2 Guar and Crop Insurance....217
12.7.3 Suitable Field Sites for Guar....218
12.7.4 Guar Planting and Establishment....218
12.7.5 Guar Fertility including Rhizobium Inoculation of Legume....219
12.7.6 Guar and Irrigation....219
12.7.7 In-Season Assessment of Growth....220
12.7.8 Guar Maturity and Harvest Management....220
12.7.9 The Year after Guar....221

12.7.10 Growing a Viable, Competitive U.S. Guar Production Industry: What Will It Take to Increase U.S. Guar to 250,000 Acres Annually? 221
12.8 Summary and Concluding Remarks 223
References 223

12.1 OVERVIEW

Guar is an important ingredient in hydraulic fracturing. While it can be produced in semi-arid and arid parts of the United States, installation of additional processing capacity is necessary for increasing its production in the United States.

12.2 INTRODUCTION

Guar gum is a galactomannan extracted from the seeds of the leguminous plant *Cyamopsis tetragonolobus* (L.)[1] (Figure 12.1a–c). Guar has been cultivated for generations as a vegetable, forage,[2] and green manure[3] crop in Asian countries especially in India and Pakistan. The discovery during World War II of guar gum as a substitute for locust bean gum in the paper industry has brought wide recognition to the importance of the crop and has garnered the interest of many industries.[3] For example, guar gum has been used in various industries, such as food, pharmaceuticals, textiles, and cosmetics. Currently, it has a diverse range of food and nonfood applications. In addition, the recent discovery that guar gum can be used in hydraulic fracturing in horizontal oil well drilling has boosted industrial interest on guar gum. The mining and mineral processing industries also use considerable amount of guar gum.[4]

Guar is an extremely drought tolerant plant and very easy to cultivate under a limited supply of resources, especially with a limited supply of water. Gresta et al.[5] reported that guar seeds have been produced with 2.65 m^3/ha average water supply. Therefore, guar is mainly cultivated in the semiarid regions of the world such as northwestern India, Pakistan,[6] and the drier part of the United States (e.g., Western parts of Texas and Oklahoma). Furthermore, the plants are well adapted

FIGURE 12.1 (a–c) Guar plants (*Cyamopsis tetragonoloba*). (Picture source: Dr. Dick Auld, Texas Tech University, Lubbock, TX.)

to mechanical planting and harvesting. Even though there are some guar varieties grown in India for human consumption, commercial scale guar cultivation is basically focused on industrial gum production and commercial scale cultivation has increased steadily since the 1950s.[3] The estimated annual world market for guar gum is around 150,000 tons, and approximately 70% of the world market of guar gum comes from India and Pakistan.[7]

Guar seeds are dicotyledonous, small, and very hard beans that are found inside bean-like pods. Figure 12.2 shows a cluster of guar beans, Figure 12.3 shows opened guar beans, and Figure 12.4 shows a guar seed sample. The seed color can vary

FIGURE 12.2 Cluster of guar beans.

FIGURE 12.3 Opened guar beans.

FIGURE 12.4 Guar seeds.

from a dull white color to a black color, and the effect of the seed coat color on the water sorption, germination, and quantity of seed components has been studied.[8] In this study, the authors reported that the seed coat color has control over the water permeability and the percent germination of the seeds. However, changes in seed coat color have no impact on the quantity of the endosperm. Endosperm, the most important part for extracting commercial guar gum, accounts for approximately 1/3 of the bean weight and is composed mainly of galactomannan, which is a heteropolysaccharide of galactose and mannose. Guar meal, the remaining 2/3 of the seed, is a good source of protein and fiber; the guar meal is composed of the hull and the germ.

Previous studies have evaluated the relationship between guar seed weight and the gum content. Seed weight showed negative correlation with the galactomannnan content but positive correlation with the protein content.[5] Thus, increasing the seed weight will not have a great advantage with respect to gum content. Hence, breeding programs can increase the gum content by increasing the number of seeds or pods per plant.

12.3 GUAR APPLICATION IN THE PETROLEUM INDUSTRY

Guar use in the petroleum industry began in the mid-1950s as a thickener for fresh water or brines mixed with hydrocarbons using various methods. Early in the next decade, it was used in the hydraulic fracturing process to create thickened fluids prepared only from water. Hydraulic fracturing is a process well described and applied since its beginning in 1949. Today, essentially all the oil and gas produced globally requires well stimulation via hydraulic fracturing, and guar gum is essential to this process. In this section, four primary areas will be discussed regarding the important usage of guar gum used in hydraulic fracturing (1) admixture with water and hydration rate, (2) rheology properties, (3) use of breakers and additives, and (4) fracture conductivity.

1. Guar gum is manufactured as a dry powder with a particle size average ranging from about 25–100 μm. Guar manufacturers have standards and proprietary specifications regarding particle size, but the primary goal of all is to provide a guar gum optimal rate of hydration coupled with maximization of viscosity. Typically, in the petroleum industry, guar viscosity is measured on a couette-type rheometer such as a Model 35 or Model 50 Fann rheometer, while other industries normally use a Brookfield rheometer to measure viscosity. A common oil industry method is to measure the viscosity of a guar solution containing 40 lb of guar mixed into 1000 gal of water. This solution would be commonly called a "40 lb gel" and at 511 s^{-1} gives a viscosity of 30–45 centipoise, depending on the quality of the guar gum.

   Early use of guar consisted of mixing the guar into the water while circulating the system into a large, typically 20,000 gal. tank. Lumping was a big problem, and in the late 1980s, guar began being premixed into an hydrocarbon such as diesel oil or more environmentally friendly oils. The guar content of this system, known as "liquid gel concentrates" (LGC) was typically 4 lb guar/gal of slurry. This method allowed formation of lump free gels and also continuous mixing processes, using a work tank, provided a method that did not require batch mixing of each tank. Today, the use of LGC is the industry standard, but a trend is growing to directly add powder to the work tank via high-energy mechanical mixers. Today, fast hydrating guar gums, known as FHG, are desired, which provide about 80% of maximum viscosity in 3 min contact time with water, and >95% hydration in 30 min.

2. Guar gum is normally not used as a simple solution. Using selected buffers, guar gum is complexed with multivalent ions made from boron, zirconium, aluminum, and titanium. When these metals react with guar, complex arrays of molecules are produced, and the viscosity of the base guar solution is multiplied at least 3–10 times. This resulting rheology is dependent on shear during formation, shear during pumping, temperature, other additives, and other variables. The purpose of this cross-linked guar solution is to transport solid, round particles such as sand, ceramics, or sintered bauxite into the hydraulic fracture. These materials are called proppants, and after the hydraulic fracturing treatment "prop" the created hydraulic fracture open, hence creating a highly permeable pathway for fluid flow compared to the native rock strata containing the produced hydrocarbon. Since the temperature of many geologic formations is hot—up to 400°F, fracturing fluids are critically evaluated such that adequate viscosity for a time–temperature profile is obtained to provide adequate proppant transport properties of the fracturing fluid.
3. As mentioned, each metal crosslinker used with guar has a specific optimum pH, so buffers are almost always used in the fracturing fluid composition along with the guar gum and crosslinker. Also present are chemicals called breakers, which depolymerize the guar polymer at a predetermined and desired rate. This breaking is a critical balancing process. One must depolymerize the guar gel after the pumping operation is complete, but not prematurely and compromise the proppant transport ability of the fracturing fluid. Breakers fit into two categories: enzymes and oxidizers. The enzymes are from the enzyme family called mannosidases, which break the mannose polymer backbone of guar. The oxidizers are generally from the persulfate family such as ammonium, sodium, or potassium persulfate. At higher temperatures, breakers from chlorine-based chemicals such as sodium chlorite may be employed. At temperatures above about 120°C, gel stabilizers are used to prevent the latent oxygen in the water from oxidizing the guar gum. Sodium thiosulfate is the industry standard gel stabilizer. The fracturing fluid additives include surfactants for emulsion prevention, foaming agents, bacteria control agents, corrosion inhibitors, scale inhibitors, and perhaps other additives for specific function.

4. The hydraulic fracturing process creates a pathway to the producing borehole for fluid flow. This pathway is desired to have great flow capacity compared to the porous media, which produces the hydrocarbon. Measurement of this conductivity is not a fully developed art/science. But regardless of the exact methodology, the overall idea is that the fracturing fluid solution is admixed with the intended proppant type, then all placed between two rock plates forming a sandwich-like system. Then, these rock plates are pressured together using pressures expected in the down-hole environment. Pressures used typically range from 2,000 to 10,000 psi. At this point, the guar fracturing fluid solution can only damage the ideal flow capacity; it cannot improve it. Damage from the observed fracture conductive compared to ideal, no guar present conductivity may show a reduction of up to 90%, but the desired level of damage is less than 50%, and ideally less than 10%.

So, where does this damage come from? The answer is complex and not fully understood. First, guar is not a pure, soluble material. Non-galactomannan gum material from other cellular debris is found in guar. If a guar solution, uncrosslinked, is broken using breakers, the remaining insoluble material may be as high as 10%. This material is known as residue and found using tests called residue after break (RAB) tests. Also potentially contributing to the fracture conductivity damage is incomplete breaking of the fluid. This incomplete breaking problem is enhanced by a gel concentrating effect called filter cake.

At high pressures used in hydraulic fracturing, some of the guar solution is forced out of the fracture at the fracture face and into the formation rock. But the guar polymer may be actually too large to enter the rock pores, and is physically filtered off at the formation face. This filtered material may contain guar at a concentration 2–10 times the original fracturing fluid, and hence be much more difficult to completely break (depolymerize). Current industry needs include methods to manufacture guar which contains no insoluble residue, breakers that can effectively break down the guar filter cake, additives for minimization of the formation of guar filter cake, and higher molecular weight guar polymers that would require less polymer to be used for obtaining equivalent fracturing fluid rheology.

12.4 GUAR BREEDING

Guar is of particular agronomic importance in the Lower Great Plains of North America because it has been identified as one of a small handful of "desert crops" that could be grown on the Lower Great Plains to help mitigate the hotter, drier, and more erratic weather expected to be caused by global warming and climate change.[9,10] Foliar diseases have historically limited guar production in the more humid areas of the United States.[3,11] The high protein co-products derived from processing guar seed to extract the gum will support the continued viability and sustainability of the livestock feeding industry of this area. Having a drought tolerant legume that has historically been used as a forage crop that is well adapted to arid regions may also provide significant benefits to livestock growers.

Our central hypothesis is that the development of high-yielding guar cultivars adapted to the Lower Great Plains of North America is critical to the development of sustainable guar production in the United States. Our long-term plan is to develop cultivars and production systems that will allow guar to compete on an economic basis with supplemental irrigated cotton.[12] These cultivars will need to produce high levels of gums with specific mannose/galactose (M/G) ratios tailored for products in both edible and industrial markets.[9] We believe that characterization of the physical and chemical properties of the gum extracted from seed of our elite lines at the Texas Tech University guar breeding program and the USDA-ARS guar germplasm collection will provide ample sources of unexploited genetic variation for these traits.[13] Perhaps one of our more pressing goals is to find a guar gum with a composition that allows hydraulic fracturing to use either fresh or saline water resources to make this process more environmentally benign.

Because guar is an obligatorily self-pollinated diploid (2n = 14), it is an ideal species to target for rapid genetic improvement using conventional genetics.[13] Prior to our work in cooperation with Halliburton (1998–2007), there had been limited recent efforts on the genetic improvement of guar in the United States with the latest cultivars, "Santa Cruz" and "Lewis," released in 1985.[14] Texas Tech University in cooperation with Halliburton Energy Service conducted a highly successful guar breeding effort from 1998 to 2007. Dr. Ellen Peffley (retired) conducted a conventional plant improvement in this program starting with an evaluation of almost 200 plant introductions from the USDA-ARS collection at Griffin, Georgia. This program invested ~$880,000 in research and generated two improved guar cultivars (Matador and Monument) and several hundred highly diverse breeding lines. Matador had an incremental yield increase compared to earlier U.S. cultivars; moderate gum content; and because of its highly branch growth habit produced high seed yield when grown in 1 m rows.[15] Monument had significantly higher gum contents but because it produces no lateral branches, it had modest seed yields in 1 m rows when compared to Matador.[16] However, Monument has a higher potential seed yield at higher plant population due to the production of a larger and more complex raceme.

Dr. Peffley also developed segregating populations from which several hundred F2 and F3 progeny rows plants were evaluated for agronomic potential in replicated field trials and harvested seeds were analyzed for viscosity to estimate gum content.[8,17] Additional crosses were made between types with divergent M/G ratios. Dr. Richard Dixon at the Samuel Roberts Noble Foundation in Ardmore, Oklahoma, also conducted some of the first molecular biology work on guar during this period.[18]

In 2007, corporate leadership at Halliburton Energy Services (HES) decided that production and distribution of agricultural products such as guar and guar gum lay outside the core competency of their company. Dr. Lewis Norman of Halliburton helped increase breeder seed of both Matador and Monument and funded the systematic storage of the 400 breeding lines. Dr. Dick Auld served as a Co-PI in the earlier breeding program at TTU and as the current curator of stored guar lines developed in this initial program. During this program, we will try to identify lines capable of producing seed yields in excess of 2000 kg/ha of seed with only 150 mm of supplemental irrigation. Segregating lines will be advanced using pedigree selection of three plants per entry in the F4 and F5 generation. Texas guar producers have historically planted in late June following a failed cotton crop with little or no inputs.[12] Often the guar has been planted in 1 m rows with plant populations of only ~80,000 plants/ha and grown under completely dryland conditions with no fertility and weed control. We anticipate that a plant population of ~500,000 plants/ha will be needed to provide yields consistently exceeding 2000 kg/ha.[9] Guar has approximately 33,000 seeds/kg, which would require only ~6.8 kg of planting seed per hectare.

12.5 GUAR PRODUCTION AND PROCESSING IN THE THAR DESERT REGION OF INDIA AND PAKISTAN

Clusterbean (*Cyamopsis tetragonoloba* L.), popularly known as guar, is a drought-resistant arid legume crop that has been grown for centuries in the Indian subcontinent. Domesticated spp is hypothesized to be evolved from a drought tolerant African relative in India.[19] The word guar is derived from the ancient Sanskrit word *Gau Aahaar*, which means "cow food." The crop is also known by various local names including *Guari*, *Khutti*, and *Darari* in India.[20] Considerable natural variability of this arid legume is present in drier regions of the country.

Guar has been traditionally grown as protein rich and highly palatable forage for centuries and is still a popular cow feed in the Indian subcontinent. Young pods of guar are also harvested and used for cooking as one of the protein sources for the vegetarian populations of the subcontinent. Guar pods are also a good source of vitamin A, calcium, iron, phosphorus, and ascorbic acid.[21] Guar is known for health benefits like reducing cholesterol levels in the body, controlling diabetes, and maintaining regular bowel movements in human beings.[19] Currently, with increased demand of guar seed and its products such as guar gum for various industries, more and more area is cultivated for seed production.

Traditionally, guar has been a poor man's crop in the Indian subcontinent, particularly for marginal farmers in the arid areas. Because of its high adaptation to poor and erratic rains, low inputs and lower management needs, soil building properties, integration in multiple cropping systems and industrial importance, guar has high economic and social significance in arid areas.[20] Significant amounts of guar are produced on marginal land under dryland conditions.

India is the world leader for production of guar, contributing 80% of the total production in the world.[22] The desert bordering region between India and Pakistan called "Thar Desert" is the home for the major guar production center in the world. The region produces more than 90% of guar in the world.[19] Guar crop is cultivated mainly during *Kharif* (monsoon) season in India in the northwestern parts of the country, encompassing the states of Rajasthan (Churu, Nagaur, Banner, Sikar, Jodhpur, Ganganagar, Sirohi, Dausa, Bikaner, Hanumangarh, and Jhunjhunu districts), Gujrat (Kutch, Banaskantha, parts of Mehsana, Sabarkantha, Vadodara, and Ahmedabad districts), Harayana (Bhiwani, Gurgaon, Mahendragrh, and Rewari districts), and Punjab (Bhatinda, Ferozpur, Muktsar, and Mansa districts).[22] Within India, Rajasthan state including the Thar Desert has more than 75% of total area cultivated under guar, but faces high fluctuation in production because of erratic and insufficient rains.[22]

During the 1970s, guar was also grown regularly in the northern state of Uttar Pradesh, mid-Indian state of Madhya Pradesh, and southern state of Orissa. As processing facilities have been closed down in Uttar Pradesh and Madhya Pradesh, the cultivation in these states is currently negligible and currently, no guar cultivation takes place in the Orissa state of India.[22]

The area, production, and yield of the crop are inconsistent due to its overdependence on weather and on production confined to a limited geographical area largely made up of arid regions. Based on a 10-year average from 2001 to 2011, guar was cultivated on about 2.96 million hectares of area and produces 1.22 million tons of guar seed in the country with an average national productivity of 0.41 ton/ha with a large year-to-year variation especially in years of drought.[23] Its cultivation extended to new regions and higher areas after high speculative prices in 2011. Average productivity of guar varies from state to state with 1096 kg/ha in Haryana, 769 kg/ha in Punjab, 582 kg/ha in Gujarat, and 319 kg/ha in Rajasthan.[23] The annual production of guar during the last 3 years ranged from 1,100,000 to 1,287,000 metric tons.[22] Total production of guar bean in India is estimated to have crossed 2.7 million metric tons during the agricultural year 2013–2014.[24]

Pakistan, with 15% of world production, is next to India with major guar production in the arid and semi-arid region of the Thar Desert in the east and southeast part of the country.[22] Before the 1990s, about 80% of the guar in Pakistan was grown under irrigated conditions; therefore, yields were higher.[22] During that period, guar was grown in Punjab, Multan, Muzaffargarh, Mianwali, and Sargodha.[22] Other areas included Bahawalpur, Banawalnagar, and Sindh Province. The annual production of guar in Pakistan during this period ranged between 180,000 and 250,000 metric tons annually.[22] Guar and guar products exported from Pakistan earned just $29 million in 2006–2007.[34] It shot up to a record high of $152 million in 2011–2012 before coming down to $139 million, still more than many other items. According to the Pakistan Bureau of Statistics, guar was cultivated over 154,821 ha in 2008–2009.[34] If weather remains conducive for the crop, it produces nearly 0.13 million tons. During 2012, around 0.11 million tons of guar seed were produced in Pakistan and in 2013, production was expected to touch the level of 0.1 million tons.[23]

India and Pakistan provide better agro-climatic conditions for the cultivation of guar and high-quality gum, though it has also been grown in the United States, South Africa, Brazil, Zaire, and Sudan.[22] The United States is the largest importer of guar and its derivatives from India. Due to the expansion of the shale oil and gas industry, demand and price for guar seed and guar gum has jumped. For instance, the value of guar product exports from India to the United States shot up to a billion dollars in 2011, further to $3.4 billion in 2012 and then came down to $1.6 billion in 2013.[24]

12.6 PROCESSING

Guar seeds are processed for the production of guar gum and the by-products, including husk and germ, are collectively called *Guar meal.* Processing involves two stages: the processing of guar seed to guar splits and then guar splits to powder or guar gum. There are more than 150 split units in India, mostly located in Jodhpur, Barmer, Sri Ganganagar, and Bikaner districts of Rajasthan state. Additional processing industries also operate in Bhiwani and Sirsa districts of Haryana state and Deesa and Ahmadabad regions of Gujarat state.[25]

Guar was introduced from India to the Oklahoma Plains and North Texas in 1903.[26] In the 1940s, researchers in the United States explored its industrial use as an alternative to locust bean gum.[19] After a breakthrough in the use of guar gum in industry by General Mill and Stein Hall and Company in 1950, guar gum production was first started commercially in 1955 in India by the Bhiwani Guma Guar Factory in collaboration with Stein Hall.[22]

Ever since the industrial importance of guar was realized, increasing guar production and productivity have been the focus of Indian agriculture research. The Indian Council of Agricultural Research (ICAR) established an all-India coordinated project in the 1970s to increase cooperation and funding for research on arid region legumes to develop higher yielding cultivars and to develop agronomic as well as pest management practices to improve productivity.[27] Within 50 years, guar became an important industrial crop in India and Pakistan.[28] What had been a minor crop with limited business interest and virtually no need for analysis became, in 2 years, India's largest agricultural export to the United States. Higher prices for guar have resulted in production expanding in the traditional states of Rajasthan, Punjab, Haryana, and Gujarat as well as into new areas such as Uttar Pradesh and Andhra Pradesh. While the United States is the major buyer, demand for Indian guar gum from other gas- and oil-producing countries, notably Russia, Venezuela, and Mexico, is growing.[24]

Indian policymakers are assessing potential ways to increase guar production in India to meet global demand and maintain the dominance of the region in world guar supply. However, lack of quality seeds, large proportion of acreage on marginal soils, uncertainty of rainfall, lack of money for research, market uncertainty, lack of value added products, etc., are all limiting rapid growth in the guar industry and are threatening India's dominance in the industry.[19]

12.7 GUAR PRODUCTION IN THE SEMI-ARID U.S. SOUTHWEST

Guar was initially grown in Texas around 1903 but only for vegetable, green manure (which remained about half of the planted crop through the 1970s), and some cattle feeding.[12] Commercial production began in 1950. Guar acreage has fluctuated considerably in the U.S. Southwest based on demand, market price, and drought severity. The latter factor may prevent planting in some years, and although guar is as drought tolerant and heat tolerant as any agricultural crop grown in the United States, without moisture to establish the crop there can be little expected from cropping. Recent acreage since 2000 has ranged from about 5,000 to 110,000 acres, the former acreage in 2011 in record-breaking low rainfall that prevented almost all planting, while the latter acreage was driven higher in 2013 following significant failure of cotton leading to replanting opportunities. Though guar typically yields better when planted as a primary crop largely due to earlier planting and more time for maturation, it is an inexpensive replacement crop that can generate income.

In addition to Texas and southwestern Oklahoma where 95%–100% of the crop has been produced, guar may be occasionally grown in New Mexico and Arizona.[12] Guar is limited by late plantings, soil types, shorter season, and potentially detrimental humid conditions from consistent reliable production in other U.S. states. Contract guar prices paid to farmers since 2000 have fluctuated from $0.125 to $0.35/lb. Since 2007, only one mechanical splitter has been operating in the United States so this caps the potential acreage until more infrastructure may be installed. With the recent demand from the oil and gas industry for guar gum, several new companies are exploring the merits of installing U.S. splitting facilities from which guar materials can then be shipped to numerous powder and gum manufacturing facilities.

In addition to the potential improvement in guar discussed elsewhere, cropping management is key even for a low-input crop like guar.[12] Guar, in general, realizes its best relative advantage to other crops in semi-arid environments, which are typically defined as 10″–20″ of annual rainfall (though the heat conditions in the U.S. Southwest associated with this lower rainfall make effective production even more challenging). Guar also edges east in Oklahoma, North Texas, and the lower Texas Gulf coastal region into areas that may approach 30″ of annual rainfall. However, this represents more humid conditions for which guar is not as well adapted and has increasing issues with plant health including foliar fungal and bacterial diseases like *Alternaria*, bacterial blight, etc. Arid regions (<10″ annual rainfall) of the U.S. Southwest—far west Texas, southern New Mexico, and Arizona will require supplemental irrigation.

12.7.1 Guar Production Basics for the U.S. Southwest

Guar genetics and guar varieties available to farmers are discussed elsewhere in this chapter. With only two exceptions, guar varieties used in almost all of the U.S. Southwest are older public lines released in the mid-1980s and earlier.[12] These lines perform reasonably well but do not have the needed disease resistance to be recommended for humid regions nor do they represent available advances in plant breeding and genetics. Production of guar seed for planting must maintain high-quality seed (especially germination) with good vigor, and seed supplies must ensure the exclusion of potential contaminant morning glory, which is the same size seed as guar and cannot be cleaned out.

12.7.2 Guar and Crop Insurance

Guar currently is not a program crop of the USDA for crop insurance purposes.[12] Thus, this valuable risk management tool is not available to guar growers. This may preclude obtaining agricultural loans for financing farming operations as loan terms often require the farmer to insure the crop. This is a potential bottleneck for many growers though, ironically, the amount of operating capital required with guar production is much lower than most other crops.

12.7.3 Suitable Field Sites for Guar

Guar grows best on medium to coarse (sandy) soils with adequate drainage.[12] Foremost, guar is not suitable for fields that are already or are expected to be weedy. Though mechanical cultivation is used to help control weeds, there are only three currently labeled active herbicide ingredients for guar: trifluralin (applied and incorporated in the soil before planting to control grasses and small-seeded broadleaves), carfentrazone (using hooded sprayers that protect from contact with the guar plant for control of broadleaf weeds), and clethodim (grass control in established guar fields). If the expected weeds cannot be controlled with this combination of mechanical cultivation and herbicides, then guar is not recommended. Other herbicides are labeled for pre-plant burndown of preexisting weeds, but these chemicals have minimal carryover to a newly planted crop. For purposes of herbicides and crop tolerances among a myriad of minor crops, the IR-4 Project specialty crop project of the U.S. Environmental Protection Agency classifies guar in crop subgroup 6-C, dried shelled pea and bean (except soybean).[29] The IR-4 program facilitates the evaluation and potential registration of additional pest management for specialty and minor use crops. In addition to guar, other subgroup 6-C crops include lupines; other peas and dried beans including kidney, navy, pinto, mung, and cowpea; chickpea (garbanzo); lablab; and pigeon pea.

Most field conditions in the U.S. Southwest, provided the crop is established and has minimal rainfall or better, can provide a guar yield potential in dryland conditions of 400–1200 lb/A.[12] To reiterate, this is greatly dependent on rainfall and also deep subsoil moisture at planting, but typical dryland yields for guar fall in the 600–1000 lb/A range. With limited irrigation of 3″–6″ the expected yield potential rises to about 1000–1500 lb/A. Recent guar contractor yield goals for guar in the region are 900 lb/A dryland and 1500 lb/A irrigated.

12.7.4 Guar Planting and Establishment

Guar requires warmer soil temperatures, at least ~70°F (21°C), than almost all other crops for optimal seeding and establishment.[12] Seeding depth is best into moisture at 1.0″–1.5″ deep, whereas guar may have difficulty emerging from deeper seeding due to lack of push, or fragile seedlings, with deeper planting. The heart of the guar production region, the Texas Rolling and lower South Plains and southwest Oklahoma, should target seeding by mid-May to the end of June for best production. The lower Texas Gulf coast can target seeding April 15–May 31 (and aim for harvest before tropical storm season), whereas Arizona and southern New Mexico have warmer early season temperatures and a potentially broader planting window.

Field conditions should be free from weeds at planting, and due to the difficulty of harvesting guar pods on lower branches it is favorable to seed guar on a slightly raised bed at least 2″–4″ higher than the row middles.[12] Guar is suitable in dryland conditions on row spacings up to 40″. However, higher production potential with irrigation or higher rainfall may find benefit in seeding on rows as little as 15″ wide. With guar averaging about 13,000–15,000 seeds/lb, seeding rates from 2 to 10 lb an acre have shown little difference in yield,[30] but currently practices raise the seeding rates on dryland above 5–8 lb/A or more in order to stimulate inter-plant competition and more likely upward basal stem branching for easier combine harvest of low-set pods.

12.7.5 Guar Fertility including Rhizobium Inoculation of Legume

As a legume, guar has been widely used as a "plow-down" or green manure crop with the assumption that infection of roots by a guar-specific *Rhizobium* bacteria is capturing and fixing atmospheric N for plant use.[12] This does require actual nodulation of the roots, which in Texas observations, is usually not the case. Nodulation is sporadic at best so the commonly assumed nitrogen credit attributed to guar may not be merited.

Currently, there is only one commercial company that markets any material that contains guar-specific *Rhizobium*.[12] Other manufacturers cite development costs and lack of potential acres, for example, limited sales, in order to develop inoculant products like those that commonly exist for other legumes grown on larger acreage. Evaluation of different inoculants is currently ongoing for not only different products or *Rhizobium* strains that may nodulate guar but also the means of how the inoculant is applied, whether on the seed or placed in-furrow at planting (usually preferred on other regional legumes). Historically, seedbox powders have sometimes been available for guar, but these products have relatively low bacterial counts and lower bacterial survivability in hotter soils, common to bare soil when guar is planted, is compromised.

Because guar is a legume, it is not normally fertilized with nitrogen and it is not recommended that growers consider N fertilizer applications in dryland cropping.[12] In the absence of sufficient N nodulation in higher yield environments (irrigated), it is possible that guar production could become limited by low N. Phosphate is most likely the nutrient of interest for fertilizing guar, and state Extension services in Texas and Oklahoma[31] provide soil test recommendations for guar based on soil testing. Potassium also has fertility recommendations in TX and OK for Southwest guar, but soil test K levels in the guar production region are inherently high and a response to applied K is not expected provided soil test levels are typical.

12.7.6 Guar and Irrigation

At most, only a few percent of U.S. guar acres are irrigated. Farmers historically use irrigation first on high-yield, high-value crops.[12] In these production environments, on a relative basis, guar is less competitive economically. Nevertheless, guar can be effectively grown with limited irrigation (4″–6″ in typical rainfall years) though it is not essential, which is one advantage that guar frequently has versus other crops in hotter, drier environments. Farmer records suggest that West Texas guar yields approximately 100–150 lb/A per 1″ of irrigation. Individual fields that have received higher levels of irrigation, more than 12″, often have higher disease incidence. Field observations suggest that, after initial irrigation and when guar starts flowering in 30–40 days after planting, over-the-top sprinkler irrigation should be avoided.

In cases of heavy irrigation, the plants turn nearly black with disease and yields are even reduced relative to adjacent nonirrigated land.

12.7.7 In-Season Assessment of Growth

Insect damage is rare in guar, and the guar midge, also commonly known as the alfalfa midge, is the only insect that has been noted for treatment in the recent past.[12] The insect infests the branch terminals and retards new growth. Foliage feeding worms have not normally been a problem.

As noted earlier, no one disease predominates in the U.S. Southwest. Guar, and disease issues that limit guar production are uncommon in the drier, low humidity regions of optimum guar adaptation.[12] When foliar and stem disease issues like *Alternaria* are observed on guar, the crop is usually far into the production season (mid-August or later) and the potential effect on guar yield is minimal and does not reach economic thresholds for treatment. The use of fungicides is rare to nonexistent and the low-input approach to guar production precludes willingness on the part of producers to treat in part because fungicide applications, if they were needed frequently, would reduce the potential economic advantage that guar may offer.

Occasional diseases in guar that have been documented recently include bacterial wilt (which is seed-borne), *Alternaria*, as well as *Fusarium* and *Myrothecium* blight.[12] The former two diseases are relatively weak and though they may affect significant portions of the foliage, their late appearance has little impact on the overall crop. Under more moist conditions in central Texas and the lower Texas Gulf coast, either rainfall or common higher humidity, foliar diseases are potentially much more common and can significantly diminish yield potential. But due to the low-input approach to guar or the unwillingness to spend, crops are likely left untreated.

12.7.8 Guar Maturity and Harvest Management

Guar maturity depends on the indeterminant nature of the plant (it will keep growing if possible), but current varieties can reach suitable maturity in about 90–120 days.[12] Historically, guar has been left in the field until well after physiological and harvest maturity often due to the difficulty in getting the crop dry enough to combine. Even after a moderate freeze, the guar stem may remain green and tough for up to a month. This makes combine harvest and threshing more difficult. Thus, much guar is not harvested until well into November and December. Prolonged field exposure to environmental factors makes the guar pod prone to shattering off the plant.

Some custom harvesters have added air reels to their combine to shoot a blast of air at the cutter bar to push shattering pods onto the header and reduce harvest losses as much as 50 lb/A.[12] Excessive rain after maturity may increase the proportion of black seed that is *perceived* to have lower quality (either reduce gum content or lower molecular weight gum). This has not been verified under controlled conditions, and, in fact, Liu et al.[8] noted that black seed has higher germination. This may be due to increased cracking of the seed coat, which thus allows the seed to more easily imbibe water, but in turn these cracks may lead to reduced ability to remove the seed coat in two uniform halves.

Current research is investigating the use of harvest aids, similar to how cotton is managed for earlier harvest,[32] to hasten harvest and potentially preserve higher quality seed and gum. Currently, paraquat dichloride and sodium chlorate are labeled for harvest aid use in guar. A particular advantage of harvest aids is their application when the bulk of the guar seed in a field is mature with little additional yield potential present. Harvest aid applications will terminate any further growth and could potentially move the harvest forward a month or more relative to natural drying.

Current USDA grain standards for grade #1 guar are moisture content ≤13.5% and a minimum test weight of 60 lb/bushel.[12] Contract prices will be based on the grade of guar delivered, and as moisture increases and test weight decreases, the pay price for the seed declines. This furthers the interest of using harvest aids to facilitate drydown to remove moisture and avoid potential loss of test weight in the field. Guar prices will likely discount for black/dark seed above 50% as dark seed may indicate possible lower gum quality (again, as noted earlier, this may be based solely on perception).

12.7.9 The Year after Guar

Guar works well in rotation with other crops such as cotton, grain sorghum, and wheat. Early crop rotation research in the Texas Rolling Plains documented up to 15% higher cotton yields the year after guar in dryland farming.[30] Some of this may be attributed to improved soil nitrogen status if the guar nodulated well and fixed significant quantities of nitrogen. Numerous growers have also commented that they like the condition of their soil after farming guar. One management issue that may arise, however, is the presence of widespread volunteer guar the next year. This is best controlled by the use of glyphosate in subsequent cotton production, particularly as current cotton varieties tolerant to this herbicide allow multiple applications.

12.7.10 Growing a Viable, Competitive U.S. Guar Production Industry: What Will It Take to Increase U.S. Guar to 250,000 Acres Annually?

Guar production technologies have not advanced at a rate comparable to the principle crops (cotton, wheat, grain sorghum) in the likely production regions of the U.S. Southwest, giving the crop a disadvantage as farmers consider their cropping alternatives.[12] Guar varietal improvement will overcome some of these disadvantages, and when coupled with improved farming methods and guar's heat and drought tolerance, a stable, reliable market makes guar a viable profit potential crop for many producers when prices are good. Many producers expect the current demand for guar gum in oil and gas operations to provide that stable demand at a good price. Guar in the U.S. faces competition on many fronts including:

- Guar from other sources, primarily India and Pakistan, normally available in massive amounts
- Other crop options for U.S. farmers, for which there is more production information, herbicide control options, etc.

- Domestic infrastructure is lacking to process increased U.S. guar production
- Agricultural financing gravitates toward established and insurable crops
- Minor crops like guar are at a disadvantage in competing for research funds or capital for improving the crop

Current U.S. guar production at 25,000–50,000 ac does not significantly impact domestic supplies (though it can for individual users) since most current guar is either dryland or planted as a "catch crop" after primary crop failure (normally cotton).[12] This leads to a relatively low yield environment subject to major drought in the U.S. Southwest (e.g., 2011–2013).[33] To achieve a more reliable crop, U.S. guar supply growers, contractors, and research/education staff believe 250,000 ac is a meaningful target to impact guar supplies and have some stabilizing influence on prices of guar products among U.S. users.[12] Though the acreage could be grown, the U.S. guar industry must install additional processing capacity to handle the crop and be willing to support domestic U.S. guar. Without that support, farmers have other attractive cropping options with crops that have less heat and drought tolerance (hence some level of higher risk). The following are forthcoming crop production goals in U.S. guar, though these will require private investment to fund needed work:

- A guar breeding program with a full-time university faculty or private company guar breeder.
- The development of some guar production on irrigated land to smooth out sharp reductions in domestic supply that will occur during widespread drought either from less planting or lower yields.
- The development of a meaningful crop insurance product—either a federal program crop or via the private sector. This should unlock agricultural loans for prospective guar producers otherwise loan entities are reluctant to loan money on uninsured crops.
- Improve public documentation of the quality and chemical properties of U.S. guar gum to address the perception that some U.S. guar may be of inferior quality.
- In addition to guar breeding, other research priorities identified among agricultural universities in the U.S. Southwest include
  - Identification of existing pre-emergent and post-emergent herbicides that have acceptable minimal injury levels on guar.
  - Continued development of harvest aid research to potentially reduce field shattering losses at harvest and enable earlier harvest so guar gum quality or quantity does not potentially degrade in the field due to delayed harvest.
  - Control of volunteer guar in subsequent crops, including improved harvesting to minimize seed losses in the field at the combine header.
  - Optimum irrigation timing for limited irrigation (≤6″) guar.
  - Limited use of N fertilization in combination with improved experimental *Rhizobium* inoculant products to stimulate increased yield.

12.8 SUMMARY AND CONCLUDING REMARKS

In this chapter, we focus specifically on the use of guar in the petroleum industry and discuss issues related with global supply of this commodity and breeding and cultivation in the United States.[9,33] Guar is a drought-tolerant crop and as such well suited for cultivation in arid and semi-arid parts of the United States. However, it is often planted as a catch crop after the failure of a primary crop (cotton), which often results in poor yields. It is widely recognized in the industry that 250,000 acres is a meaningful target to impact guar supplies and have some stabilizing influence on prices of guar products among U.S. users. While this acreage could be grown, installation of additional processing capacity is necessary to support the U.S. guar industry. In conclusion, guar gum is a vital element needed for hydraulic fracturing processes, and hydraulic fracturing is an underpinning technology for production of the global energy needed by society. Both hydraulic fracturing and guar gum has been used for over 50 years, and much incremental advancement has been made in their optimization, but no direct replacement technology has been found. Future work in guar breeding, manufacturing, agronomic practices, and global guar growing diversity should be done in order to provide continual availability of guar. It is quite likely guar will remain the water thickener of choice for hydraulic fracturing for decades to come.

REFERENCES

1. Miyazawa, T. and T. Funazukuri. Noncatalytic hydrolysis of guar gum under hydrothermal conditions. *Carbohydrate Research* 341(7) (2006): 870–877.
2. Chudzikowski, R. J. Guar gum and its applications. *Journal of the Society of Cosmetic Chemists* 22 (1971): 43–60.
3. Whistler, R. H. and T. Hymowitz. *Guar: Agronomy, Production, Industrial Use and Nutrition*. West Lafayette, IN: Purdue University Press (1979).
4. Ma, X. and M. Pawlik. Intrinsic viscosities and Huggins constants of guar gum in alkali metal chloride solutions. *Carbohydrate Polymers* 70(1) (2007): 15–24.
5. Gresta, F., O. Sortino, C. Santonoceto, L. Issi, C. Formantici, and Y. M. Galante. Effects of sowing times on seed yield, protein and galactomannans content of four varieties of guar (*Cyamopsis tetragonoloba* L.) in a Mediterranean environment. *Industrial Crops and Products* 41 (2013): 46–52.
6. Reddy, T. T. and S. Tammishetti. Free radical degradation of guar gum. *Polymer Degradation and Stability* 86(3) (2004): 455–459.
7. Gong, H., M. Liu, B. Zhang, D. Cui, C. Gao, B. Ni, and J. Chen. Synthesis of oxidized guar gum by dry method and its application in reactive dye printing. *International Journal of Biological Macromolecules* 49(5) (2011): 1083–1091.
8. Liu, W., E. B. Peffley, R. J. Powell, D. L. Auld, and A. Hou. Association of seedcoat color with seed water uptake, germination, and seed components in guar (*Cyamopsis tetragonoloba* (L.) Taub). *Journal of Arid Environments* 70(1) (2007): 29–38.
9. Auld, D. L., T. D. Miller, E. Peffley, B. R. Hendon, and T. Witt. Production of guar for high gum and seed yields. In: McMahan, C. M. and Berti, M. T. (eds.), Paper presented at the *24th Annual AAIC Meeting, Industrial Crops: Developing Sustainable Solutions: Program and Abstracts*, Sonoma, CA, November 12–15, 2012.
10. Saunders, N. *Energy for the Future and Global Warming: Nuclear Energy*. Pleasantville, NY: Gareth Stevens Pub (2008).

11. Undersander, D.J., D.H. Putman. A.R. Kaminski, K.A. Kelling, J.D. Doll, E.S. Oplinger, and J.L. Guunsolus. Guar—Alternative Field Crops Manual. University of Wisconsin. Cooperative Extension. (1991). www.hort.purdue.edu//newcrop/afcm/guar.html (accessed February 9, 2015).
12. Trostle, C. D. *Guar in United States—Prospects for Domestic Seed Supply and Impact on Gum Supplies*. Lubbock, TX: Texas Agrilife Research & Extension Center (2012).
13. Morris, J. B. Morphological and reproductive characterization of guar (*Cyamopsis tetragonoloba*) genetic resources regenerated in Georgia, USA. *Genetic Resources and Crop Evolution* 57(7) (2010): 985–993.
14. Ray, D. T. and R. E. Stafford. Registration of 'Santa Cruz' guar. *Crop Science* 25(6) (1985): 1124–1125.
15. Texas Tech University and Halliburton Energy Services. Guar—Matador. USDA-AMS, PVPO 200400235 (2004). http://apps.ams.usda.gov/CMS/CropSearch.aspx (accessed February 9, 2015).
16. Texas Tech University and Halliburton Energy Services. Guar—Monument. USDA-AMS, PVPO 200400301 (2010). http://apps.ams.usda.gov/CMS/CropSearch.aspx (accessed February 9, 2015).
17. Liu, W., A. Hou, E. B. Peffley, D. L. Auld, and R. J. Powell. The inheritance of a basal branching type in guar. *Euphytica* 151(3) (2006): 303–309.
18. Naoumkina, M., I. Torres-Jerez, S. Allen, J. He, P. X. Zhao, R. A. Dixon, and G. D. May. Analysis of Cdna libraries from developing seeds of guar (*Cyamopsis tetragonoloba* (L.) Taub). *BMC Plant Biology* 7(1) (2007): 62.
19. Mudgil, D., S. Barak, and B. S. Khatkar. Guar gum: Processing, properties and food applications—A review. *Journal of Food Science and Technology* 51 (2014): 409–418.
20. Kumar, D. and N. B. Singh. Guar—An introduction. In: *Guar in India*, Kumar, D. and Singh, N. B. (eds.), pp. 1–10. Jodhpur, India: Scientific Publishers (2003).
21. Roy, S. K. and A. K. Chakarborti. Vegetables for temperate climate. In: *Encylopaedia of Food Science, Food Technology and Nutrition*, Macrae, R., Robinson, R. K., and Sadler, M. J. (eds.), p. 4715. London, U.K.: Academic Press (1993).
22. DSIR. Technology status study on guar based industry in India. Department of Scientific and Industrial Research. http://www.dsir.gov.in/reports/tmreps/guar.pdf (accessed on July 14, 2014) (2003).
23. NRAA. Potential of rainfed guar (cluster beans) cultivation, processing and export in India, Policy Paper No. 3. New Delhi, India: National Rainfed Area Authority (2014).
24. NIAM. An analysis of guar crop in India. Global agricultural information network report: US Department of Agriculture Foreign Agricultural Service, Network Report IN4035. Jaipur, Rajasthan: National Institute of Agricultural Marketing (2014).
25. Sharma, P. and K. C. Gummagolmath. Reforming guar industry in India: Issues and strategies. *Agricultural Economics Research Review* 25(1) (2012): 37–48.
26. Hymowitz, T. and R. S. Matlock. Guar in the United States. *Oklahoma Agricultural Experimental Station Bull*, B611 (1963): 3–14.
27. NIAM. *Guar Industry Outlook 2015*. New Delhi, India: National Institute of Agriculture Marketing (2015). http://ccsniam.gov.in./GUAR_Booklet_03122013.pdf (accessed on February 10, 2015).
28. Yadav, N. S. and R. L. Ray. Industrial aspect. In: *Guar in India*, Kumar, D. and Singh, N. B. (eds.), pp. 201–216. Jodhpur, India: Scientific Publishers (2003).
29. EPA. U.S. Environmental Protection Agency Crop Groups—Federal Register. Environmental Protection Agency (1995). http://www.ir4.rutgers.edu/Other/CropGroups.htm (accessed February 9, 2015).

30. Tripp, L. D., D. A. Lovelace, and E. P. Boring. *Keys to Profitable Guar Production*. College Station, TX: Texas A&M AgriLife Research—Texas Agricultural Experimentation Station (1977). www.lubbock.tamu.edu/files/2011/10/keysguard-prod1977-1.pdf (accessed February 10, 2015).
31. Zhang, H., B. Raun, and B. Arnall. *OSU Soil Test Interpretations*. Stillwater, OK: Oklahoma State Cooperative Extension (2014).
32. Kelley, M., Keeling, W., and G. Morgan. *High Plains and Northern Rolling Plains Cotton Harvest Aid Guide*. College Station, TX: Texas A&M AgriLife Extension Service (2013).
33. Trostle, C. and D. Auld. Agronomy of guar (*Cyanopsis Tetragonoloba*) in the southwest U.S. In: *Agronomy Abstracts*, American Society of Agronomy, Madison, WI (2013).
34. Hassan, S. Pakistan's guar is as good as India's, says agribusiness investor. http://tribune.com.pk/story/657018/pakistans-guar-is-as-good-indias-says-agribusiness-investor/ (accessed February 9, 2015).

13 Characterization of the Properties of Guar Gum to Improve Hydraulic Fracturing Efficiencies

Noureddine Abidi and Sumedha Liyanage

CONTENTS

13.1 Overview 228
13.2 Introduction 228
13.3 Extraction of Guar Gum 228
13.4 Chemical Structure of Guar Gum 230
13.5 Solubility of Guar Gum 231
13.6 Hydrolysis of Guar Gum 232
13.7 Importance of the Characterization of Guar Gum 233
13.8 Determination of Mannose to Galactose Ratio 233
13.9 Determination of Degree of Polymerization and Molecular Weight of the Gum 236
13.10 Study of Rheological Properties of the Gum 237
13.10.1 Effect of Concentration on Viscosity 237
13.10.2 Effect of Temperature on Viscosity 238
13.10.3 Effect of pH on Guar Gum Viscosity 238
13.10.4 Effect of Impurities on Guar Gum Viscosity 238
13.11 Study of the Thermal Stability of the Gum 240
13.12 Application of Fourier Transform Infrared Spectroscopy for Guar Gum Characterization 242
13.13 Investigation of Crystallinity of Guar Gum 245
13.14 Investigation of the Effect of Salt on the Viscosity of Guar Gum 246
13.15 Modification of Guar Gum 247
13.16 Summary and Closing Remarks 248
References 248

13.1 OVERVIEW

Analytical methods to characterize the material properties of Guar are reviewed in this chapter. This understanding is critical to guide improvements in hydraulic fracturing operations.

13.2 INTRODUCTION

In the last chapter, large-scale issues related to cultivating guar and increasing its production in the southwestern United States were discussed. While guar is a drought-tolerant crop, its yields and material characteristics vary widely due to genetic diversity and is affected by a variety of factors including rainfall amounts, soil fertility, and cropping practices. Understanding and characterizing material characteristics of guar is critical to improving the performance of guar-based gels used for hydraulic fracturing. A variety of analytical methods can be used for chemical characterization of guar in order to understand its structure, strength, and rheological properties. This fundamental understanding can in turn be used to guide improvements in hydraulic fracturing engineering operations. This chapter introduces the chemistry of guar and discusses various state-of-the-art analytical methods for characterizing its properties.

13.3 EXTRACTION OF GUAR GUM

Guar seed is mainly composed of three different components, the endosperm, the hull, and the germ. Figure 13.1a and b shows a cross section of a guar seed and endosperm powder.

In general, guar splits or endosperms are obtained after separation of the hull and the germ. Several other processes have been developed to extract guar gum. In the laboratory, this can be done by soaking the raw guar seed in water for 12 h and then separating the germ and the hull manually. The endosperms are then collected, dried at 105°C for 20 min, and ground into the required particle size.[1] The difference in the hardness of the seed components after water sorption is used in this

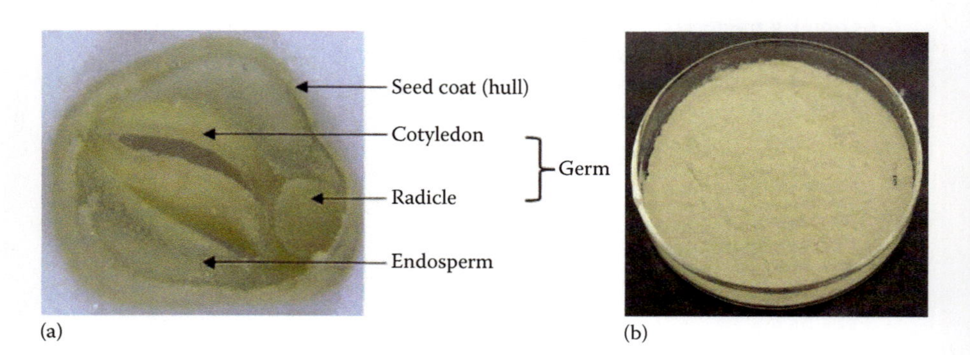

FIGURE 13.1 (a) Cross section of guar seed, (b) endosperm powder.

method. Otherwise, seeds can be loosened by sending them rapidly through a flame and then the loosened hull and germ particles are removed by scouring and milling operations, respectively.[2] Some commercial scale productions first roast the seeds in a furnace to loosen the seed coat and then break the seeds using specific mills. The hull particles remaining in the endosperms are removed by further milling operations.[3]

Crushed endosperm powder can be used in industrial applications without any purification. However, insoluble particles coming from the germ and hull residues of the seeds and proteins may have an adverse impact on the properties of guar gum solutions. Thus, a final heat treatment is required to inactivate any remaining enzymes in the germ fragments. Moreover, the endosperm is composed of several other compounds and pure galactomannan accounts only for 86% of the endosperm. This galactomannan is composed approximately of 36.6% of D-galactose anhydride and 63.1% of mannose anhydride.[2] Therefore, purification is required to isolate the galactomannan fraction from the endosperm.

Numerous methods have been used to purify guar gum. The most commonly reported procedure is based on ethanol/alcoholic precipitation of guar galactomannan from an aqueous solution of crushed guar seeds, guar endosperm powder, or commercial guar gum (Figure 13.2a and b). However, filtration or centrifugation at high speed is required to remove insoluble particles prior to ethanol precipitation to achieve high purity. When guar gum is dispersed in water, the galactomannan molecules are surrounded by water molecules and the polymer is dispersed in water through hydrogen bonding. The addition of ethanol breaks hydrogen bonds between the water and the galactomannan molecules. Then, the ethanol molecules link to water molecules, making a precipitate of the polymer.[4] Repetitive ethanol precipitation can be used to achieve a high level of purity.[5] Lubambo et al.[4] reported that the addition of ethanol to aqueous guar solution and the addition of guar solution to pure ethanol result in different cumulative molecular weights.

Complexation of the gum with Ba^{2+} and Ca^{2+} salts or dialysis and membrane filtration can also be used as purification methods.[3] However, the chemicals used in the

(a)

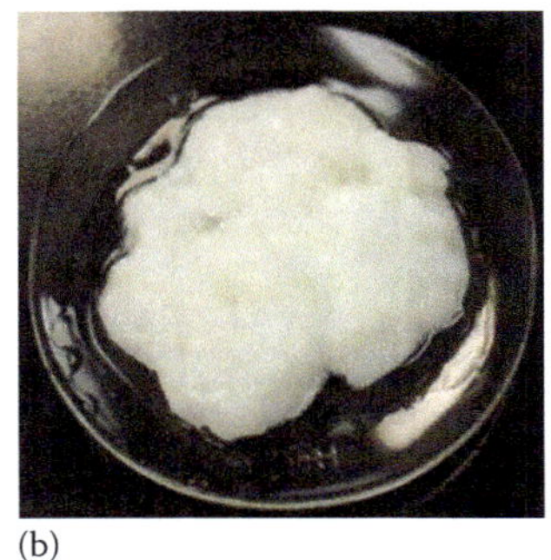

(b)

FIGURE 13.2 (a and b) Purification of guar gum using ethanol precipitation.

purification may contaminate the final product. The purified sample may contain Cu(II) when Fehling solutions are used for purification.[6] Even though copper is an essential nutrient for all living beings, larger doses may be poisonous. Therefore, care should be taken when selecting a method for purification to minimize contamination and to optimize the yield. The purity and the properties of the purified guar gum, such as intrinsic viscosity and Huggins' coefficient, can be greatly influenced by the method of purification. Furthermore, Chudzikowski[2] states that the properties of natural guar gum, such as purity, particle size, viscosity potential, rate of hydration, and dispersion properties depend on their commercial grades. Preservatives such as sodium azide are added to guar solutions to minimize the microbial and enzymatic degradation of the polymer. The preservatives are selected according to the final use of the gum.

13.4 CHEMICAL STRUCTURE OF GUAR GUM

Galactomannan, a common polysaccharide found in most legume seeds (such as alfalfa, fenugreek, and locust bean), serves as a food reserve for germination.[2,7] Galactomannans extracted from different sources differ in mannose to galactose ratio,[3] galactose residues distribution along the mannose backbone, molecular weight, and molecular weight distribution (MWD).[8] Coffee (*Coffea arabica*) and coconut (*Cocos nucifera*) are two common examples for nonleguminous galactomannan sources.[3] Galactomannan-type hemicelluloses are said to be deposited inside the primary cell walls of the storage tissues, and the proposed pathway of galactomannan synthesis is reported by Buchanan et al.[9]

The structure of guar galactomannan has been identified by acid hydrolysis followed by paper chromatography, methylation and oxidation, and enzyme hydrolysis (Figure 13.3). Chemically, guar gum is a heteropolysaccharide consisting of a linear chain of (1 → 4)-linked β-D-mannopyranosyl sugar units (M) with (1 → 6)-linked α-D-galactopyranosyl sugar (G) as single unit side chains.[5,8,10] Even though, β-D-mannopyranosyl units are bonded through (1 → 4) linkages, Mahammad et al.,[11] considered that there are three types of β-(1 → 4) bonds along the mannose backbone. Depending on the α-D-galactopyranosyl substitution, there are β-(1 → 4) linkages

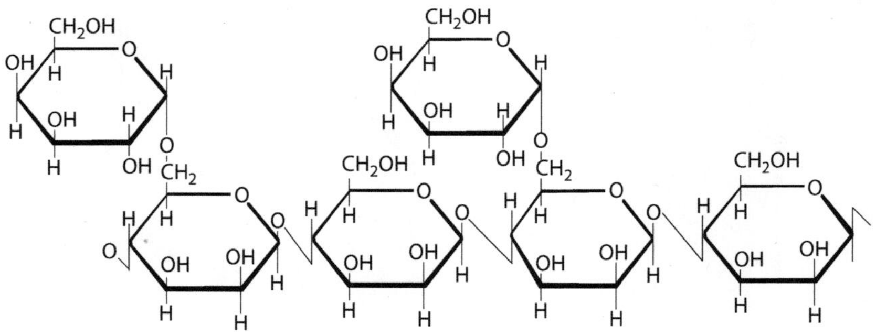

FIGURE 13.3 Chemical structure of guar gum.

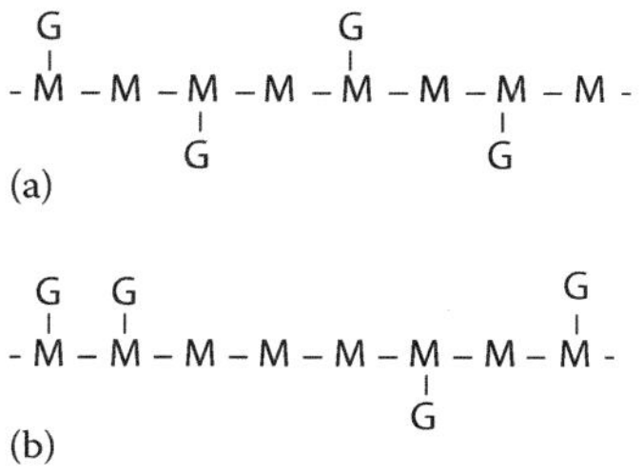

```
 G               G
 |               |
-M - M - M - M - M - M - M - M -
         |               |
         G               G
(a)

 G   G                       G
 |   |                       |
-M - M - M - M - M - M - M - M -
                     |
                     G
(b)
```

FIGURE 13.4 (a) Regular and (b) random distribution of D-galactose (G) on D-mannose (M) backbone in galactomannan molecule.

between two unsubstituted mannose units, between two substituted mannose units, and between a substituted and an unsubstituted mannose unit.

Guar gum has an overall ratio of mannose to galactose in the range of 1.6–1.8.[6,12] It was believed that galactose side groups are attached at regular intervals (Figure 13.4a). However, recent studies showed rather random distribution (Figure 13.4b). Many studies showed that the ratio of mannose to galactose has an important impact on the solubility of the guar gum,[13] chain stiffness, and spatial extension of the polymer.[14] The galactose poor regions of galactomannan molecules form partially crystalline regions due to inter- and intramolecular associations.[14]

13.5 SOLUBILITY OF GUAR GUM

Galactomannans are insoluble in most organic solvents but soluble even in cold water due to the presence of a large number of hydroxyl groups in the molecule. When dispersed in water, guar gum hydrates rapidly forming a colloidal solution with unusually high-viscosity characteristics even at very low concentrations[2] It attains full viscosity in cold water and hydrates rapidly in warm water. However, improperly dispersed lumps of guar encapsulate within the hydrated outer layer, avoiding further water penetration. This reduces the viscosity of the solution. Therefore, vigorous stirring of guar powder sprinkled in water is required in order to get a uniform solution. Guar flour can be mixed with additives like sugar or pre-retardants such as alcohol and glycerin before it is mixed with water to obtain better dispersion.[2] According to Gittings et al.,[15] dissolution by hand shaking of the solution immediately after adding guar powder to water followed by rotating on a spinning wheel at higher speeds, is preferable over the solution preparation using a magnetic stirrer. The authors noted that the aggregation of the gum particles seems to be facilitated by the magnetic stirring, ending up with larger and more numerous aggregates.

In general, the solubility of the galactomannan is decreased with lesser amounts of galactose side branches. The approximate galactose to mannose ratio in guar gum is 1:2[16] and 1:4 in locust bean gum. It makes guar gum readily soluble even in cold water and locust bean gum less soluble in water.[2] The pure polymannose without galactose side branches is completely insoluble in water and identical to polyglucose chains in cellulose having a stiff helical structure that is stabilized by hydrogen bonding. Galactose branches keep the polymer chains far apart and prevent the formation of helical structures or covalent bonding between the main polymer chains and, therefore, facilitate the solubility by weakening the intermolecular interactions.[3,13] It has been reported that the fully substituted (1:1) galactomannan chain is stiffer than the (1:2) chain, but it is more flexible than the unsubstituted chain. The distribution of galactose subunits along the backbone has great influence on the conformational characteristics and flexibility of the chain; the substitution pattern and the mannose to galactose ratio depend on the plant source.[14]

The very high viscosity attained at low concentrations makes guar gum an excellent thickener in the food industry, for example. Guar is a nonionic gum, which is not affected by ionic strength[16] or stable over wide range of pH.[2] However, the highest rate of hydration is observed between the pH value of 6–9 and the lowest hydration is observed at pH 3.5. Hence, guar can be easily hydrated in cold water at pH 3.5 and the rate of hydration can be increased by changing the temperature to 80°C and pH 8.[2] Even though guar gum can be hydrated in solutions with lower pH values, the acetal linkages may be hydrolyzed due to the acidity. Guar gum shows higher acid stability compared to locust bean gum or most of the other natural galactomannan gums due to the presence of a large number of 1–6 glycosidic bonds that are easily hydrolyzed by acids. It has been reported that salts influence the solubility of guar gum. Detailed information on the influence of salt on the behavior of guar gum is discussed in Section 13.14.

An aqueous guar solution is composed of both fully dispersed guar molecules and partially dispersed small clusters of molecules, which increase in size with time.[17] Those partially dispersed particles can impact the light scattering and rheological properties of the solution. Therefore, the effect of different pressure and temperature conditions was investigated to maximize the solubility of three different purified commercial food grade guar gum solutions. Samples were analyzed using viscometer and light scattering data. The results show that the intrinsic viscosity of the guar solutions seems to decrease with time at a given temperature when there is no added pressure on the samples. Similarly, the increase in temperature results in the reduction of intrinsic viscosity with time when there is no applied pressure. It has been clearly observed that the intrinsic viscosity of aqueous solutions of guar gum is significantly affected by the solvent pressure.[17] The authors reported that the best dissolution of guar gum can be obtained when guar is dissolved at 130°C with the presence of 4–12 bar pressure for 10–40 min. The pressure disrupts the polymer aggregation in the solution.

13.6 HYDROLYSIS OF GUAR GUM

Hydrolysis of guar gum is often required before characterizing with different analytical techniques and for several industrial applications. Acid hydrolysis, enzymatic hydrolysis, and noncatalytic hydrolysis under hydrothermal conditions have been reported. Trifluoroacetic acid (TFA), sulfuric acid, and hydrochloric acid are the main acids used in the hydrolysis process. When hydrolyzing guar gum, the cleavage of $\alpha(1 \rightarrow 6)$ acetal linkage liberates galactose monomers to the system and it may reduce the solubility without showing significant reduction in the molecular weight.

On the contrary, cleavage of $\beta(1 \rightarrow 4)$ acetal linkage between mannose sugar units in the mannose backbone shows sudden decrease in the molecular weight and change in the viscosity of the solution without showing significant change in the solubility.[13] Furthermore, each cleavage of $\alpha(1 \rightarrow 6)$ acetal linkage liberates one galactose unit. Therefore, the galactose concentration in the solution increases significantly just after starting the hydrolysis. However, the cleavage of $\beta(1 \rightarrow 4)$ acetal linkage does not necessarily release a mannose unit to the system except from the chain scission at the far most ends of the polymer chains. Thus, there is no sudden increase in the mannose content of the solution at the beginning of the reaction. The enzymatic hydrolysis of guar gum may provide highly variable mixture of products compared to acid hydrolysis due to the site specificity of the enzymes. Enzymatic degradation is a time-consuming process compared to hydrothermal degradation and concentrated acid hydrolysis. Free radical degradation of guar gum using hydrogen peroxide and potassium persulfate (KPS) under both thermal and microwave conditions has been reported.[18]

13.7 IMPORTANCE OF THE CHARACTERIZATION OF GUAR GUM

There is an increased interest in guar gum and its application particularly in food and hydraulic fracturing industries. Therefore, numerous efforts have been made during the last few decades to characterize guar gum using different analytical tools and to understand the chemistry of the guar galactomannan, which govern the vital properties of guar gum. The chemical and physical characterization of the guar gum involves the determination of the mannose and galactose contents; determination of the molecular weight, MWD, and degree of polymerization; determination of the viscosity and other rheological properties; investigation of the thermal stability; and finally investigation of the impact of salt on the viscosity of guar gum solution.

13.8 DETERMINATION OF MANNOSE TO GALACTOSE RATIO

As mentioned earlier, the galactose to mannose ratio, which is a very important determinant of galactomannan behavior, can be determined using chromatographic methods; the galactose to mannose ratio has great impact on the solubility of guar gum and a dramatic decrease in solubility can be observed with lower numbers of galactomannan side chains. Several chromatographic methods can be used such as gas chromatography and liquid chromatography. High-performance liquid chromatography (HPLC) and gel permeation chromatography (GPC) or size exclusion chromatography come under the liquid chromatography category. In chromatographic methods, a mobile phase carries a sample through a stationary phase. In this process, the molecules in the sample are separated into their components while moving through the column. Those compounds are sensed by the detector located at the end of the separation column. The plot of the detector's response and the retention time is known as a chromatogram. Then, the molecules are identified by their retention time in the column because different molecules have different retention time.

HPLC is a widely used chromatographic method due to its speed and separation power. Sample preparation includes several steps depending on the nature and composition of the sample to be analyzed. HPLC is composed of finely divided stationary phase and liquid mobile phase. The cylindrical shaped separation column is generally packed with porous spherical particles. Generally, porous silica, where each pore is covered with the stationary phase such as C_{18} groups,[19] is found inside the separation column. The liquid mobile phase is highly pressurized through the stationary phase to maintain a satisfactory flow. The sample molecule can flow with the mobile phase and may go through the pores while moving through the column. The stationary phase will interact with the materials that pass through the stationary phase differently. Different types of molecules may differ in time to pass through the column depending on the reactivity of the molecule with the stationary phase. Sample molecules reach the detector depending on the reactivity of the molecule with the stationary phase. Generally, the same type of molecules take similar elution time. Both identification of molecules by the retention time and quantification using a well-developed calibration curve are possible with HPLC. Figure 13.5 shows typical chromatogram obtained when analyzing a sugar solution by HPLC. Clearly separated peaks indicate that HPLC can be used for a reliable identification and quantification of galactose and mannose in galactomannan.

In chromatographic methods, a compound has to be fully hydrolyzed before analysis. Acid hydrolysis using TFA, sulfuric acid, or hydrochloric acid has been reported. There are few studies related to chromatographic analysis of guar galactomannan.[8] The authors used different stationary phases, mobile phases, and hydrolyzing techniques. An aqueous solution of guar gum was completely hydrolyzed using 0.5 M sulfuric acid at 100°C for 3 h and analyzed using high-performance anion exchange chromatography (HPAEC).[8] Sodium hydroxide solution 0.02M and

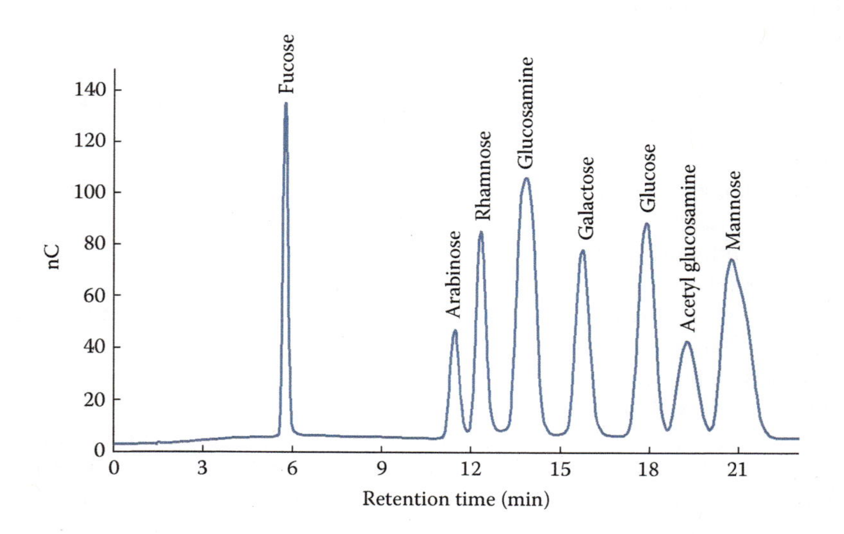

FIGURE 13.5 Typical chromatogram obtained when analyzing a solution containing different carbohydrates using HPLC. Mannose and galactose are well separated.

0.02 M sodium hydroxide solution/1 M sodium acetate solution were used as eluents A and B, respectively, at 35°C. The authors used a pulsed amperometric detector and CarboPac PA1 column (4 mm I.D. × 250 mm long) and obtained mannose to galactose ratio of 1.44 for a commercial guar gum sample.

The effect of the purification method on the characteristics of guar gum was studied by Cunha et al.[6] The authors purified commercial guar gum using four different methods. All purified samples and nonpurified samples were separately hydrolyzed using TFA at 100°C for 5 h and then analyzed using a gas liquid chromatography. The authors used Hewlett Packard chromatograph (model 5890A series II), BAScientific DB-225 column (0.25 mm I.D. × 30 m), acetone as the eluent, and nitrogen as the flow gas. The results show that the corresponding mannose to galactose ratios vary between 1.61 and 2.53. The commercial guar sample without any purification had 1.67 of mannose to galactose ratio.

Hansen et al.[20] analyzed the composition of guar gum using HPLC after hydrolyzing 0.2 g of guar galactomannan samples using 2 mL of 2 M TFA in a pressure cooker (approximately 105 kPa) for 1 h. Bio-Rad HPX-87P column at 85°C with water as the mobile phase (60°C) at a flow rate of 0.7 mL/min was used in this study. The gum extracted from several discolored samples due to wet weather during maturation was analyzed. Samples showed mannose to galactose ratios varying from 1.27 to 1.71. Hansen et al.[20] reported that the extraction methods of guar gum affect the gum content. Furthermore, the method of purification has significant impact on mannose to galactose ratio.[6] Similarly, mannose to galactose ratios of galactomannans from five different plant sources, including commercial guar gum, were estimated using gas chromatography. The authors reported a ratio of 1.80 for commercial guar gum. The study was accomplished with flame ionization detector and 30 m × 0.25 mm DB-225 column.[21]

Previously, it was believed that there was a galactose side branch at every other mannose unit and the mannose to galactose ratio was 2:1. However, recent studies confirmed that the mannose to galactose ratio is less than 2, and galactose molecules show random distribution on the mannose backbone. The ratio is below 2 and most probably is around 1.6. As discussed in the previous paragraph, the mannose to galactose ratio is influenced by the method of purification. However, there are several other factors that affect the mannose to galactose ratio, such as type and concentration of hydrolyzing agent, reaction temperature, and reaction time and most importantly the variation in guar germplasm. Therefore, the mannose to galactose ratio, which is an important determinant of the guar gum behavior, should be compared using several cultivars. Complete characterization with HPLC will help establish a relationship between the mannose to galactose ratio and other vital behaviors of guar galactomannan solutions. Since commercial guar gum does not come with varietal information, a comprehensive study with known guar varieties should be carried out to compare their existing properties and mannose to galactose ratio. New varieties can be developed through breeding programs to obtain improved properties. Furthermore, HPLC can be used to study guar gum degradation kinetics under different conditions, such as acid, enzymes, and thermal conditions and shear rates by measuring variation of the mannose and galactose content in the solution with time.

13.9 DETERMINATION OF DEGREE OF POLYMERIZATION AND MOLECULAR WEIGHT OF THE GUM

The molecular weight (MW) and the MWD of guar gum, two highly important parameters in the characterization of this polysaccharide, are studied using GPC. As discussed in the previous section, chromatography is an analytical technique in which different types of molecules of a sample are separated based on the rate at which they are carried through the stationary phase by the mobile phase. Unlike HPLC, the sample molecules do not interact with the stationary phase in the GPC and the separation is based only on the size of the molecule. The small molecules have higher retention time as they go through most of the small pores in the stationary phase, while larger molecules have lower retention time as they skip the small pores in the stationary phase. The detector senses how many molecules come out of the stationary phase at a given time. It is possible to determine the molecular weight of a polymer depending on the elution time and a proper calibration curve. The average degree of polymerization of a polymer with known monomer unit can be calculated by dividing the total weight of the polymer by the molecular weight of the monomer. Similarly, the MWD of an unknown sample can be investigated using GPC. GPC is also considered as the most important technique for studying the degradation kinetics of a polymer under different conditions. There are several studies reported on molecular weight and MWD of guar gum. We will discuss those studies briefly in the next few paragraphs.

Mudgil et al.[22] reported that the average degree of polymerization of natural guar gum estimated using intrinsic viscosity measurements is around 3295. The partially hydrolyzed guar gum prepared using cellulase enzyme has shown lower average degree of polymerization compared to the original guar gum (~29). Three different acid hydrolysis methods and enzymatic hydrolysis method were used and compared to study the variation of molecular weight and MWD of guar gum due to hydrolysis.[12] The authors developed an acid hydrolysis procedure to obtain larger quantities of various molecular weight guar gum solutions with consistent molecular weight. The GPC was used to determine the molecular weight variation in order to study the effectiveness of the hydrolysis method. The three columns used in the GPC were TSK G3000PWXL, TSK G6000PWXL (TosoHaas, Montgomeryville, PA), and GDNA-PW as the guard column. The columns were calibrated using both guar and pullulan samples. Degraded guar solutions (0.05% [w/w]) were sent through the columns at 40°C using 55 mM sodium sulfate (Na_2SO_4) and 0.02% (w/w) sodium nitrite (NaN_3) as the mobile phase at a flow rate of 0.6 mL/min. The weight average molecular weight and number average molecular weight of the nondegraded guar sample were 1.98×10^6 and 1.2×10^6 g/mol, respectively.

Cunha et al.[6] analyzed crude guar sample and different purified guar samples obtained from the same crude guar sample using GPC. The authors concluded that the molecular weight is greatly influenced by the method of purification. The molecular mass estimated using intrinsic viscosity measurements also confirmed that the molecular weight is reduced from 1.88×10^6 to 0.51×10^6 g/mol with increasing level of purification. The GPC curves shifted toward lower molecular weight region. The high molecular weight compounds in a solution travel rapidly through the stationary phase and give peaks in the lower elution volume region and vice versa. This study used water as the eluent and pullulan sample was used as a standard. Furthermore, the study was carried out using sample concentration of 0.1% (w/v) at a flow rate of 0.5 mL/min using an ultrahydrogel linear column at room temperature.

Miyazawa and Funazukuri[8] obtained the MWD of hydrothermally degraded guar sample at 200°C after 10, 20, and 25 min of reaction period using a GPC on a HPLC system. The GPC column was a SB-803 HQ (Shodex, Tokyo, Japan) and 0.05 M sodium nitrate solution at 40°C with flow rate of 0.5 mL/min was used as the eluent. The shift of the molecular weight toward the lower molecular weight with the degradation time was clearly observed in this study. Barth and Smith[23] characterized guar gum with size exclusion chromatography using 100–4000 Å SynChropac column, 0.7 ionic strength acetate buffer as the mobile phase at a flow rate of 0.5 mL/min. Mahammad et al.[11] used a GPC to evaluate their newly developed mathematical model for predicting the kinetics of MWD during enzymatic de-polymerization of the guar galactomannan. Similar procedures can be applied to study the guar galactomannan stability under different conditions such as heat, flow, shear stress, and other commonly used conditions in the food and hydraulic fracturing industry.

13.10 STUDY OF RHEOLOGICAL PROPERTIES OF THE GUM

Guar galactomannan possesses remarkable rheological properties, which allow it to be used as a thickener to control viscoelastic properties in food, personal care, and oil recovery industries.[14] Rheological characterization of guar gum is critical to determine its linear and nonlinear rheological properties over an extended range of thermodynamic properties such as deformation, concentration, and temperature. Rheological measurements such as viscosity, elastic, and storage moduli can be carried out using properly calibrated rheometers. However, the viscosity of guar gum solution decreases when the gum undergo microbial degradation under favorable conditions. Therefore, guar gum solutions should be analyzed within 24 h of preparation unless some preservatives such as sodium benzoate, benzoic acid, formaldehyde and substituted phenols, etc., are added.[2] Tiwari et al.[24] noted that guar gum solution can be fermented and enzymatically hydrolyzed by microorganisms. Therefore, guar solutions should be treated with preservatives depending on the ultimate use of the product.

Most of the guar gum applications benefit from the remarkable rheological properties of the gum. Guar gum can acquire very high viscosity at low concentration in water. Viscosity is widely used as a performance index for comparing different grades of guar gum. In addition, viscosity comparison is normally conducted at 1.0% (w/v) concentration.[2] Concentration, dispersion, temperature, pH, and presence of impurities are the major determinants of viscosity development of guar gum.

13.10.1 Effect of Concentration on Viscosity

When the concentration increases up to about 1.5%, an exponential increase in the viscosity is observed, and further increase in the concentration up to 3% showed a gradual increment in viscosity.[2] The author reported that the increase in the concentration to more than 3% results in a gel-like paste, which does not have the properties of a true structural gel. Furthermore, it was shown that guar gum requires low concentration to develop particular viscosity compared to most of the other hydrocolloids such as tragacanth, sodium alginate, and sodium carboxymethyl cellulose (high viscosity) at the same temperature.

13.10.2 Effect of Temperature on Viscosity

The rate of hydration and viscosity development can be accelerated by increasing the temperature up to an optimal level. Particular guar grade which takes 5 h to develop full viscosity may take about 10 min to reach maximum viscosity at 80°C.[2] Guar sols show gradual decrease in viscosity as the temperature increases. Furthermore, the author mentioned that the guar gum has comparatively higher retention of viscosity compared to most of the other gums at high temperature. Furthermore, guar gum can be kept at high temperature for a long period of time without much effect on the ultimate viscosity at room temperature. Similarly, Ma and Pawlik[25] observed a reduction in guar solution viscosity with increasing temperature. The solutions gained the initial viscosity when cooling down to room temperature. The intrinsic viscosities of guar solution were 10.6 and 10.3 dL/g at 70°C and 90°C, respectively. According to Moser et al.,[26] guar gum solution showed pseudoplastic rheological behavior with good stability upon freezing and thawing. The modification of cationic guar gum (CGG) with fluorinated monomer units showed better heat resistance compared to the initial material.[27]

Moreover, variation in viscosity with temperature is of a great concern in both food and hydraulic fracturing. Cunha et al.[6] studied the variation of viscosity with increasing temperature within 20°C–40°C at constant shear rate of 20 S^{-1}. The activation energy of flow, the energy required for a particle to move from one equilibrium position to the next equilibrium position, was calculated from the slope of the graph drawn between the viscosity and the reciprocal of temperature. Flexibility and interaction between molecules are the main factors that affect the activation energy. The purification methods showed a great influence on the activation energy of guar gum solutions. The activation energy was 16.6 and 25.8 kJ/mol for unpurified and highly purified guar gum sample, respectively.

13.10.3 Effect of pH on Guar Gum Viscosity

Guar gum is stable over the entire usable range of pH (1–10.5) due to its nonionic nature, as discussed earlier. The maximum viscosity is reached at pH 7–9 range and practically stable range of 4–10.5.[2] However, the viscosity decreased with the presence of strong bases or strong acids.[16] Galactomannan is depolymerized in mild acidic conditions.[13]

13.10.4 Effect of Impurities on Guar Gum Viscosity

When a compound with strong affinity to water is present in guar gum solution, the gum molecule and the aforementioned dissolved substance will compete for water in the solution and form a galactomannan precipitate. That is why galactomannan can be precipitated by adding water miscible solvent to guar solutions such as alcohol, acetone, or glycerin in the purification process. Guar solutions are unaffected by electrolytes, and there is no effect of hard water. However, galactomannan molecules can be electrically charged by absorbing ions from the solution. Thus, it can affect the ultimate properties such as viscosity and rate of hydration. Sodium benzoate increases the rate of hydration and viscosity while sodium sulfate inhibits the hydration and the viscosity.[2] Proteins, fibers, and pentosans found as impurities in the crude guar gum[3] may also influence viscosity development of guar gum. Similarly, the presence of additives such as salts and co-solvents in the guar gum solution can affect the viscoelastic and structural properties. Hence, cone and plate rheometer were used to study the rheological properties of guar gum solutions with different concentrations at different temperatures without any additives.[14] The apparent viscosity of guar gum can be positively changed by increasing the concentrations of polyols that are the main category of additives commonly used to improve texture, nutritional value, and flavor of foods.

Guar gum solutions are used under high shear conditions in both hydraulic fracturing and food industry. Therefore, viscosity and rheological behavior should be studied under different shear rates. It has been reported that the viscosity of guar gum solutions has strong dependency on shear rate.[2] However, it is preferable to have constant viscosity with the increase of shear rate in industrial applications. The effective viscosity of non-Newtonian fluids varies with the shear rate where the effective viscosity of Newtonian fluids does not vary with the shear rate. Intrinsic viscosity measurement of 0.02%–0.08% aqueous solution of food grade guar gum was published by Mudgil et al.[22] The intrinsic viscosity of natural guar gum at 25°C is 9 dL/g, and it drastically changes to 0.28 dL/g due to enzymatic hydrolysis of the polymers. Mudgil and coworkers[22] further studied the flow behavior index (FBI) of 1% guar solution at 25°C. They reported a value of 0.3094, where FBI below 1.0 represents non-Newtonian behavior. These authors reported that 1% aqueous solution of guar gum has pseudoplastic behavior and pseudoplastic behavior decreases as the guar concentration decreases. However, low molecular weight partially hydrolyzed guar gum showed a Newtonian behavior. The authors stated that these differences in rheological behavior may be due to the vast difference in the molecular weight in the two samples. Furthermore, the increase in shear rate could remove the entanglement of the molecules and reduce the viscosity of the solution. Cunha et al.[6] reported on the effect of shear rate and temperature on viscosity of aqueous solutions with 0.1% and 1% (w/v) guar concentrations. Even though 1% solution of guar gum can have a pseudoplastic (shear thinning) behavior, only three out of five samples showed pseudoplastic behavior at concentration of 1% (w/v). Newtonian behavior was reported for the other two samples at the same concentration. This study showed that the rheological behavior of guar solution is greatly affected by the method of purification. However, at 0.1% (w/v) concentration all samples showed the expected Newtonian behavior; no variation in the viscosity was found with the increase of shear rate at 25°C for all samples. The intrinsic viscosity reported for 1% guar solutions in this study was 12.5 dL/g, and there is a significant reduction in viscosity with increasing level of purification.

Rheological study with storage time is a very important parameter since guar gum is used in various industries where the chemical and physical stability of the gum are of concern. When guar gum is used in chemically active environments such as textile, ink, or pharmaceutical industries, the stability of the gum changes as it reacts with other compounds in the solution. Two types of aqueous solutions of commercial gum, including guar gum, were studied for 37 days in 1-week intervals to investigate the change in rheological properties with the time. The aqueous solutions were preserved with sodium azide. The apparent viscosity decreased continuously up to 20 days of storage time at low shear rate ($<10\ S^{-1}$). After that, the apparent viscosity remained practically constant. Flow curves practically coincided at shear rate more than 10 S^{-1} without the influence of storage time, indicating that the shear thinning behavior is not affected during the storage time.[28]

The interaction between polymers in a solution can be determined by the Huggins coefficient, which is determined using the slope of the intrinsic viscosity graph. The Huggins coefficient of a normal polymer in a good solvent is around 0.3–0.4, and the increase in the value indicates a high interaction between polymer chains. The aqueous solution of guar gum showed Huggins coefficient of 0.7–0.8. Therefore, the interaction between galactomannan chains is very high compared to other polymers. A dramatic decrease in the Huggins coefficient was observed with the hydrolysis of the gum indicating a loss of intermolecular interactions.[12] According to Cunha et al.,[6] the Huggins coefficient of commercial guar gum sample was 0.94, and it also confirms that guar gum has higher intermolecular interactions. Furthermore, three purification methods out of four have resulted in a lower Huggins coefficient, which indicates that the purification method has greater impact on rheological behavior of guar gum. Ma and Pawlik[25] reported that, based on the Huggins coefficient values, guar gum chains can aggregate in distilled water and in dilute solutions of lithium chloride, potassium chloride, sodium chloride, and cesium chloride.

A kinetic study on guar gum degradation demonstrated that guar gum hydrolysis is far more complex than expected and the degradation continues long after the initial loss of viscosity and it produces a mixture of different molecular weight and mannose to galactose ratio.[13]

13.11 STUDY OF THE THERMAL STABILITY OF THE GUM

Thermal stability is a major factor in selecting water-soluble polymers. Therefore, it is important to evaluate the thermal properties of the galactomannans using thermogravimetric analysis (TGA). TGA is the most widely employed thermal technique for the characterization of materials.[29] TGA thermogram provides weight loss in the material with increasing temperature (Figure 13.6). Weight loss is directly related to the composition of the material, because the constituents are removed from the system differently as they reach their degradation temperature. The weight loss could be associated with the evaporation of volatile substances, thermal decomposition with the formation of gaseous product in an inert environment, or oxidative decomposition of material in air, etc.[30] The thermal behavior of the polymer can be determined using the percent weight loss and the peaks of decomposition. Moreover, the thermograms of different material can be compared using the derivative thermogram, which shows

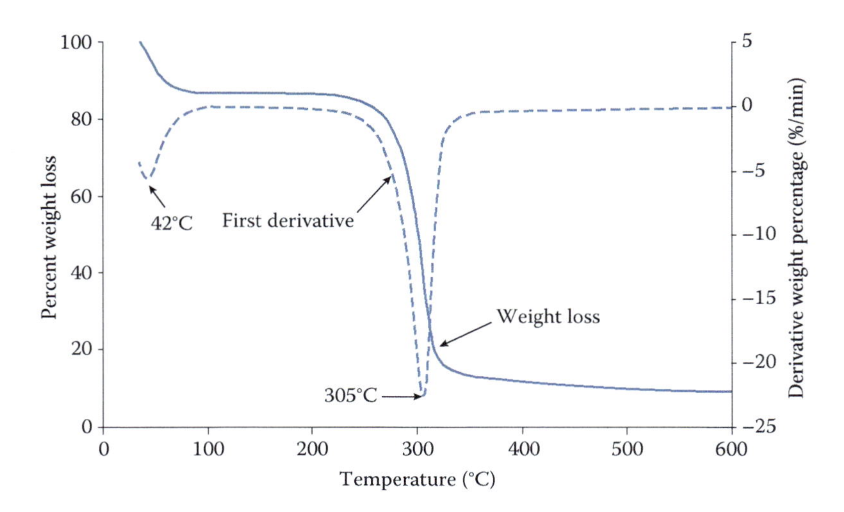

FIGURE 13.6 TGA thermogram of GGM.

the rate at which the weight changes. However, the sample preparation should gain additional attention in order to obtain uniformity in particle size and sample size to have a better comparison of the thermal behavior of different materials.[30]

The weight loss can be recorded by maintaining the sample at a constant temperature for a period of time. TGA helps determine characteristics of polymer such as degradation temperature, amount of absorbed water, and, most importantly, the thermal stability of materials. Differential scanning calorimetry (DSC) is used to study the phase transition of molecules such as melting, glass transition, and crystallization. It measures the heat flow absorbed or released by a sample as a function of time or temperature. This change in heat flow is used to identify thermal transitions in the material as it is cooled down or heated up.

Thermal stability of aqueous solution of guar gum is of critical concern in both food and hydraulic fracturing industries, which employ a wide range of temperatures. The thermal degradation is dependent on the availability of oxygen.[31] Since the elimination of oxygen is not feasible in the field or industrial situations, the behavior of the polymer under a wide range of thermal conditions should be studied to determine possible alternatives to prevent thermal degradation of the material. Variations of viscosity and rheological properties are widely measured in addition to the variation in molecular weight of the polymer under different thermal conditions. Bradley et al.[31] studied the variation in viscosity with temperature under constant shear stress. However, the authors were unable to determine the changes in viscosity at high concentration due to heat variation because polymers degrade not only due to the heat but also to the flow.

The TGA thermograms of guar samples were obtained with a heating rate of 10°C/min over temperature range of 25°C–900°C.[6] The authors reported that the first dramatic change is observed at 70°C, which is due to the removal of moisture from the sample. Guar gum purified by different methods showed different thermal behaviors and different residual mass at 900°C.[6] Therefore, the methods of purification should be selected according to the expected thermal stability of guar gum. A thermal analysis carried out using three different galactomannan sources confirmed that the thermal behavior of galactomannan is affected by the molecular weight and the mannose and galactose content of the polymer.[32]

The thermal behavior of oxidized guar galactomannan was studied using thermogravimetry (TGA), differential thermogravimetry (DTG), and differential thermal analysis (DTA) in air and nitrogen atmospheres at a heating rate of 10°C/min.[31] The authors reported that the temperatures at the onset of degradation under air and nitrogen atmospheres are the same. However, the final degradation is lower in air atmosphere. The final degradation temperature under air atmosphere is closer to 500°C while the final degradation temperature is >600°C under nitrogen atmosphere. This finding confirms previous reports, which indicated that the thermal degradation is especially dependent on the availability of oxygen.[31] Furthermore, the percent weight loss in the air atmosphere is greater than that in nitrogen atmosphere. The authors observed variations in the peaks of DTG thermograms depending on the atmosphere and the level of oxidation in the guar gum.[33]

A thermogravimetric study was conducted to investigate the insertion of the carboxymethyl group to a guar gum sample. The crude guar gum and purified guar gum samples showed two distinct weight loss regions in the temperature ranges of 25°C–115°C and 230°C–350°C. These weight losses correspond to water evaporation and galactomannan degradation, respectively. The presence of one additional weight loss region in the carboxymethylated guar gum due to the degradation of carboxymethyl group confirms that the guar gum sample was successfully modified with carboxymethyl group.[34] This confirms that the TGA is a good tool for confirming the effectiveness of grafting chemical groups on the polymer or to identify the effectiveness of the chemical modifications. Yadav et al.[35] characterized guar gum graft copolymer using TGA where the partially carboxymethylated guar gum was modified through graft copolymerization of vinyl-sulfonic acid. The modified product showed three distinct regions of degradation outside of the expected weight loss due to the evaporation of the water molecules; the peak temperatures of weight loss located at 233°C, 345°C, and 737°C are attributed to the elimination of CO_2, SO_2 and are due to the removal of hydroxyl groups from the pendent chains attached to the partially carboxymethylate guar gum, respectively. The modified product showed better thermal stability compared to the starting material.

13.12 APPLICATION OF FOURIER TRANSFORM INFRARED SPECTROSCOPY FOR GUAR GUM CHARACTERIZATION

Infrared spectroscopy is a widely used technique to characterize both organic and inorganic material. It utilizes the interaction between the material and the infrared light for the characterization irrespective of the state of matter. When an infrared light interacts with the matter, the chemical bonds absorb infrared radiation in a specific wavenumber range, regardless of the structure of the rest of the molecule, causing the chemical bonds in the material to vibrate. IR studies reported specific wavenumbers related to most of the chemical functional groups found in nature.

Chemical functional groups can have slightly different peak position according to their chemical environments. The wavenumbers that give the peaks in the IR spectra are used to identify the chemical groups found in the unknown material and help determine the chemical structure of the material. Fourier transformation makes this IR spectrum digital and Fourier transform infrared spectrometer (FTIR) is the most commonly used infrared spectrometer. Furthermore, FTIR is considered as the most powerful tool for determining the structure of organic and inorganic compounds. The new generation of FTIR, the attenuated total reflectance (ATR)-FTIR, requires minimal sample preparation and has eliminated the use of traditional potassium bromide (KBr) pellet technique.[36] However, some of the FTIR studies reported in this paper include KBr pellet techniques in sample preparation. Furthermore, FTIR can be used to identify the structural and compositional changes due to the chemical modification of guar gum and for the identification of unknown materials found in guar samples. Figure 13.7 and Table 13.1 show representative IR spectra of guar galactomannan and the corresponding main vibration assignments.

When guar gum is used in chemically active environments, there can be some structural changes in the polymer. Therefore, FTIR can be used to identify those changes. The concentration of the functional groups and the structure of the food material are widely identified using FTIR spectroscopy in the food industry.[26] The authors used FTIR to evaluate the structural changes of the guar gum in frozen foods with the presence of polyols or alcoholic sugars that are commonly used in food industry. The spectra were collected after freezing and thawing cycles with a spectral resolution of 4 cm^{-1}, in the wavenumber range of 3000–1200 cm^{-1}.

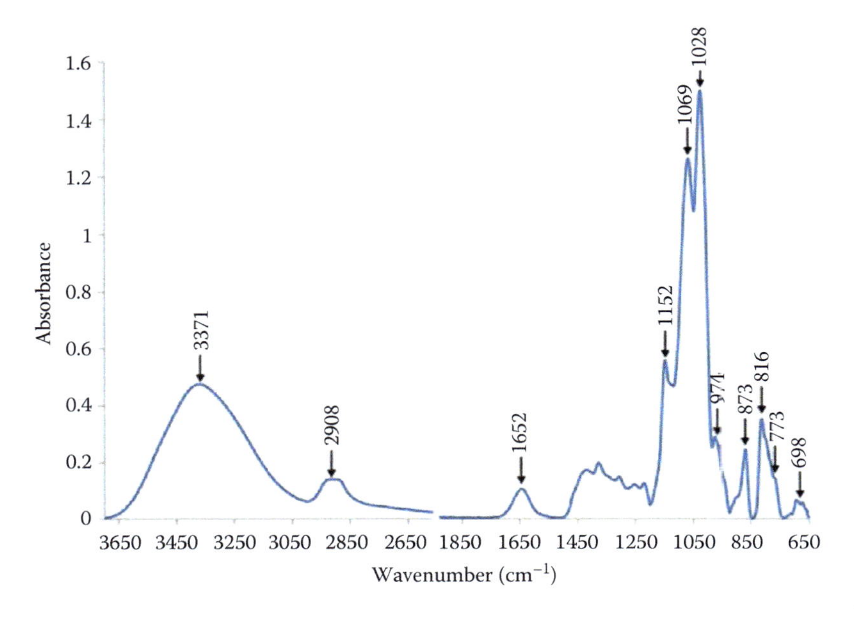

FIGURE 13.7 FTIR spectrum of guar galactomannan obtained using Universal Attenuated Total Reflectance.

TABLE 13.1
IR Assignment of the Main Vibrations in the IR Spectra of Guar Galactomannan

Wavenumber (cm^{-1})	Tentative Assignment
~3371	O–H stretching (intermolecular hydrogen bonding)
~2908	CH group stretching
~1633	Adsorbed water
~1152	C–OH
~1069	Primary alcoholic CH_2–OH stretch
~1028	CH_2 twisting vibration
~873	Galactose and mannose
~773	$\alpha(1 \rightarrow 6)$ linkage

Sources: Dodi, G. et al., *Cell. Chem. Technol.*, 45(3), 171, 2011; Mudgil, D. et al., *Int. J. Biol. Macromol.*, 50(4), 1035, 2012.

It was reported that partially hydrolyzed guar gum using enzymes do not show any functional group modifications and, therefore, the FTIR spectra from both natural guar gum and partially hydrolyzed guar gum are almost superimposable.[37] This study clearly showed that enzymatic hydrolysis significantly affects the physiochemical and rheological characteristics of guar gum without a significant change in the basic structure, hence it can be used as a good source of dietary fiber without changing the viscosity and texture of the food product.

Reddy and Tammishetti[18] used FTIR spectroscopic method to study the functional group modification of guar gum after free radical degradation of the guar gum. FTIR study of guar gum, purified guar gum, and carboxymethyl guar gum showed that the properties of crude guar gum can be improved by purification and carboxymethylation.[34] The authors observed that the intensity of the peak at 3418 cm^{-1}, which is attributed to OH stretching, was reduced with the carboxymethylation, which is an indication of the OH group substitution with carboxymethylation. The study confirmed that FTIR spectroscopy is a good research tool to identify the effectiveness of chemical modifications of guar gum. Similarly, FTIR and NMR studies of fluorinated cationic guar gum (FCGG) confirmed that the fluorinated reactive monomers were successfully introduced.[27] Kumar et al.[38] also used FTIR to investigate the effectiveness of carboxymethylation of guar gum and to study the interaction between the $–COO^-$ groups in carboxymethyl guar gum and $–NH_3^+$ groups in chitosan in interpolymer complex films of chitosan and carboxymethyl guar gum. The availability of additional vibration at 1184 cm^{-1}, which is attributed to $–SO_2\text{-}O^-$, in the partially carboxymethylated guar gum-*g*-poly vinylsulfonic acid confirmed the successful grafting of the vinylsulfonic acid on the partially carboxymethylated guar gum. The shift in the peak at 3450 cm^{-1}, which is attributed to –OH stretching vibration might be due to the grafting of the acid on OH groups.[35]

Many studies carried out using IR spectroscopy confirmed that FTIR study is a very important analytical tool for the characterization of guar gum. However, the applicability of the FTIR spectroscopy should be further studied to characterize and quantify different properties of guar gum obtained from guar gum genotypes since this is a nondestructive and fast analytical technique.

13.13 INVESTIGATION OF CRYSTALLINITY OF GUAR GUM

X-ray diffraction (XRD) is a widely used technique for analyzing crystallinity of samples. The source (usually copper) generates the x-ray, which is focused through a filter on the sample. X-rays are diffracted according to the orientation of atoms in the sample and the intensities of the diffracted radiation are collected by the detector. Then, a recorder plots the intensity against the diffraction angle. This method is widely employed to estimate the percent crystallinity, crystallite size, orientation of crystallites, and structural variation under different conditions.

The degree of galactose subunits along the main backbone influences the properties of the galactomannan molecules. The regions in the galactomannan molecules with less number of galactose side branches are less soluble in water. Furthermore, those areas can associate intra- and inter-molecular bonds to form partially crystalline complexes.[14] The crystallinity index measured with x-ray diffractometer demonstrated very low percent crystallinity for native guar gum (3.86%) and comparatively higher percent crystallinity for partially hydrolyzed guar gum (13.2%).[22] This slight increment in percent crystallinity may be the result of higher reactivity of the hydrolyzing agent (enzyme, acids) in the less organized amorphous regions of the guar sample leaving more crystallized areas. However, the enzymatic hydrolysis of guar gum resulted in negligible changes in the XRD pattern. Chunha et al.[39] reported little increment in the percent crystallinity of cross-linked guar gum. High levels of crystallinity have also been observed with acetylated guar gum.[40] The crystallinity showed a positive relationship with the degree of substitution of acetyl group. The authors reported that natural guar gum forms a large number of random hydrogen bonds due to the availability of hydroxyl groups. Therefore, natural guar gum shows lower levels of organization. However, the random hydrogen bond formation reduces with acetylation and increases the order of molecules. Therefore, acetylated guar gum shows higher levels of crystallinity. Wang et al.[27] reported that the crystallinity of the CGG is reduced from 3.01% to 2.5% with the introduction of fluorinated reactive monomers. Further reduction of crystallinity was observed (up to 0.67%) with increasing levels of the degree of substitution. According to the authors, the crystallinity reduction could be attributed to the replacement of the hydroxyl groups that create hydrogen bonds between free hydroxyl groups of the polymer chains.

CGG has lower degree of crystallinity compared to the native guar gum. This could be due to the replacement of hydroxyl groups, which creates hydrogen bonds with the cationic groups. Pal et al.[41] reported that CGG shows lower levels of crystallinity compared to native guar gum. According to the authors, lower levels of crystallinity of CGG could be due to the disruption of crystalline structure by the cationic group. Even though different modifications result in slight changes in the present crystallinity, natural guar gum exhibits very small overall crystallinity.[22,41]

13.14 INVESTIGATION OF THE EFFECT OF SALT ON THE VISCOSITY OF GUAR GUM

The energy requirements of the world increase due to the manufacturing of new machineries and vehicles with higher energy consumption. Therefore, efficient and effective ways should be discovered to harvest petroleum and natural gas trapped beneath the rock layers deep in the earth. Hydraulic fracturing, the method normally used to break the rock layers to reach the energy sources, uses highly pressurized water/fracturing fluids to break the rocks. The hydrocolloids mixed with fracturing fluids, such as guar gum that has remarkable rheological properties, play a vital role in minimizing the friction of the hydraulic fracturing system. However, a huge amount of fresh water is required for hydraulic fracturing and fresh water resources have depleted considerably with time. Hence, it is worth testing the applicability of sea water, a mixture of salt and other minerals, as an alternative to fresh water. However, the impact of salts on the performance of hydrocolloid or guar gum should be first evaluated to optimize the process. Furthermore, most of other industrial applications make use of guar gum with different additives such as salts and alcohols in varying concentrations. Those additives can alter the viscoelastic and structural properties of guar solution by changing the chain conformation and intermolecular association.[42] Hence, the stability of guar gum under different salts should be evaluated by correlating changes in the polymer structural and intermolecular association with the changes in rheological behavior for proper characterization of guar gum.

Few studies were performed on evaluating the performance of guar gum under salty conditions. It has been reported that the presence of salt can affect the rheological and thermal properties of the gum.[16] The guar gum concentration showed a great impact on the viscosity measured at 37°C. Guar gum (0.25% [w/w]) in Krebs bicarbonate buffer has significantly lower viscosity compared to a guar solution without buffer at the same concentration. On the contrary, guar solution (0.5% [w/w]) showed high viscosity with the presence of the buffer. Gittings et al.[42] reported that salts have different influences on guar gum aqueous solutions using sodium chloride (NaCl), sodium carbonate (Na_2CO_3), and sodium thiocyanate (NaSCN). The authors stated that all salts studied have similar moderate effect on the rheological properties of guar gum. There was only a slight decrease in the complex viscosity without quantitative change in the rheological properties. However, the level of viscosity reduction showed great dependency on the type of salt. Salts affect the guar on a local scale leading to a distorted chain configuration and, therefore, reduced viscosity of the solution. Furthermore, the solubility of the polymer has also been affected by salts and sodium chloride has not shown considerable influence on the solubility of guar galactomannan. The solubility decreased as the salt ions disrupt the water layer, which surrounds the galactomannan molecules. The influences of lithium chloride, potassium chloride, sodium chloride, and cesium chloride on the viscosity of the guar solution were studied by Ma and Pawlik.[25] Only saturated solutions of lithium chloride, sodium chloride, and cesium chloride exerted a significant effect on the viscosity of the guar solutions. All four salts, up to the concentration of 4.1 mol/L, have not given statistically different viscosity results when considering the experimental errors. However, some other reports indicated that the viscosity of nonionic polymers such as guar is marginally affected by the presence of high concentrations of monovalent salts.

According to the literature review, it is clear that the guar gum is being evaluated using different analytical tools. It clearly demonstrates that it is important to characterize guar gum completely with those analytical tools to evaluate the relationships between the chemical structure and the rheological behavior. Then, the chemical structure, for example, the galactose to mannose ratio and the distribution of galactose units along the mannose backbone, can be modified through available high technology breeding programs to accomplish advanced rheological properties. Furthermore, those approaches could also be used to optimize the extraction, purification, and other processing steps of guar gum.

13.15 MODIFICATION OF GUAR GUM

A number of chemical modifications have been applied to guar gum to make it versatile and useful in a diverse range of industrial applications. The abundant hydroxyl groups in galactomannan molecules allow chemical reactions such as esterification[2] and alkoxylation[2], oxidation,[43] etherification[43], and grafting[43]. Guar esterification results in a variety of compounds having different characteristics; for example, guar triacetate is a water insoluble, strong, and flexible film whose properties are comparable with cellulose triacetate. A number of other compounds can be produced by various functional modifications and some of them are strongly stable under alkaline solutions. Galactomannan molecules can make three-dimensional networks through complexing reaction with salts of copper, calcium, aluminum, or chromium that can form a gel according to the pH value of the solution. The complex formed between galactomannan and copper salt is not reduced by Fehling's solution even at prolonged boiling. However, it produces an insoluble gel-like substance. Four hydroxyl groups in two chains of galactomannan can be cross-linked with Borate ion (sodium borate) via hydrogen bonding. This reaction occurs even at very low concentrations of both borate ions and galactomannan molecules. Under alkaline conditions ($pH \geq 8$), it forms a gel and it solubilizes in acidic media ($pH \leq 7$).

The properties of guar gum and partially hydrolyzed guar gum were analyzed by Mudgil et al.,[37] using DSC, TGA, and XRD. Guar gum is hydrolytically degraded under hydrothermal conditions without any catalyst. The effect of reaction temperatures and time on the product distribution was studied. The hydrolyzed sample did not show viscous properties compared to crude guar gum solution. According to the MWD obtained by GPC at different reaction times, the decomposition of guar gum showed gradual conversion of guar polysaccharide to oligosaccharides and then to monomers with longer reaction time.[8] Free radical degradation of guar gum was studied by Reddy and Tammishetti[18] in order to degrade guar gum into smaller fragments using KPS and hydrogen peroxide (H_2O_2). Two different modes of degradation, acid and enzymatic hydrolysis, were conducted and compared using GPC and dilute solution viscometry by Cheng et al.[12]

Oxidized guar gum was produced for reactive dye printing and the product was characterized using FTIR, GPC, differential scanning calorimetric, and scanning electron microscope.[43] Furthermore, guar gum can be modified to form CGG by replacing the hydroxyl groups with the quaternary ammonium groups.[27] The authors modified CGG with fluorinated groups in order to obtain excellent solubility with special rheological behavior. Guar gum hydrolyzate or partially hydrolyzed guar gum modified with *n*-octenyl succinic anhydride and oleic acid was successfully used as a wall material in microencapsulation of mint oil, where microencapsulation is used to stabilize essential oils.[44] Guar gum hydrogel prepared by reacting guar gum and glutaraldehyde was suggested as a matrix for modified drug delivery.[7]

Guar derivatives are usually produced by reacting guar with the required ingredient under high temperature and pressure and then is dried and ground to meet the required quality. The degree of substitution provides an idea on the average number of hydroxyl groups substituted per sugar unit. Guar has maximum possible nine hydroxyl groups for making additional bonds if we assume one repeating unit of guar has one galactose sugar and two mannose sugar units. Therefore, guar gum can be modified with several other compounds to obtain different guar gum derivatives and those materials also have huge interests in the industry.

13.16 SUMMARY AND CLOSING REMARKS

From the literature review, it is clear that there is a great interest for the characterization of guar gum using different analytical tools. However, most of the studies used food-grade guar gum or other commercial guar gum samples. The characterization of gum extracted from the seeds of currently available guar germplasms is still limited. Therefore, a detailed characterization of guar gum extracted from known cultivars is required to identify the best-performing guar varieties. Moreover, the chemical characterization of guar gum will allow us to (1) establish the carbohydrate profiles of the commercially available guar gum, (2) optimize the method for guar gum extraction from the guar seeds, (3) establish the carbohydrate profiles (mannose/galactose ratio) of the guar gum produced in the United States, (4) determine the rheological and thermal properties of the gum, and (5) establish the effect of salt on the structure and the rheological properties of the gum.

REFERENCES

1. Sabahelkheir, MK, AH Abdalla, and SH Nouri. Quality assessment of guar gum (endosperm) of guar (*Cyamopsis tetragonoloba*). *Journal of Biological Sciences* 1(1) (2012): 67–70.
2. Chudzikowski, RJ. Guar gum and its applications. *Journal of Society of Cosmetic Chemists* 22 (1971): 43–60.
3. Srivastava, M and VP Kapoor. Seed galactomannans: An overview. *Chemistry and Biodiversity* 2(3) (2005): 295–317.
4. Lubambo, AF, RA de Freitas, M-R Sierakowski, N Lucyszyn, GL Sassaki, BM Serafim, and CK Saul. Electrospinning of commercial guar-gum: Effects of purification and filtration. *Carbohydrate Polymers* 93(2) (2013): 484–491.

5. Dea, ICM, and A Morrison. Chemistry and interactions of seed galactomannans. *Advances in Carbohydrate Chemistry and Biochemistry* 31 (1975): 241–312.
6. Cunha, PLR, RCM Paula, and JPA Feitosa. Purification of guar gum for biological applications. *International Journal of Biological Macromolecules* 41(3) (2007): 324–331.
7. Sandolo, C, P Matricardi, F Alhaique, and T Coviello. Dynamo-mechanical and rheological characterization of guar gum hydrogels. *European Polymer Journal* 43(8) (2007): 3355–3367.
8. Miyazawa, T and T Funazukuri. Noncatalytic hydrolysis of guar gum under hydrothermal conditions. *Carbohydrate Research* 341(7) (2006): 870–877.
9. Buchanan, BB, W Gruissem, and RL Jones. *Biochemistry & Molecular Biology of Plants*, Vol. 40. American Society of Plant Physiologists, Rockville, MD, 2002.
10. McCleary, BV, AH Clark, I Dea, and DA Rees. The fine structures of carob and guar galactomannans. *Carbohydrate Research* 139 (1985): 237–260.
11. Mahammad, S, RK Prud'homme, GW Roberts, and SA Khan. Kinetics of enzymatic depolymerization of guar galactomannan. *Biomacromolecules* 7(9) (2006): 2583–2590.
12. Cheng, Y, KM Brown, and RK Prud'homme. Preparation and characterization of molecular weight fractions of guar galactomannans using acid and enzymatic hydrolysis. *International Journal of Biological Macromolecules* 31(1) (2002): 29–35.
13. Weaver, J, R Gdanski, and A Karcher. Guar gum degradation: A kinetic study. Paper presented at the *International Symposium on Oilfield Chemistry*, Houston, TX, 2003.
14. Dressler, M, P Fischer, and EJ Windhab. Rheological characterization and modeling of aqueous guar gum solutions. Paper presented at the *International Symposium on Food Rheology and Structure*, Zürich, Switzerland, 2003.
15. Gittings, MR, L Cipelletti, V Trappe, DA Weitz, M In, and C Marques. Structure of guar in solutions of H_2O and D_2O: An ultra-small-angle light-scattering study. *The Journal of Physical Chemistry B* 104(18) (2000): 4381–4386.
16. Srichamroen, A. Influence of temperature and salt on viscosity property of guar gum. *Naresuan University Journal* 15(2) (2007): 55–62.
17. Picout, DR, SB Ross-Murphy, N Errington, and SE Harding. Pressure cell assisted solution characterization of polysaccharides. 1. Guar gum. *Biomacromolecules* 2(4) (2001): 1301–1309.
18. Reddy, TT and S Tammishetti. Free radical degradation of guar gum. *Polymer Degradation and Stability* 86(3) (2004): 455–459.
19. Snyder, LR, JJ Kirkland, and JW Dolan. *Introduction to Modern Liquid Chromatography*. John Wiley & Sons, New York, 2009.
20. Hansen, RW, SM Byrnes, and AD Johnson. Determination of galactomannan (gum) in guar (*Cyamopsis tetragonolobus*) by high performance liquid chromatography. *Journal of the Science of Food and Agriculture* 59(3) (1992): 419–421.
21. Azero, EG and CT Andrade. Testing procedures for galactomannan purification. *Polymer Testing* 21(5) (2002): 551–556.
22. Mudgil, D, S Barak, and BS Khatkar. Effect of enzymatic depolymerization on physicochemical and rheological properties of guar gum. *Carbohydrate Polymers* 90(1) (2012): 224–228.
23. Barth, HG and DA Smith. High-performance size-exclusion chromatography of guar gum. *Journal of Chromatography A* 206(2) (1981): 410–415.
24. Tiwari, A, D Terada, and H Kobayash. Polyvinyl modified guar-gum bioplastics for packaging applications. In S Pilla, ed., *Handbook of Bioplastics and Biocomposites Engineering Applications*, p. 177. Wiley, New York, 2011.
25. Ma, X and M Pawlik. Intrinsic viscosities and Huggins constants of guar gum in alkali metal chloride solutions. *Carbohydrate Polymers* 70(1) (2007): 15–24.

26. Moser, P, ML Cornelio, and VRN Telis. Influence of the concentration of polyols on the rheological and spectral characteristics of guar gum. *LWT-Food Science and Technology* 53(1) (2013): 29–36.
27. Wang, C, X Li, B Du, P Li, and H Li. Associating and rheological behaviors of fluorinated cationic guar gum in aqueous solutions. *Carbohydrate Polymers* 95(2) (2013): 637–643.
28. Chenlo, F, R Moreira, and C Silva. Rheological behaviour of aqueous systems of tragacanth and guar gums with storage time. *Journal of Food Engineering* 96(1) (2010): 107–113.
29. Abidi, N, E Hequet, and D Ethridge. Thermogravimetric analysis of cotton fibers: Relationships with maturity and fineness. *Journal of Applied Polymer Science* 103(6) (2007): 3476–3482.
30. Gabbott, P. *Principles and Applications of Thermal Analysis*. John Wiley & Sons, New York, 2007.
31. Bradley, TD, A Ball, SE Harding, and JR Mitchell. Thermal degradation of guar gum. *Carbohydrate Polymers* 10(3) (1989): 205–214.
32. Cerqueira, MA, BWS Souza, J Simões, JA Teixeira, M Rosário, M Domingues, MA Coimbra, and AA Vicente. Structural and thermal characterization of galactomannans from non-conventional sources. *Carbohydrate Polymers* 83(1) (2011): 179–185.
33. Varma, AJ, SP Kokane, G Pathak, and SD Pradhan. Thermal behavior of galactomannan guar gum and its periodate oxidation products. *Carbohydrate Polymers* 32(2) (1997): 111–114.
34. Dodi, G, D Hritcu, and MI Popa. Carboxymethylation of guar gum: Synthesis and characterization. *Cellulose Chemistry and Technology* 45(3–4) (2011): 171–176.
35. Yadav, M, A Srivastav, SK Verma, and K Behari. Graft (partially carboxymethylated guar gum-*g*-poly vinyl sulfonic acid) copolymer: From synthesis to applications. *Carbohydrate Polymers* 97(2) (2013): 597–603.
36. Abidi, N, L Cabrales, and CH Haigler. Changes in the cell wall and cellulose content of developing cotton fibers investigated by FTIR spectroscopy. *Carbohydrate Polymers* 100 (2013): 9–16.
37. Mudgil, D, S Barak, and BS Khatkar. X-ray diffraction, IR spectroscopy and thermal characterization of partially hydrolyzed guar gum. *International Journal of Biological Macromolecules* 50(4) (2012): 1035–1039.
38. Kumar, V, AK Tiwary, and G Kaur. Investigations on chitosan-carboxymethyl guar gum complexes interpolymer complexes for colon delivery of fluticasone. *International Journal of Drug Delivery* 2(3) (2010): 242–250.
39. Cunha, PLR, RR Castro, FAC Rocha, RCM Paula, and JPA Feitosa. Low viscosity hydrogel of guar gum: Preparation and physicochemical characterization. *International Journal of Biological Macromolecules* 37(1) (2005): 99–104.
40. Shenoy, MA and DJ D'Melo. Synthesis and characterization of acryloyloxy guar gum. *Journal of Applied Polymer Science* 117(1) (2009): 148–154.
41. Pal, S, D Mal, and RP Singh. Synthesis and characterization of cationic guar gum: A high performance flocculating agent. *Journal of Applied Polymer Science* 105(6) (2007): 3240–3245.
42. Gittings, MR, L Cipelletti, V Trappe, DA Weitz, M In, and J Lal. The effect of solvent and ions on the structure and rheological properties of guar solutions. *The Journal of Physical Chemistry A* 105(40) (2001): 9310–9315.
43. Gong, H, M Liu, B Zhang, D Cui, C Gao, B Ni, and J Chen. Synthesis of oxidized guar gum by dry method and its application in reactive dye printing. *International Journal of Biological Macromolecules* 49(5) (2011): 1083–1091.
44. Sarkar, S, S Gupta, PS Variyar, A Sharma, and RS Singhal. Hydrophobic derivatives of guar gum hydrolyzate and gum arabic as matrices for microencapsulation of mint oil. *Carbohydrate Polymers* 95(1) (2013): 177–182.

14 Communicating Fracturing Impacts and Technologies

Assessment, Public Understanding, and Theoretical Linkages

Shawna R. White, R. Glenn Cummins, Melanie Sarge, and Erik P. Bucy

CONTENTS

14.1 Overview 251
14.2 Introduction 251
14.3 Public Perception of Fracturing Impacts and Technologies 252
14.4 Media Influence on Public Perception 261
 14.4.1 Agenda-Setting Influence of the Media 261
 14.4.2 Media Framing of Issues 262
 14.4.3 Mental Models of Issues 263
 14.4.4 Selective Processes in Message Consumption 264
 14.4.5 Toward an Integrated Model of Studying Media Influence and Public Perception of FIT 265
14.5 Summary and Concluding Remarks 267
References 268

14.1 OVERVIEW

A review of public opinion and communications literature is used to develop an integrated framework for studying media influence and public perceptions of fracturing impacts and technologies.

14.2 INTRODUCTION

As demonstrated by other chapters in this volume, hydraulic fracturing is a highly technical endeavor. However, unconventional production technologies also have important social and cultural dimensions: what different constituencies understand about the issues surrounding energy production; how the public, particularly affected residents and landowners, view the oil and gas industry; the amount of risk people consider acceptable with unconventional production technologies; and, how the fracturing issue is covered or framed in the news and entertainment media. Despite the contentious nature of public debate surrounding fracturing impacts and technologies (FIT) and the accompanying media coverage (e.g., Onishi, 2014), little scholarly work has examined the role of mass media in shaping public perceptions or even served as an information source in this context. In addition, research on hydraulic fracturing as an important social, scientific, and environmental issue is just now beginning to appear in the communication research and public opinion literature (e.g., Shen et al., 2014). Thus, this chapter outlines a theoretically informed research agenda so that a coherent research program can begin to examine the media's impact on public understanding of fracturing, as well as the attitudes toward this rapidly expanding form of energy production.

In this chapter, we perform a general review of the literature to assess public perceptions of fracturing impacts and technologies, and identify the conceptual frameworks that investigators are employing to study the area. We begin by noting how the press has framed fracturing, as well as how members of the public understand the issue in terms of potential benefits or detriments. We then describe the relevance of media agenda-setting, framing, and mental model approaches to understanding media influence on public perception. Finally, drawing on these classic theories of news influence, we propose an integrated framework for studying how the press portrays and how the public perceives fracturing impacts and technologies.

14.3 PUBLIC PERCEPTION OF FRACTURING IMPACTS AND TECHNOLOGIES

Some scientific developments, such as biofuels, genetically modified foods, or climate change, necessitate that the public have a degree of technical understanding (Cacciatore et al., 2012; Williams, 2011; Wohlers, 2013). Given uneven levels of knowledge, public perceptions of fracturing impacts and technologies are decidedly mixed, fraught with uncertainty, and depend upon a variety of individual and contextual variables. As shown by a review of the literature, theoretical insights from communication research bring considerable explanatory power to understanding this diversity of attitudes and how they may be shaped by media coverage. Extant empirical evidence suggests an overall lack of familiarity with the hydraulic fracturing process, and the potential impacts of the process create ambivalence among the public toward the issue (Boudet et al., 2014).

Surveys and interviews with affected populations have revealed a basic dichotomy that characterizes responses to many new technologies and policies (e.g., Anderson and Theodori, 2009; see Table 14.1 for summary). On the one hand, the benefits of hydraulic fracturing are understood as potential gains to impacted communities and the wider economy. Locally, benefits are articulated in terms of expanded employment opportunities, services, and community development (Ladd, 2013; Wolske et al., 2013). Regionally and nationally, benefits from fracturing are expressed in

TABLE 14.1
Summary of Communication Research on Hydraulic Fracturing

Study	Step in Integrated Model of Communication Research	Method	Key Variables Studied	Overall Findings
Anderson and Theodori (2009)	Examination of systematic/structural factors	Interview	Perceived consequences of hydraulic fracturing; social impacts	Local leaders and citizens recognize economic benefits but also express concern for public health and safety, environmental impacts, and temporary disturbances.
Boudet et al. (2014)	Examination of public understanding/perception	Survey	Familiarity with issue; levels of support/opposition; media use; education; political party identification/ideology; gender; world views	Public has limited familiarity with the process and impacts of HF and are generally uncertain of their position on the issue. Several individual characteristics are associated with attitudes toward HF.
Brasier et al. (2011)	Examination of systematic/structural factors	Case study	Level of development; previous extractive history; population density; awareness and perceptions of natural gas impacts	Valence of perceptions about HF impact vary according to stage of energy development as well as experience with extractive industries.
Brown et al. (2013)	Examination of public understanding/perception	Telephone survey	Positive and negative views of overall impact; support/opposition of regulation and taxation; perceptions of risk and reward; political party; opinions of state lawmakers' actions; direct experience; media exposure	Respondents from Pennsylvania and Michigan both appear uncertain about the level of risk and are more likely to believe an expert assessment of high risk than low. Responses by Pennsylvania residents are more informed by direct experiences and media exposure than Michigan respondents.
Davis and Fisk (2014)	Examination of public understanding/perception	Survey	Fracking attitudes; support/opposition of associated policies; political party identification; environmental attitudes; issue perception; gender; urban versus rural areas	Demographics (female, urban residence) are slightly associated with HF opposition and support for regulation, while Democratic identification and pro-environmental attitudes are strongly associated.

(Continued)

TABLE 14.1 (*Continued*)
Summary of Communication Research on Hydraulic Fracturing

Study	Step in Integrated Model of Communication Research	Method	Key Variables Studied	Overall Findings
Evensen et al. (2014)	Examination of media coverage and content	Content analysis; interview	Social representations of impacts; environmental and economic impacts; newspaper coverage; geographical contexts; journalist perspectives	Effects on water quality most prevalent environmental representations in newspaper coverage. Economic representations focus on jobs, leases, and royalties, but vary substantially across geographical contexts. Representations of social impacts are relatively rare. Journalists offer explanations for frequency of various HF impacts.
Ferrar et al. (2013)	Examination of public understanding/perception	Interview	Concerns for health and social impacts; mental and physical perceived stressors	Participants attributed 59 unique health impacts and 13 stressors to Marcellus Shale development. Stress was the most frequently reported symptom. Over 2 years, health impacts increased while stressors remained constant.
Heuer and Lee (2014)	Examination of public understanding/perception	Survey	Stakeholders' perceptions of environmental, social, and economic impacts; business sectors (nonprofit, government, and private)	Responses by the three sectors are relatively similar across four categories of perceptions: economic opportunity, protecting health and safety, preserving communities, and energy security.
Jacquet (2012)	Examination of public understanding/perception	Survey	Landowners' attitudes; environmental attitudes; leasing development; employment experience; proximity to development	Landowners are generally more polarized and negative toward gas development than wind farm development. These attitudes are highly dependent on environmental attitudes and industry leasing, development, or employment experience.

(*Continued*)

TABLE 14.1 (*Continued*)
Summary of Communication Research on Hydraulic Fracturing

Study	Step in Integrated Model of Communication Research	Method	Key Variables Studied	Overall Findings
Jacquet and Stedman (2011)	Examination of systematic/structural factors	Interview	Landowner coalitions; member motivations and benefits; public benefits; collective action	Coalitions are primarily concerned with the advancement of private member benefits, while public benefits of collective action are poised to accrue indirectly.
Jacquet and Stedman (2013)	Examination of public understanding/perception	Survey	Landowner perceptions of impacts; wind farming; natural gas drilling; positive and negative impacts; resident attitudes toward energy developments; place attachment; length and type of residency	Impact perception explained a large portion of variation and residents' place meanings explained some variation in attitudes, while place attachment and length and type of residency had little or no effect on attitudes.
Jacquet and Stedman (2014)	Examination of systematic/structural factors	Review	Community acceptance; social-psychological values (attachment and community/place-based identity); oppositional behavior; land use planning	This collection of literature demonstrates the threat of disruption to place-based identities may spur oppositional behavior.
Kriesky et al. (2013)	Examination of public understanding/perception	Telephone survey	County support; perceptions of economic opportunity and environmental threats; lease holders	County-based differences in support of Marcellus Shale drilling are due to Washington County residents seeing more of an economic opportunity in the MS and their greater likelihood of having a family-held lease.
Ladd (2013)	Examination of public understanding/perception	Interview	Community stakeholder perceptions; positive economic benefits; negative socioeconomic/environmental costs	Most Haynesville residents believe that the socioeconomic benefits of development outweigh the collective socioeconomic/environmental costs to the region. A substantial minority of respondents are also skeptical or disagree that the benefits to date had been worth the risks.

(*Continued*)

TABLE 14.1 (*Continued*)
Summary of Communication Research on Hydraulic Fracturing

Study	Step in Integrated Model of Communication Research	Method	Key Variables Studied	Overall Findings
Rahm (2011)	Examination of systematic/structural factors	Case study	Shale gas production; Texas shale gas regions; controversy	Texas is a major player in shale gas developments. Of the eight states and coastal areas that account for the bulk of U.S. gas, Texas has the largest proved reserves.
Schafft et al. (2013)	Examination of public understanding/perception	Secondary data analysis	Perceptions of risk and opportunity; Pennsylvania Marcellus Shale region school districts; amount of local drilling	Strong positive association between perceptions of risk and opportunity associated with gas extraction. Intensity of perception of both risk and opportunity is directly associated with the amount of local drilling.
Schafft et al. (2014)	Examination of public understanding/perception	Survey; interview	Educators' and educational administrators' perceptions; community impacts	Respondents in areas with high levels of drilling are significantly more likely to perceive the effects of local economic gains, but also report increased inequality, heightened vulnerability of disadvantaged community members, and pronounced strains on local infrastructure.
Shen et al. (2014)	Examination of public understanding/perception	Experimental	News framing; narrative versus informational; environmental versus economic; issue attitude; cognitive responses; empathy	Both news formats (narrative, informational) and frames (environmental, economic) significantly affect issue attitudes and other responses. Narrative environmental news reports have a significantly greater impact than informational environmental news, including delayed impacts, on issue attitudes. Cognitive responses and empathy act as mediators of narrative impact.

(*Continued*)

TABLE 14.1 (*Continued*)
Summary of Communication Research on Hydraulic Fracturing

Study	Step in Integrated Model of Communication Research	Method	Key Variables Studied	Overall Findings
Stedman et al. (2012)	Examination of public understanding/perception	Natural experiment; survey	Support or opposition; trust in the industry; public participation; perceived knowledge; perceived impacts	New York residents are more opposed to development, characterize the industry in a more negative light, and are more likely to engage in various forms of public participation than Pennsylvania residents. Residents' perceived knowledge did not differ by state.
Theodori (2009)	Examination of public understanding/perception	Survey	Counties in Barnett Shale region of Texas; levels of energy development; negative and positive perceptions; social, environmental, economic, and service-related issues	Residents of a mature natural gas industry county (Wise County) are significantly more likely than residents of a less established county (Johnson County) to view social or environmental issues negatively and economic and/or service-related issues positively. Residents in Johnson County are more likely to view two social and/or environmental issues as getting worse.
Theodori (2012)	Examination of public understanding/perception	Survey	Counties in Barnett Shale region of Texas; perceptions of natural gas industry; mineral rights ownership	Residents of a mature natural gas industry county (Wise County) exhibit more negative perceptions of the energy industry than residents of a less established county (Johnson County). Mineral rights ownership is a relatively strong factor associated with public perceptions of the natural gas industry.

(*Continued*)

TABLE 14.1 (*Continued*)
Summary of Communication Research on Hydraulic Fracturing

Study	Step in Integrated Model of Communication Research	Method	Key Variables Studied	Overall Findings
Theodori (2013)	Examination of public understanding/perception	Survey	Public perception of natural gas industry; energy industry actions; social, environmental, economic, and service-related issues	Members of the public in Tarrant County dislike certain potentially problematic social or environmental issues perceived to accompany natural gas development but view less negatively the economic or service-related benefits that often result from such development. The social/environmental perceptual variable is a key factor in explaining past behaviors and predicting future behaviors.
Theodori et al. (2013)	Examination of public understanding/perception	Survey	Perceptions of hydraulic fracturing process; views on and familiarity with the management of, disposal of, and technologies related to flowback wastewater; information sources; trust in sources; factual knowledge about topic; level of agreement	Respondents in high-density drilling counties are more familiar with processes than counterparts in low-density counties. Trustworthy sources of information were identified and natural gas industry is viewed as the least trustworthy. Both the natural gas industry and regulatory agencies are proactively dispensing information in high-density drilling locations. Additional findings discussed.
Willits et al. (2013)	Examination of public understanding/perception	Secondary data analysis	Knowledge of impacts; opinions of costs and benefits; concerns about negative environmental impacts	2009–2012 saw a decline in the proportion of residents knowing very little or nothing about impacts of gas drilling. Residents formed more opinions about costs/benefits and opposition/support and showed increased concern about negative environmental impacts. Most residents supported the development of the industry, but there was little change over time.

(*Continued*)

TABLE 14.1 (*Continued*)
Summary of Communication Research on Hydraulic Fracturing

Study	Step in Integrated Model of Communication Research	Method	Key Variables Studied	Overall Findings
Wolske et al. (2013)	Examination of public understanding/perception	Secondary data analysis	Public awareness; risk perceptions; desire for regulation; concerns about water quality and other community impacts	In Michigan, a clear majority (82%) of residents have heard at least "a little" about fracking and nearly half report that they follow debates about fracking in the state "somewhat" to "very closely." Consistent with other polls, a slight majority of residents (52%) believe that benefits of HF outweigh the risks, but concerns remain about the impacts on water quality.
Wynveen (2011)	Examination of public understanding/perception	Open-ended survey responses; thematic analysis	Perceptions about impacts; positive and negative; economic, social, and environmental	Open-ended survey responses from residents of two Texas counties were coded; themes ranged from positive to negative, and reflected economic, social, and environmental impacts accompanying unconventional natural gas development.

terms of energy independence and economic growth. On the other hand, detriments are expressed in terms of perceived economic and environmental losses from fracturing. Economic detriments include inflated housing prices, "boom/bust" cycles, and uneven benefits to communities due to mineral rights leases, while environmental harms are identified as potential groundwater contamination, excessive water use, artificially induced seismic activity, and quality-of-life issues such as noise pollution and degraded roads (Ladd, 2013; Wolske et al., 2013). This bifurcated understanding and articulation of the issue pervade both media coverage and citizen understandings.

Research on fracturing impacts and technologies is showing that individual differences are consequential in accounting for variations in issue perceptions. Level of trust in the industry, perceived knowledge, and perceived impact of energy development, including the aforementioned environmental, economic, and social impacts, are significantly associated with attitudes toward energy operations (Jacquet and Stedman, 2013; Stedman et al., 2012; Wynveen, 2011). Furthermore, these attitudes are dynamic, as changes in the perceived impacts of hydraulic fracturing are shown to occur according to the stage of energy development for affected communities (Brasier et al., 2011). Gilmore (1976) advances four attitudinal stages that residents experience when undergoing rapid community change: first, initial *enthusiasm* for the prospect of new economic opportunities transitions into *uncertainty* as community changes occur or potential detriments become increasingly visible; *uncertainty* then progresses to *near panic* in some cases as residents no longer recognize long-held ways of life; finally, *adaptation* occurs either through acceptance of the new reality as progress or rejection of the new reality with the decision to move away.

Recent research exemplifies this attitude transformation. In a 2009 survey of the residents in two counties, Theodori uncovered paradoxical attitudes toward unconventional natural gas development, such that community members in both areas perceived the possibility for both positive and negative outcomes. At that time, residents of the county in which the industry was more established were more likely to hold more positive attitudes. Three years later, Theodori (2012) noted that members of that same community began expressing more negative attitudes, stating concerns over proximity of drilling to residential areas, the political power of the industry, lack of consideration for the environment, and disbelief in the long-term positive benefits for local residents. Willits et al. (2013) provide additional support for such changes by comparing the results of a 2009 and 2012 survey of 21 counties in one shale region. Over time, respondents reported higher levels of knowledge regarding potential impacts of shale drilling, formed more solid opinions about drilling, communicated more concern for environmental impacts, and expressed higher levels of opposition to drilling. The authors suggest various reasons for these attitudinal changes, not the least of which is media coverage.

In addition to the maturity of the industry within a community, the sheer amount of local drilling is linked to the perception of both risks and opportunities (Schafft et al., 2013, 2014). Although community members with high drilling levels are more inclined to recognize economic benefits, they also perceive heightened inequality, vulnerability of underprivileged sectors, and local infrastructure strains as direct results of drilling (Schafft et al., 2014).

Survey research has demonstrated that certain individual characteristics are associated with attitudes toward energy development. Opposition to hydraulic fracturing is related to being female, residing in an urban area, upholding egalitarian beliefs, being more familiar with the process, attributing detrimental environmental impacts to the process, and reading newspapers more than once a week (Boudet et al., 2014; Davis and Fisk, 2014). Additionally, identifying with the Democratic party and attitudes consistent with pro-environmental policy are associated with opposing hydraulic fracturing, as well as endorsing current or increased regulation of the practice (Davis and Fisk, 2014). On the other hand, characteristics linked with support for hydraulic fracturing include being older, more educated, politically conservative, endorsing the economic benefits of the practice, and watching televised news more than once per week (Boudet et al., 2014; Kriesky et al., 2013).

Despite theoretical perspectives that suggest the potential influence of media coverage on public attitudes, little research has systematically examined the nature of news surrounding specific energy developments. Moreover, the scarce research available suggests that the different news sources mentioned earlier (i.e., newspapers and television news) are associated with differing positions on hydraulic fracturing (e.g., Boudet et al., 2014; Davis and Fist, 2014; Kriesky et al., 2013). Such findings call for in-depth examination of media coverage. Evensen et al. (2014) conducted a content analysis of how newspapers presented the social, environmental, and economic impacts of FIT in one shale region over a 4-year span. The prevalence and valence of each of the impact topics were found to vary across newspapers and geographic contexts, and journalists reported that their coverage was largely influenced by a local, public perspective (Evensen et al., 2014).

14.4 MEDIA INFLUENCE ON PUBLIC PERCEPTION

As previously noted, research has documented a diversity of attitudes, knowledge, and understanding of hydraulic fracturing operations. But the media's role in shaping these attitudes and serving as a source of information has largely gone unexplored, despite the potential relevance that theoretical perspectives might supply, particularly from the field of mass communication. Although many perspectives could be brought to bear, three have obvious applicability: agenda setting, framing, and mental models.

14.4.1 Agenda-Setting Influence of the Media

Agenda setting is a decades-old and robustly supported theory that is often summarized by the pithy hypothesis that the mass media "may not be successful much of the time in telling people what to think, but it is stunningly successful in telling its readers what to think about" (Cohen, 1963, p. 13). More formally, the theory asserts that news media have the ability to influence the public's perception of the salience of news topics through selection and frequency of topic coverage (McCombs and Reynolds, 2009). Therefore, an agenda-setting effect is detected when audience member's beliefs about the prominence of certain news issues resembles the representation of those issues by news media outlets (McCombs and Reynolds, 2009).

The seminal agenda-setting study conducted by researchers in Chapel Hill, North Carolina, simultaneously surveyed undecided voters on their beliefs about the most important issues of the time and analyzed content produced by major news outlets in the area to determine key issues covered in the news (McCombs and Shaw, 1972). Results from both methodologies were found to correlate, providing evidence of an agenda-setting effect (McCombs and Shaw, 1972). Since then, similar effects have been demonstrated in a variety of contexts (e.g., presidential election coverage, Weaver et al., 1981; civil rights issue coverage, Winter and Eyal, 1981; portrayals of crime in entertainment media, Holbrook and Hill, 2005). Ader (1995) examined media coverage of environmental pollution over a 20-year period and demonstrated a correlation between increased media coverage and public perception of issue salience. Moreover, no relationship emerged between public perception and real-world conditions, further demonstrating the powerful role of the media in raising an issue's public profile.

14.4.2 Media Framing of Issues

Although agenda setting suggests tremendous potential to raise the perceived salience of an issue, one limitation of the theory is that it fails to shed insight into the nature of these perceptions. Thus, *framing* has emerged as a useful perspective shows how key attributes of media coverage can influence how audiences view an issue or event. The process of framing refers to the intentional or unintentional selection of words and/or images by media producers that can influence the interpretation and evaluation of the message or news story (Entman, 1993; Tewksbury and Scheufele, 2009). Through this emphasis, framing makes certain aspects of a topic more salient in the message recipient's mind, and excludes or diminishes the importance of other aspects, thereby suggesting a certain perspective or view on the topic (Lecheler and de Vreese, 2011; Tewksbury and Scheufele, 2009). Among other things, frames consist of recurring characteristics, language, and consistent visual portrayals assigned by journalists in coverage of issues or events. Although framing's relevance in a news context is obvious as a function of journalistic practice, it has also emerged as an effective means of influencing attitudes in public relations campaigns (e.g., Hallahan, 1999; Knight, 1999) as well as primetime dramas (e.g., Holbrook and Hill, 2005). Media scholars have also increasingly called for the study of news visuals (Grabe and Bucy, 2009) first through systematic content analysis and then with studies, that demonstrate how different visual frames influence audiences.

Although the story-specific elements emphasized within coverage of an issue or event are seemingly infinite, mass communication researchers have identified several global typologies of news frames. Iyengar (1994) developed the categories *thematic framing*—providing information on a topic in a broad, general context—and *episodic framing*—presenting an individual or group's story, as an exemplar, pertaining to a particular topic. Experimental studies have found that use of specific frame types can influence audience response. For example, episodic frames that emphasize individuals tend to elicit emotional reactions from audience members (Aarøe, 2011; Yu et al., 2010), and can lead to placing the attribution of responsibility on the affected individuals as opposed to society (Iyengar, 1994).

Given the previous acknowledgment that individual attitudes toward hydraulic fracturing are influenced by whether one views the technology as having positive attributes (e.g., economic benefits) or negative attributes (e.g., concerns over water use) (Ladd, 2013; Wolske et al., 2013), the global typology of *gain* or *loss* frames has clear import. Kahneman and Tversky's (1979) work on prospect theory introduced the typologies of gain frames, constructing a message to emphasize the positive aspects of a situation, and loss frames, constructing a message to emphasize the negative aspects of a situation. Research in both health (Nan, 2012) and environmental (Newman et al., 2012) communication has shown that loss frames tend to be more persuasive than gain frames. Preliminary analysis points to the existence of an overarching gain/loss frame that can be subdivided into at least four, more specific frames in the context of hydraulic fracturing: *energy independence*, *environmental harm*, *economic benefits*, and *detriments*.

One recent study provides the most direct evidence for the impact of gain versus loss frames in this context. Shen et al. (2014) examined the differences in effects produced by an article about shale drilling framed from both an economic benefits perspective and an environmental consequences perspective. The news article framed from the environmental perspective produced more negative immediate and delayed attitudes toward drilling than did the economic perspective. Additionally, the researchers found that combining an environmental frame with a narrative format (rather than an informational format) produced more negative immediate and delayed issue attitudes, whereas the economic frame placed in the same formats did not produce significantly different attitudinal effects.

14.4.3 Mental Models of Issues

While agenda setting explains how the media can steer the public's attention to certain issues and framing explains how the media can shape public attitudes about an issue, the *mental models* approach explains how media coverage shapes cognitive representations of an issue. A mental model is a cognitive representation of how a person thinks about something that exists or works in the world (Garnham, 1997; Gentner and Stevens, 2014; Johnson-Laird, 1983; Roskos-Ewoldsen et al., 2009). This cognitive representation may include physical aspects of an issue, situations or events pertaining to an issue, or concepts associated with an issue. Referencing the work of Glenberg and Langston (1992) as well as Wyer and Radvansky (1999), Roskos-Ewoldsen and colleagues (2002) suggested that given the nature of media with their ability to blend linguistic and pictorial features, media presentations "should be particularly effective at influencing the construction of mental models" (p. 114). Considering the public's limited understanding and ambivalent position on issues surrounding hydraulic fracturing (Boudet et al., 2014), research examining the construction of hydraulic fracturing mental models could be extremely beneficial to message designers and stakeholders, since they work to shape public perceptions and attitudes (e.g., Hallahan, 1999). Indeed, various media sources, such as news articles, social media discussions, television specials or documentaries and even films, can influence the mental models of those who are not yet familiar with hydraulic fracturing.

Previous research has examined media influence on the construction of short-term mental models, known as *media priming*, about such issues as stereotypes (e.g., Dixon, 2006), violence (e.g., Anderson, 1997; Josephson, 1987), and politics (e.g., Krosnick and Kinder, 1990). These studies typically only examine the impact of single media encounters on creating short-term associations or the early formation of a mental model (e.g., Josephson, 1987). However, a longer lasting impact can be obtained through multiple encounters or heavy exposure. *Cultivation theory* coined this idea as the "drip effect" (Gerbner et al., 1980). This theory posits that over time, people will be exposed to consistent, repetitive depictions or themes within the media that help create mental models of particular issues or perceptions of reality. For example, much empirical work testing cultivation theory has demonstrated that heavy exposure to violent media can lead people to perceive the world as more violent and increase the expectation they will experience more crime (Gerbner et al., 1980; Shananhan and Morgan, 1999) when compared to people who consume less violent media. As coverage of hydraulic fracturing technology becomes more common, and achieves some consistency across media outlets, cultivation effects could shape mental models of the issue. As we will argue, by yoking studies of media content and public perception, research employing a mental models approach could also explore how misperceptions may be generated.

Researchers examining risk perception have attempted to reconstruct the mental models of stakeholders using a method referred to as *concept mapping* (Atman et al., 1994; Morgan et al., 1992). This approach focuses on determining the mental model that influences decision-making processes when a risk is present. Most of this research explores relatively new risks meaning that the affected public has limited preexisting mental models or concepts (Atman et al., 1994). For example, previous studies have been conducted on topics such as nuclear energy resources (Maharik and Fischhoff, 1992) and climate change (Bostrom et al., 1994). Given the aforementioned ambiguity surrounding hydraulic fracturing risks, mental models approaches regarding hydraulic fracturing are an appropriate place to apply such risk perception investigations.

14.4.4 Selective Processes in Message Consumption

Preexisting perceptions, attitudes, and mental models are also called upon when people are introduced to new information about an issue or topic. In other words, perceptions, attitudes, and mental models developed through these theoretical processes (agenda setting, framing, and media priming) create a lens through which individuals select and understand new information. *Selective exposure* refers to biased media selection determined by personal or situational characteristics (Zillmann and Bryant, 1985). One type of selective exposure motivated by attitudes is the confirmation bias or reinforcement hypothesis (for review, see Frey, 1986). This perspective holds that people prefer media content that is consistent with their preexisting attitudes, which, in turn, strengthen or reinforces already existing attitudes. In addition, this reinforcement drive creates a spiral effect between attitudinal outcomes and media selection (Slater, 2007). In contrast, *selective perception* is the phenomenon by which people "see what they want to see" (Balcetis and Dunning, 2006;

Vidmar and Rokeach, 1974). As people use existing attitudes and mental models to understand new concepts, this understanding is slanted to fit or relate to preexisting attitudes and models. A comprehensive examination of the public's perception of FIT should not only study the creation of perceptions, attitudes, and mental models through media exposure, but also the influence these preexisting individualities have on subsequent message selection, attention, and comprehension.

Based on these theoretical perspectives, news coverage and media portrayals clearly have the ability to influence the public thought about hydraulic fracturing (i.e., agenda setting), to present and influence a particular perception or attitude about hydraulic fracturing (i.e., framing), and to paint a vivid picture of hydraulic fracturing in the minds of the public (i.e., mental models). Once these perceptions, attitudes, and models are established, they are difficult to change because of their significant influence on interpreting and understanding subsequent information (i.e., selective processes). Insights from this research can then be employed in the service of more effective message development and placement.

14.4.5 Toward an Integrated Model of Studying Media Influence and Public Perception of FIT

Clearly, the growth of hydraulic fracturing activities presents a tremendous opportunity to examine the dynamic process of attitude formation and its relationship to media coverage of such developments. With respect to potential framing effects, a comprehensive approach to analyzing media influence would thus identify (1) salient image frames used in television coverage of hydraulic fracturing, (2) the constituent elements of those frames, and (3) the persuasive influence of different issue framing on viewers and consumers of news. A fourth stage of research would involve analysis of the public opinion environment and message-shaping strategies employed by key stakeholders who have an interest in the promotion and acceptance of specific frames in the news. Here, discourse analysis would be a useful technique to examine the "frame wars" that take place between the oil and gas industry and activist groups that seek to reframe the debate in such a way (namely, as an imminent threat to public health) as to spark collective action. Indeed, from an issue perspective, frames have been defined as "conceptual tools which social movements, media, and individuals rely on to convey, interpret, and evaluate information" (Neuman et al., 1992, p. 60). Thus, communication concepts and methods of analysis have an important role to play in illuminating how hydraulic fracturing is understood and received by members of the public across different levels of issue awareness and involvement.

Perhaps the best way to probe these issues is the use of multiple methods, or triangulation, to provide a many-sided view of a process or phenomenon. Triangulating findings with multiple methods and measures involves integrating diverse methods such as content analysis, panel studies, focused interviews, and other data collection techniques, into a single comprehensive project. Mixed methods approaches may also use different sets of measures, such as integrating the results of a survey design with detailed observations from a set of in-depth interviews and case study analyses to help explain research outcomes.

In mixed methods research, an organization may convene a focus group as a prelude to conducting a survey, or perform a framing analysis of election news and run an experiment to test the effects of frames on cognition. Multiple-method studies enable researchers to document content or opinion trends in the first step of data collection and then test hypotheses around these elements in the second step to generate fuller findings and more complete conclusions. One increasingly popular approach to examining the relationship between media and public perception is the use of hybrid survey-experiments.

"Population based" survey experiments embed an experimental design within an online survey, allowing researchers to randomly assign participants to conditions and treatments, and then generalize beyond a narrow pool of convenience sample volunteers (Mutz, 2011). Typically "large *N*" survey experiments can better establish causality while retaining the advantages of representative probability surveys because researchers can assign larger and more diverse samples to experimental conditions. The versatility of hybrid studies is likely to encourage "more successive studies comprising a larger research agenda, and research that speaks to real world events and policies" (Mutz, 2011 p. 14).

In terms of explanatory power, the most illuminating means of examining the relationship between media and public perception is through an integrated model: this would comprehensively examine (1) the systemic organizational, contextual, or structural elements that precede media content; (2) the actual content itself; and, (3) resultant outcomes such as attitudes, mental models, and behaviors (Riffe et al., 2014).

First, examination of community characteristics should be assessed in order to document the cultural milieu surrounding proposed fracturing developments. Both survey research and qualitative interviews or focus groups could be employed to amass benchmark data that assembles community characteristics and individual response (e.g., Theodori, 2009, 2012). Furthermore, corporate communication activities such as public engagement efforts—either through traditional media channels or via social media efforts—could also be examined to document message strategies surrounding fracturing operations (Roper, 2005). In addition, qualitative examination of journalistic practices could also explore the interplay between corporate communication efforts, community response, and media coverage (e.g., Evensen et al., 2014).

A second and central element of this research agenda should be the comprehensive and systematic study of media coverage about fracturing developments over time. Evensen et al.'s (2014) work provides a basic model that could be expanded to assess message content characteristics. For example, differences in content could be documented as a function of the size of the media organization (e.g., local, weekly newspapers vs. larger metropolitan dailies vs. papers with national circulation). Moreover, the global nature of fracturing activities also permits international comparisons (e.g., U.S. coverage vs. European coverage). Again, these content characteristics could be tied to aforementioned community or news organization variables.

A final phase of this integrated model, which could be guided by any of the preceding theoretical perspectives, is to examine individual perceptions of fracturing activities explicitly as a function of message characteristics. Although survey-based research can provide some evidence tying together general media content properties with public perception, the strongest means of linking content with individual perception is through controlled experimental research (e.g., Shen et al., 2014) or, as previously noted, hybrid survey-experiments. In a sample study design, select attributes of media messages can be systematically altered to determine differences in selection of and/or response to the messages, either within a laboratory setting or in more naturalistic field settings.

Although individual self-report measures are the most commonly used outcome variables, alternate means of assessing response may provide unique insights into message processing. One form of measuring the dynamic or changing nature of individual response is through continuous response measurement (CRM) (Biocca et al., 1994). This technique provides an online measure of response to audio-visual stimuli (typically through hand-held dials) where changes in message evaluation are linked to specific message elements. For example, an experiment could assess how individuals respond to television news sources within stories concerning hydraulic fracturing, where the nature of the source is systematically altered (e.g., corporate spokesperson vs. government official).

An additional means of shedding light into individual perception of media messages is the use of psychophysiological or biometric measures. One advantage of psychophysiological measures of cognitive and emotional processing is that they can assess responses that are free of bias compared to traditional self-report measures (Potter and Bolls, 2012). Furthermore, like CRM, such measures provide an online assessment of individual response that can be synchronously linked to specific message elements. In the context of hydraulic fracturing, an experiment could examine individual response to television news content where spokespersons employ various framing strategies or where study participants vary in knowledge of the technology. Not only could individual response be assessed after message consumption via traditional paper and pencil measures, but both CRM and psychophysiological measures of emotional response could be employed to study viewers' positive or negative affect to alternative framing strategies (e.g., gain or loss frames). Moreover, the use of specific visual frames (e.g., imagery, shot composition, or perspectives) could be linked to positive versus negative affective response using this technique. Again, a primary benefit of this approach is to ellicit responses that are less susceptible to conscious biases.

14.5 SUMMARY AND CONCLUDING REMARKS

In this chapter, a general review of the literature regarding public perceptions of fracturing impacts and technologies has been presented to help identify conceptual frameworks that investigators are employing to study the area. Ultimately, the greatest understanding of the media's role in shaping perceptions of fracturing developments would be gleaned through the integrated combination of information about the cultural environment surrounding fracturing developments, media coverage of such developments, and individual perceptions of both the issue in general and the responses to specific message strategies.

REFERENCES

Aarøe, L. (2011). Investigating frame strength: The case of episodic and thematic frames. *Political Communication*, *28*(2), 207–226. doi:10.1080/10584609.2011.568041.

Ader, C. R. (1995). A longitudinal study of agenda setting for the issue of environmental pollution. *Journalism & Mass Communication Quarterly*, *72*, 300–311.

Anderson, B. J. and Theodori, G. L. (2009). Local leaders' perceptions of energy development in the Barnett Shale. *Southern Rural Sociology*, *24*(1), 113–129.

Anderson, C. A. (1997). Effects of violent movies and trait hostility on hostile feelings and aggressive thoughts. *Aggressive Behavior*, *23*(3), 161–178.

Atman, C. J., Bostrom, A., Fischhoff, B., and Morgan, M. G. (1994). Designing risk communications: Completing and correcting mental models, Part I. *Risk Analysis*, *14*, 779–788.

Balcetis, E. and Dunning, D. (2006). See what you want to see: Motivational influences on visual perception. *Journal of Personality and Social Psychology*, *91*(4), 612–625.

Biocca, F., David, P., and West, M. (1994). Continuous response measurement (CRM): A computerized tool for research on the cognitive processing of communication messages. In A. Lang (ed.), *Measuring Psychological Responses to the Media* (pp. 15–64). Hillsdale, NJ: Erlbaum.

Bostrom, A., Fischhoff, B., and Morgan, M. G. (1992). Characterizing mental models of hazardous processes. *Journal of Social Issues*, *48*(4), 85–100.

Bostrom, A., Morgan, M. G., Fischhoff, B., and Read, D. (1994). What do people know about global climate change? 1. Mental Models. *Risk Analysis*, *14*, 959–970. doi: 10.1111/j.1539-6924.1994.tb00065.x.

Boudet, H., Clarke, C., Bugden, D., Maibach, E., Roser-Renouf, C., and Leiserowitz, A. (2014). Fracking controversy and communication: Using national survey data to understand public perceptions of hydraulic fracturing. *Energy Policy*, *65*, 57–67.

Brasier, K. J., Filteau, M. R., McLaughlin, D. K., Jacquet, J., Stedman, R. C., Kelsey, T. W., and Goetz, S. J. (2011). Residents' perceptions of community and environmental impacts from development of natural gas in the Marcellus Shale: A comparison of Pennsylvania and New York cases. *Journal of Rural Social Sciences*, *26*(1), 32–61.

Brown, E., Hartman, K., Borick, C., Rabe, B. G., and Ivacko, T. (2013). Public opinion on fracking: Perspectives from Michigan and Pennsylvania. University of Michigan. http://closup.umich.edu/files/nsee-fracking-fall-2012.pdf.

Cacciatore, M. A., Binder, A. R., Scheufele, D. A., and Shaw, B. R. (2012). Public attitudes toward biofuels: Effects of knowledge, political partisanship, and media use. *Politics and the Life Sciences*, *31*(1–2), 36–51.

Cohen, B. C. (1963). *The Press and Foreign Policy*. Princeton, NJ: Princeton University Press.

Davis, C. and Fisk, J. M. (2014). Energy abundance or environmental worries? Analyzing public support for fracking in the United States. *Review of Policy Research*, *21*(1), 1–16. doi: 10.1111/ropr.12048.

Dixon, T. L. (2006). Psychological reactions to crime news portrayals of black criminals: Understanding the moderating roles of prior news viewing and stereotype endorsement. *Communication Monographs*, *73*(2), 62–187.

Entman, R. M. (1993). Framing: Toward clarification of a fractured paradigm. *Journal of Communication*, *43*, 51–58.

Evensen, D. T., Clarke, C. E., and Stedman, R. C. (2014). A New York or Pennsylvania state of mind: Social representations in newspaper coverage of gas development in the Marcellus Shale. *Journal of Environmental Studies and Sciences*, *4*(1), 65–77.

Ferrar, K. J., Kriesky, J., Christen, C. L., Marshall, L. P., Malone, S. L., Sharma, R. K. et al. (2013). Assessment and longitudinal analysis of health impacts and stressors perceived to result from unconventional shale gas development in the Marcellus Shale region. *International Journal of Occupational and Environmental Health*, *19*(2), 104–112.

Frey, D. (1986). Recent research on selective exposure to information. In L. Berkowitz (ed.), *Advances in Experimental Social Psychology* (Vol. 19, pp. 41–80). New York: Academic Press.

Garnham, A. (1997). Representing information in mental models. In M. A. Conway (ed.), *Cognitive Models of Memory* (pp. 149–172). Cambridge, MA: The MIT Press.

Gentner, D. and Stevens, A. L. (eds.). (2014). *Mental Models*. East Sussex, England: Psychology Press.

Gerbner, G., Gross, L., Morgan, M., and Signorielli, N. (1980). Aging with television: Images on television drama and conceptions of social reality. *Journal of Communication*, *30*(1), 37–47.

Gilmore, J. S. (1976). Boom towns may hinder energy resource development: Isolated rural communities cannot handle sudden industrialization and growth without help. *Science*, *191*(4227), 535–540.

Glenberg, A. M. and Langston, W. E. (1992). Comprehension of illustrated text: Pictures help to build mental models. *Journal of Memory and Language*, *31*, 129–151.

Grabe, M. E. and Bucy, E. P. (2009). *Image Bite Politics: News and the Visual Framing of Elections*. New York: Oxford University Press.

Hallahan, K. (1999). Seven models of framing: Implications for public relations. *Journal of Public Relations Research*, *11*, 205–242.

Heuer, M. A. and Lee, Z. C. (2014). Marcellus shale development and the Susquehanna River: An exploratory analysis of cross-sector attitudes on natural gas hydraulic fracturing. *Organization & Environment*, 1–18.

Holbrook, R. A. and Hill, T. G. (2005). Agenda-setting and priming in prime time television: Crime dramas as political cues. *Political Communication*, *22*, 277–295.

Iyengar, S. (1994). *Is Anyone Responsible?: How Television Frames Political Issues*. Chicago, IL: University of Chicago Press.

Jacquet, J. B. (2012). Landowner attitudes toward natural gas and wind farm development in northern Pennsylvania. *Energy Policy*, *50*, 677–688.

Jacquet, J. B. and Stedman, R. C. (2011). Natural gas landowner coalitions in New York State: Emerging benefits of collective natural resource management. *Journal of Rural Social Sciences*, *26*(1), 62–91.

Jacquet, J. B. and Stedman, R. C. (2013). Perceived impacts from wind farm and natural gas development in northern Pennsylvania. *Rural Sociology*, *78*(4), 450–472.

Jacquet, J. B. and Stedman, R. C. (2014). The risk of social-psychological disruption as an impact of energy development and environmental change. *Journal of Environmental Planning and Management*, *57*(9), 1285–1304.

Johnson-Laird, P. N. (1983). *Mental Models: Towards a Cognitive Science of Language, Inference, and Consciousness* (No. 6). Cambridge, MA: Harvard University Press.

Josephson, W. L. (1987). Television violence and children's aggression: Testing the priming, social script, and disinhibition predictions. *Journal of Personality and Social Psychology*, *53*(5), 882–890.

Kahneman, D. and Tversky, A. (1979). Prospect theory: An analysis of decision under risk. *Econometrica*, *47*(2), 263–292.

Knight, M. G. (1999). Getting past the impasse: Framing as a tool for public relations. *Public Relations Review*, *25*, 381–398.

Kriesky, J., Goldstein, B. D., Zell, K., and Beach, S. (2013). Differing opinions about natural gas drilling in two adjacent counties with different levels of drilling activity. *Energy Policy*, *58*, 228–236.

Krosnick, J. A. and Kinder, D. R. (1990). Altering the foundations of support for the president through priming. *American Political Science Review*, *84*, 497–512.

Ladd, A. E. (2013). Stakeholder perceptions of socioenvironmental impacts from unconventional natural as development and hydraulic fracturing in the Haynesville Shale. *Journal of Rural Social Sciences*, *28*(2), 56–89.

Lecheler, S. and de Vreese, C. H. (2011). Getting real: The duration of framing effects. *Journal of Communication*, *61*(5), 959–983.

Maharik, M. and Fischhoff, B. (1992). The risks of using nuclear energy sources in space: Some lay activists's perceptions. *Risk Analysis*, *12*, 383–392.

McCombs, M. E. and Reynolds, A. (2009). How the news shapes our civic agenda. In J. Bryant and M. B. Oliver (eds.), *Media Effects: Advances in Theory and Research* (pp. 1–16). New York: Taylor & Francis Group.

McCombs, M. E. and Shaw, D. L. (1972). The agenda-setting function of mass media. *Public Opinion Quarterly*, *36*, 176–187.

Morgan, M. G., Fischhoff, B., Bostrom, A., Lave, L., and Atman, C. (1992). Communicating risk to the public. *Environmental Science and Technology*, *26*(11), 2048–2056.

Mutz, D. C. 2011. *Population-Based Survey Experiments*. Princeton, NJ: Princeton University Press.

Nan, X. (2012). Relative persuasiveness of gain- versus loss-framed human papillomavirus vaccination messages for the present- and future-minded. *Human Communication Research*, *38*(1), 72–94.

Neuman, W. R., Just, M. R., and Crigler, A. A. (1992). *Common Knowledge. News and the Construction of Political Meaning*. Chicago, IL: University of Chicago Press.

Newman, C. L., Howlett, E., Burton, S., Kozup, J. C., and Tangari, A. H. (2012). The influence of consumer concern about global climate change on framing effects for environmental sustainability messages. *International Journal of Advertising*, *31*(3), 511–527.

Onishi, N. (2014, May 15). California's thirst shapes debate over fracking. *New York Times*, p. A14.

Potter, R. F. and Bolls, P. D. (2012). *Psychophysiological Measurement and Meaning: Cognitive and Emotional Processing of Media*. New York: Routledge.

Rahm, D. (2011). Regulating hydraulic fracturing in shale gas plays: The case of Texas. *Energy Policy*, 39, 2974–2981.

Riffe, D., Lacy, S., and Fico, F. (2014). *Analyzing Media Messages: Using Quantitative Content Analysis in Research* (3rd edn.). New York: Routledge.

Roper, J. (2005). Symmetrical communication: Excellent public relations or a strategy for hegemony? *Journal of Public Relations Research*, *17*, 69–86.

Roskos-Ewoldsen, D. R., Roskos-Ewoldsen, B., and Carpentier, F. R. D. (2002). Media priming: A synthesis. *Media Effects: Advances in Theory and Research*, *2*, 97–120.

Schafft, K. A., Borlu, Y., and Glenna, L. (2013). The relationship between Marcellus Shale gas development in Pennsylvania and local perceptions of risk and opportunity. *Rural Sociology*, *78*(2), 143–166.

Schafft, K. A., Glenna, L. L., Green, B., and Borlu, Y. (2014). Local impacts of unconventional gas development within Pennsylvania's Marcellus Shale region: Gauging boomtown development through the perspectives of educational administrators. *Society & Natural Resources: An International Journal*, *27*(4), 389–404.

Shananhan, J. and Morgan, M. (1999). *Television and Its Viewers: Cultivation Theory and Research*. Cambridge, MA: Cambridge University Press.

Shen, F., Ahern, L., and Baker, M. (2014). Stories that count: Influence of news narratives on issue attitudes. *Journalism & Mass Communication Quarterly*, *91*(1), 98–117.

Slater, M. D. (2007). Reinforcing spirals: The mutual influence of media selectivity and media effects and their impact on individual behavior and social identity. *Communication Theory*, *17*, 281–303. doi: 10.1111/j.1468-2885.2007.00296.x.

Stedman, R. C., Jacquet, J. B., Filteau, M. R., Willits, F. K., Brasier, K. J., and McLaughlin, D. K. (2012). Marcellus Shale gas development and new boomtown research: Views of New York and Pennsylvania residents. *Environmental Practice*, *14*(4), 382–393.

Tewksbury, D. and Scheufele, D. A. (2009). News framing theory and research. In J. Bryant and M. B. Oliver (eds.), *Media Effects: Advances in Theory and Research* (pp. 17–33). New York: Taylor & Francis Group.

Theodori, G. L. (2009). Paradoxical perceptions of problems associated with unconventional natural gas development. *Southern Rural Sociology*, *24*(3), 97–117.

Theodori, G. L. (2012). Public perception of the natural gas industry: Data from the Barnett Shale. *Energy Sources, Part B: Economics, Planning, and Policy*, *7*(3), 275–281.

Theodori, G. L. (2013). Perception of the natural gas industry and engagement in individual civic actions. *Journal of Rural Social Sciences*, *28*(2), 122–134.

Theodori, G. L., Luloff, A. E., Willits, F. K., and Burnett, D. B. (2013). Hydraulic fracturing and the management, disposal, and reuse of frac flowback waters: Views from the general public in the Pennsylvania Marcellus Shale region (Final Research Report). Sam Houston State University. http://www.shsu.edu/~org_crs/Publications/hydraulic%20fracturing%20wastewater%20treatment%20final%20report.pdf.

Vidmar, N. and Rokeach, M. (1974). Archie Bunker's bigotry: A study in selective perception and exposure. *Journal of Communication*, *24*(1), 36–47.

Weaver, D., Graber, D. A., McCombs, M. E., and Eyal, C. H. (1981). *Media Agenda-Setting in a Presidential Election: Issues, Images, and Interest.* New York: Praeger.

Williams, A. E. (2011). Media evolution and public understanding of climate science. *Politics and the Life Sciences*, *30*(2), 20–30.

Willits, F. K., Luloff, A. E., and Theodori, G. L. (2013). Changes in residents' views of natural gas drilling in the Pennsylvania Marcellus Shale, 2009–2012. *Journal of Rural Social Sciences*, *28*(3), 60–75.

Winter, J. P. and Eyal, C. H. (1981). Agenda-setting for the civil rights issue. *Public Opinion Quarterly*, *45*, 376–383.

Wohlers, A. E. (2013). Labeling of genetically modified food: Closer to reality in the United States? *Politics and the Life Sciences*, *32*(1), 73–84.

Wolske, K., Hoffman, A., and Strickland, L. (2013). Hydraulic fracturing in the State of Michigan: Public perceptions of high-volume hydraulic fracturing and deep shale gas development (Graham Sustainability Institute Integrated Assessment Report Series, Vol. II, Report 8). Retrieved from University of Michigan website: http://pocarisweat.umdl.umich.edu/bitstream/handle/2027.42/102581/08_Public%20Perceptions.pdf?sequence=1, accessed July 8, 2014.

Wyer, R. J. and Radvansky, G. A. (1999). The comprehension and validation of social information. *Psychological Review*, *106*(1), 89–118.

Wynveen, B. J. (2011). A thematic analysis of local respondents' perceptions of Barnett Shale energy development. *Journal of Rural Social Sciences*, *26*(1), 8–31.

Yu, N., Ahern, L. A., Connolly-Ahern, C., and Shen, F. (2010). Communicating the risks of fetal alcohol spectrum disorder: Effects of message framing and exemplification. *Health Communication*, *25*(8), 692–699.

Zillmann, D. and Bryant, J. (1985). *Selective Exposure to Communication*. Hillsdale, NJ: Erlbaum.

15 Multidisciplinary Teams as Mechanisms of Accountability

Neutralizing the Emotions and Politics of Hydraulic Fracturing Research

Kay J. Tindle, Daniel Marangoni, and Anna Thomas Young

CONTENTS

15.1 Overview 273
15.2 Introduction 274
15.3 Researching Hydraulic Fracturing 274
15.4 Learning from the Past 274
15.5 Public Responsibility and the Need for Unbiased Research 275
15.6 Transition to Multidisciplinary Teams 276
15.7 Multidisciplinary Teams as Mechanisms of Accountability 277
15.8 Multiple Perspectives to Balance Politics and Biases 278
15.9 Cultivating an Institutional Culture of Research Integrity 278
15.10 Summary and Closing Remarks 281
References 282

15.1 OVERVIEW

A multiple lens approach with a focus on finding innovative solutions by utilizing multiple perspectives is argued to be the best way to study contentious issues like hydraulic fracturing. This approach provides a broad perspective, and along with associated financial incentives, accountability mechanisms, and loosely coupled teams, has the ability to neutralize politically charged issues and lead to unbiased, scientifically credible solutions.

15.2 INTRODUCTION

Hydraulic fracturing, as with any politically charged topic, is either lacking in unbiased, scientific research or is clouded by emotions and politics. Both sides of the debate pay their own researchers, who then likely produce results that favor their respective side. What should be a discussion based on science becomes a subjective debate (Hubner et al., 2013). Though industry-funded research may produce unbiased, factual results, the funding source may taint these results or encourage public suspicion, especially when researchers have a vested interest in the success of an oil and gas company. It is, therefore, the responsibility of public institutions that serve the public good to utilize the scientific process to produce accurate, unbiased results that inform public policy.

Unfortunately, federal agencies and public higher education institutions (HEIs) often fear that publishing unfavorable results may harm existing or future relationships with the oil and gas industry or worry that industry will corrupt scientific inquiries to serve their interests. Multidisciplinary teams can act as accountability agents by ensuring that strong research agreements and best practices are in place to avoid such issues. Furthermore, the synergistic nature of a multidisciplinary effort results in a more comprehensive approach to the issue. Social science researchers study the impact of the issue on populations, while engineers improve the technologies and communications experts engage the public.

Multidisciplinary teams provide an expansive view of the issue and can serve as mechanisms of accountability. Politically charged issues, such as hydraulic fracturing, require a balanced, unbiased approach, which can occur with multidisciplinary teams. The biases of each single discipline provide balance and hold each other accountable. Furthermore, multidisciplinary teams need institutional support and will thrive within a culture of research integrity. Multidisciplinary teams, being loosely coupled and not psychologically attached to the issue, approach it through multiple lenses, thereby conducting responsible research. This multiple lens approach enables an accountability mechanism and allows the team to perceive the issue in context.

15.3 RESEARCHING HYDRAULIC FRACTURING

Hydraulic fracturing is best examined through a multidisciplinary lens, one that approaches the topic with a focus on finding innovative solutions by utilizing multiple perspectives. In order for teams to be successful in the current research climate, these solutions must appeal to proponents of both sides of the issue. Research teams can patent technology that will lessen the environmental impact of hydraulic fracturing and will reduce costs for oil and gas companies. It is important for these teams to fully disclose all conflicts of interest in order to prevent a reoccurrence of past scandals.

15.4 LEARNING FROM THE PAST

Although corruption in research is not unique to the hydraulic fracturing controversy, it has received a large amount of attention from the media in recent years. With decreases in government funding of research and increases in corporate research funding, more conflicts will likely be uncovered in future years. University leaders will continue to struggle with how to balance serving the best interests of the public and financing its research studies. If they do not cultivate a culture of full disclosure and due process, their institutions and their researchers will face the consequences.

In 2012, researchers at the Energy Institute at the University of Texas at Austin released a report stating there is no correlation between hydraulic fracturing and groundwater contamination. Not only did the report appear to be a rough draft that lacked serious peer review, the lead researcher, Chip Groat, failed to disclose financial conflicts of interest as a board member of Plains Exploration and Production (PXP) and his $1.6 million of stock in the oil and gas company (Connor et al., 2012). Disclosing this information and submitting the study to peer review might have saved Groat from resigning.

Later in that same year, the State University of New York (SUNY) at Buffalo closed its Shale Resources and Society Institute amidst pressure from a petition signed by over 10,500 professors, students, and trustees regarding the institute's questionable financing (Navarro, 2012). The Public Accountability Initiative (PAI), a nonprofit watchdog group, analyzed a report issued by the institute earlier in the year and found issues of self-plagiarism and industry bias (Connor et al., 2012). In response, SUNY leadership has organized a committee to examine its current conflict of interest policies and determine how to strengthen these policies to ensure all financial conflicts of interest are disclosed (Navarro, 2012). Whether peer-reviewed or not, industry funding diminishes public trust (Hubner et al., 2013). Any ties to industry should be fully disclosed in order to prevent distrust.

15.5 PUBLIC RESPONSIBILITY AND THE NEED FOR UNBIASED RESEARCH

The previous examples are indicative of a larger systemic problem that is not isolated to only the oil and gas industry. Other research reports laced with controversy evolve out of inappropriate or undisclosed relationships with pharmaceutical, tobacco, and chemical companies (Henry, 2012). Institutions pursuing research on politically charged topics should ensure their teams disclose all conflicts of interest. Failure to disclose conflicts of interest can discount the scientific merit of a publication and limit its contribution to the field of study.

Public institutions have a responsibility to serve the public trust by conducting and openly disseminating unbiased research to inform public opinion. Public controversies of this nature contradict the mission of public higher education and cause public HEIs across the nation to lose their credibility. Without the public's trust and institutional scientific integrity, no progress can be made in determining public policy that will in turn benefit the public good.

Some federal agencies are conducting scientific inquiry into the environmental impacts of hydraulic fracturing. The U.S. Geological Survey is looking into the dramatic increase of earthquakes in central Oklahoma, finding a correlation between wastewater disposal near a fault line and increased seismicity (Ellsworth et al., 2014). Also, the Shale Gas Production Subcommittee of the Secretary of Energy Advisory

Board (2011) produced a series of reports outlining best practices for safe, responsible drilling. However, neither of these has resulted in changes to public policy or environmental regulations.

The Environmental Protection Agency (EPA) is studying possible impacts of hydraulic fracturing and related activities on air and water quality, but results are slow and regulations are even slower. The recently established air pollution standards for oil and gas companies were published to meet a court-imposed deadline (EPA, 2012), and the final draft report for drinking water impacts were open for comment in late 2014 (EPA, 2014). An undesignated amount of time will be required for peer review and public comment before publishing a final report and thus, any resulting regulations might not be enacted for a few more years. Furthermore, the nationwide study will not include the investigation into the source(s) of methane in Pavillion, Wyoming's, drinking water, which, in 2013, the EPA turned over to the state of Wyoming to complete. Public HEIs can fill the gap for federal agencies, such as the EPA.

Public HEIs can reduce the time incurred from study design to changes in policies and regulations by publishing peer-reviewed, unbiased research on open-source platforms. Federal agencies can then use these studies to build on their own research, resulting in a solid research foundation upon which they can make policy decisions. Furthermore, multidisciplinary teams can engage stakeholders, which is becoming increasingly important (Hubner et al., 2013).

With politically charged issues, the most recent or most emotionally laden story sways public opinion. Public HEIs should be the voice of reason, stating the facts, conducting peer reviews of the science, and fully disclosing any conflicts of interest. Studies show public HEIs are generally viewed as trustworthy sources of information (Hubner et al., 2013); however, when individual faculty researchers break these bonds of trust, it reflects on academia as a whole. Industry funding also discredits the results of research conducted by public HEIs.

Faculty researchers need to reflect on how their actions might compromise the unbiased work of others and reflect poorly on the institution. The integrity of the research institution is based on the integrity of its researchers. Faculty researchers at public HEIs have a distinct responsibility to the consumer of research to conduct ethical studies that produce unbiased results. The following sections show that multidisciplinary teams can serve as mechanisms of accountability and more specifically can provide a larger scope of assessment than single disciplinary work.

15.6 TRANSITION TO MULTIDISCIPLINARY TEAMS

Higher education institutions often employ rigid and difficult-to-change operating policies that can limit the implementation of more flexible and multidisciplinary research methods (Lucas, 2006). This traditional framework for research in higher education has been constructed through single disciplinary work (Wuchty et al., 2007). Separating colleges, departments, and areas to create boundaries has become a normal process within higher education—which are modernly referred to as educational silos (Stock and Burton, 2011). These boundaries protect the customary methods and processes for conducting research. Traditionally, academic disciplines were used for frames of reference, methodological approaches, theoretical bases, topics and technologies, which limited interaction and exploration into broader issues (Stock and Burton, 2011). At times, the use of single discipline lenses is relevant and important when studying specific and narrow topics.

However, today's research landscape requires a multidisciplinary approach. Over the past 50 years, teamwork is increasingly more utilized for research development and commercialization activities in higher education (Wuchty et al., 2007). Multidisciplinary research teams are providing a more expansive view of dynamic issues that singular discipline lenses cannot match (Disis and Slattery, 2010). A multidisciplinary perspective uncovers new, innovative solutions to global issues. Researchers provide insight from perspectives that build upon one another (Börner et al., 2010). Furthermore, traversing traditional disciplinary boundaries can create a culture of accountability within the team.

This process of problem solving with multiple disciplines is deemed the *Team Science* approach and solves larger-based problems from multiple lenses (Pennington, 2010). The Science of Team Science, or SciTS, concept addresses key challenges spanning macro, meso, and micro levels of analysis. The macro or population level identifies growth, patterns, and substantial changes, often noted through big data consensus; the meso or group level analyzes intra team communications, composition, and contribution to team processes and outcomes; and the micro level examines the individual and the training, education, discipline, and other personal components that influence team research (Börner et al., 2010). Impacting all levels—macro, meso and micro—multidisciplinary research proves to be the preferred method for studying comprehensive and large issues, as well as those politically charged and culturally sensitive.

15.7 MULTIDISCIPLINARY TEAMS AS MECHANISMS OF ACCOUNTABILITY

Recognizing the need for and success of multidisciplinary solutions, academic institutions are investing in diverse research teams (Milem, 2003). The medical community has also recognized multidisciplinary teams as critical to finding long-term, innovative solutions. Discoveries made by such teams, in return, have a much higher rate of long-term sustainability (Disis and Slattery, 2010). Success and sustainability for multidisciplinary research teams is measured through interactions, publications, external funding, and developed Intellectual Property (IP) or tech transfer. Through initial augmentation of ideas, teams then review potential discoveries, which can optimize breakthroughs while also limiting failures.

The team approach, or network thinking, reduces time of failure by providing immediate evaluation from multiple perspectives and also serves as a mechanism of accountability (Disis and Slattery, 2010). When an issue is studied through multiple perspectives, the time to failure can be reduced, which means if an innovation will not be successful, multidisciplinary teams can recognize it earlier than a single disciplinary researcher. Examining an issue through a single disciplinary lens could encourage a fallacy of logic that would be less likely to occur when examined through multiple angles and multiple perspectives. The biases of a single discipline can be mitigated when balanced by multiple disciplines. Likewise, unethical practices can be prevented when team members are held accountable by each other. Actions and decisions must be justified to other members of the group, and a multidisciplinary approach can balance political pressure.

15.8 MULTIPLE PERSPECTIVES TO BALANCE POLITICS AND BIASES

Faculty researchers, as public employees, should uphold the highest standards of responsible research. They cannot succumb to the expectations of industry funders or political affiliations. When studying politically charged issues, academic researchers owe it to the public interest to provide an objective and unbiased approach to the issue. Forgoing all external pressures, they should focus on the science, not the expected results.

In recent years, researchers and their unethical practices have dissuaded others from studying this politically charged topic. Other reasons for caution regarding the effects of hydraulic fracturing on the environment include the lack of funding for such a research topic, a possible "whistle-blower" reaction, and the abundance of funding for increasing hydraulic fracturing technologies. One solution could be to focus on cost-effective solutions that increase productions and decrease possible harmful effects.

Multidisciplinary teams studying hydraulic fracturing can provide a balance by improving technology efficiencies while lessening environmental effects. Because the question of whether or not hydraulic fracturing is harmful is not one that can attract much funding, multidisciplinary teams can examine practices and technologies through a mitigating lens. Multiple perspectives can satisfy both sides of the issue if they find solutions that decrease cost, increase efficiency, and decrease harm. With federal agencies avoiding the issue or taking years to complete a study, the responsibility falls to the faculty researcher to take the initiative and institutional leadership to cultivate a culture of research integrity.

15.9 CULTIVATING AN INSTITUTIONAL CULTURE OF RESEARCH INTEGRITY

Institutions of higher education and their collaborative partners need to be specifically concerned with conducting research using appropriately constructed methodologies and analyzing results using rigorous techniques. To cultivate a culture of research integrity that strives to produce accurate, ethical, and unbiased results, there needs to be a particular focus on creating multidisciplinary research teams and motivating appropriate participatory behavior by team members.

In order to create effective multidisciplinary teams, especially at institutions of higher education, a common misconception of how different areas within the organization are linked should be debunked. Research institutions that receive federal contracts and grants often focus heavily on accountability, assessment, and shared goals. This mode of thinking is important to supporting ethical and legal operating research programs, but it creates a "preoccupation with rationalized, tidy, efficient, coordinated structures which [can blind] many practitioners as well as researchers" that might limit the design of multidisciplinary research teams (Weick, 1976). Furthermore, researchers should focus on the loose coupling of entities within a single institution; in this case, that means seeing different disciplines as loosely coupled within the institution (Weick, 1976). This mode of thinking promotes the ability and possibility of bringing diverse groups together. The failure to bring together multidisciplinary teams often comes from the belief that the disciplines in question are not connected enough or are too connected to work together effectively. It is important to understand that, for the sake of multidisciplinary research, this is not the case.

Researchers in a particular discipline should understand the loosely connected nature of partnerships with other disciplines or departments. Multidisciplinary collaboration is more likely to occur when research administrators or others invested in the research help participants understand the nonbinding and nonpermanent nature of the connection. To cultivate a culture of research integrity that can stem from multidisciplinary teams, institutional leadership should clearly explain the nature of the relationships and the implications it can have on the research.

There are a few important implications of bringing together loosely coupled teams. First, multidisciplinary teams that are loosely coupled are better at sensing external influences. Weick (1976) illustrates this point by discussing the difference between tossing up a handful of sand into the air compared with tossing up a handful of large rocks. The sand is made of smaller particles that are more loosely connected than those in the rocks—as such; they are better at demonstrating wide influence. Likewise, multidisciplinary teams that are loosely coupled can sense the external environment. For example, research studies performed in a single department with researchers who have been working together for decades are less likely to identify the external climate and related issues than a research team made up of members of several departments who have not worked together frequently. While institutions should foster the healthy working relationships that come from long-term engagement between researchers, it is important that they include a diverse spectrum of participants that are more loosely connected.

One issue with sand in the wind is just how vulnerable it is to the wind. It might be concerning that loosely coupled teams are so sensitive to the external environment. There are aspects of these teams, however, that make them more resilient to negative external pressures. This has positive implications for the ethical concerns that arise with such challenging research. This can be particularly true in hydraulic fracturing. Multidisciplinary teams are not only loosely coupled internally, but also loosely coupled with hydraulic fracturing or other politically charged issues.

One major drawback to single entity or single discipline research in hydraulic fracturing is the potentially biased outputs of research given the strongly coupled relationship between proponents of the issue with their selected researchers and opponents with their selected researchers. Multidisciplinary teams are less likely to support a single bias, and as such, are more loosely connected with the issue itself. Given the politically charged nature of the topic, opinions and attitudes toward hydraulic fracturing are susceptible to rapid change. The loose associations between multidisciplinary teams and opinions about hydraulic fracturing means that multidisciplinary teams have a smaller probability of changing, or being able to change, their research each time the political environment changes. Multidisciplinary teams provide a better guarantee against the possibility of money, politics, and popular opinion tainting methods and analysis.

Not only can multidisciplinary teams better ensure research integrity, they can also produce more innovative ideas and solutions. Weick (1976) says that because these groups maintain "identity, uniqueness, and separateness of elements…[they] potentially can retain a greater number of mutations and novel solutions" that would not have otherwise been the case. With external pressures potentially affecting the integrity of research on hydraulic fracturing, institutions should insist that research teams be comprised of individuals that are given the ability to preserve a relative level of separateness from each other. If researcher contamination is possible, loosely grouped teams are less likely to be influenced by individual members and are more resilient to external pressures. Moreover, the complexity of hydraulic fracturing demands research that produces real and impactful results. Multidisciplinary teams are more likely to approach problems from a number of lenses and find more novel solutions.

Institutions of higher education and their collaborative partners, especially when researching with federal funds, should insist on ethical and unbiased research. This is a matter of research quality, responsible conduct in research, and institutional accountability. As such, researchers should be concerned that their involvement with hydraulic fracturing could be susceptible to negative external influences and public suspicion. Consequently, it is natural for institutions to expect that research on hydraulic fracturing follow strict and rigorous measures to ensure research integrity. Multidisciplinary teams are a better mechanism for ensuring this integrity due to its loosely coupled relationship with hydraulic fracturing issues.

Apart from creating multidisciplinary teams as a means of creating a culture of research integrity at institutions, it is important that teams motivate appropriate behavior and contributions of its members. One benefit to a multidisciplinary team over a team of researchers from the same discipline is that each member is likely to contribute his or her separate and unique identity. It is important that institutions support the autonomy of individuals within research groups. To create a culture of research integrity, all participants must be motivated toward high-quality outputs.

Multidisciplinary teams also run the risk of not being supportive to each teammate. Often times, Principal Investigators favor working with colleagues they have worked with in the past, and communicate poorly with new or unfamiliar people. Looking at human motivation, self-determinations theory (SDT) states that people are "inherently active, intrinsically motivated, and oriented toward developing naturally through integrative processes" (Deci and Ryan, 2012). Just as people need physical nourishment to thrive, they also need psychological support to maintain their natural disposition towards motivation. SDT identifies that the needs of autonomy, competence, and relatedness must be met for healthy motivation. Since researchers, such as university faculty, are naturally motivated to produce high-quality work, research groups should help support this drive.

With loosely coupled teams, there is a natural predisposition to support individual identity, and as such, individual autonomy on projects. This means that after tasks are assigned, loosely coupled teams are well structured to support the autonomous completion of tasks. To maintain autonomy, groups must take the time to ensure that tasks are being assigned appropriately according to ability and experience, and that individuals have input regarding deadlines.

These groups should also be structured so as to foster the competence of participants. Multidisciplinary teams should strive to support the development of new abilities and the improvement of current abilities by exposing all members to new ideologies and methods that the other disciplines bring to the table. It is important for research integrity that growth, curiosity, and the desire for quality outputs be supported. While supporting autonomy, multidisciplinary groups also need to commit to a shared vision that allows all members to develop abilities together.

The need for togetherness, or relatedness, is the "need to be close to, trusting of, caring for, and cared for by others" (Deci and Ryan, 2012). Naturally, this can be a challenge in multidisciplinary groups. On a basic level, relatedness can be supported through the absence of negative experiences during team member interaction. At the very least, institutions need to foster a culture of supportive and nonnegative interactions where team members are treated respectfully. Research integrity also depends on the trust between members of the team. While multidisciplinary team members are not familiar enough with each other to be directly trusting of individual members, the loosely coupled nature of the interaction insulates the team as a whole from corruption and deliberate error. While individuals might be slower to trust other team members, institutional leadership should remind the group that multidisciplinary teams as a whole are better sensing mechanisms to potential external threats and influences.

Research integrity is of particular concern in hydraulic fracturing given its polarizing nature. Consequently, institutions can foster a culture of research integrity by developing multidisciplinary research teams that support the psychological needs of its members. Multidisciplinary teams can come together once organizations understand the loose connection between each of the disciplines represented. The misconception that departments are either too connected or not connected enough often keeps multidisciplinary collaborations from happening. Loose connections foster a more resilient environment that insulates against ethical misgivings that can plague research in this field. Once groups are created and institutions are ensured that members feel autonomous, competent, and related; then researchers will be naturally motivated to produce high-quality work.

15.10 SUMMARY AND CLOSING REMARKS

Multidisciplinary research teams attempt to attack complex societal problems from a comprehensive perspective. Collaborative research enables positive outcomes through team building, communications, and managing personalities. These constructive modes for management are promising but do not fully dismiss all concerns. Though many multidisciplinary research teams struggle with a variety of substantive differences, positive qualities have the potential to outweigh these issues to enable the generation of a single overarching conceptual framework.

Furthermore, multidisciplinary research teams can serve as mechanisms of accountability when studying politically charged topics, such as hydraulic fracturing. Institutions that create a culture of research integrity and encourage researchers to fully disclose all conflicts of interest will begin to bring back the public trust. Multidisciplinary research teams can look for areas of study with the most impact to both sides of politically charged issues. If researchers at public HEIs focus their research on hydraulic fracturing, they should focus on tangential issues, such as water scarcity, water quality, water reuse, air quality, transportation infrastructure, public opinion, and others. Approaching issues in this manner will prevent backlash and allow for the research to move forward. Multidisciplinary research will be the norm of future research approaches due to its broad perspective, associated financial incentives, accountability mechanisms, loosely coupled teams, and ability to neutralize politically charged issues.

REFERENCES

Börner, K., Contractor, N., Falk-Krzesinski, H. J., Fiore, S. M., Hall, K. L., Keyton, J., Spring, B., Stokols, D., Trochim, W., and Uzzi, B. (2010). A multi-level systems perspective for the science of team science. *Science Translational Medicine 2*, 49cm24.

Connor, K., Galbraith, R., and Nelson, B. (2012 July). Contaminated inquiry: How a University of Texas fracking study led by a gas industry insider spun the facts and misled the public. Public accountability initiative. Retrieved from http://public-accountability.org/wp-content/uploads/ContaminatedInquiry.pdf, accessed on February 19, 2015.

Deci, E. L. and Ryan, R. M. (2012). Self-determination theory. In P. A. M. Van Lange, A. W. Kruglanski, and E. T. Higgins (eds.), *The Handbook of Theories of Social Psychology* (pp. 416–437). London, U.K.: SAGE.

Disis, M. L. and Slattery, J. T. (2010). The road we must take: Multidisciplinary team science. *Science Translational Medicine 2*(22), 22cm9.

Ellsworth, W., Robertson, J., and Hook, C. (2014, January 14). Man-made earthquakes update. United States Geological Survey. Retrieved from http://www.usgs.gov/blogs/features/usgs_top_story/man-made-earthquakes/, accessed on February 19, 2015.

Environmental Protection Agency (EPA). (2012, April 4). EPA issues updated, achievable air pollution standards for oil and natural gas/Half of fractured wells already deploy technologies in line with final standards, which slash harmful emissions while reducing cost of compliance. Retrieved from http://yosemite.epa.gov/opa/admpress.nsf/d0cf6618525a9efb85257359003fb69d/c742df7944b37c50852579e400594f8f%21OpenDocument, accessed on February 19, 2015.

Environmental Protection Agency (EPA). (2014). EPA's study of hydraulic fracturing for oil and gas and its potential impact on drinking water resources. Retrieved from http://www2.epa.gov/hfstudy, accessed on February 19, 2015.

Henry, T. (2012, December 7). Why the UT fracking study controversy matters. State impact. Retrieved from http://stateimpact.npr.org/texas/2012/12/07/why-the-ut-fracking-study-controversy-matters/, accessed on February 19, 2015.

Hubner, A., Horsfield, B., and Kapp, I. (2013). Fact-based communication: The shale gas information platform SHIP. *Journal of Environmental Earth Science 70*, 3921–3925. doi: 10.1007/s12665-013-2504-y.

Lucas, C. J. (2006). *American Higher Education: A History*. New York: Palgrave Macmillan.

Milem, J. F. (2003). The educational benefits of diversity: Evidence from multiple sectors. In: Chang, M., Witt, D., Jones, J., and Hakuta, K. (eds.), *Compelling Interest: Examining the Evidence on Racial Dynamics in Higher Education* (pp. 126–169). Stanford, CA: Stanford Education.

Navarro, M. (2012, November 19). SUNY buffalo shuts down its institute on drilling. *The New York Times.* Retrieved from http://www.nytimes.com/2012/11/20/nyregion/suny-buffalo-will-end-controversial-fracking-institute.html, accessed on February 19, 2015.

Pennington, D. (2010). The dynamics of material artifacts in collaborative research teams. *Computer Supported Cooperative Work 19*(2), 175–199. doi: 10.1007/s10606-010-9108-9. Retrieved on July 18, 2014 from http://www.springerlink.com/openurl.asp?genre=articleandid=doi:10.1007/s10, accessed on July 18, 2014.

Secretary of Energy Advisory Board (SEAB). (2011, November 18). The SEAB shale gas production subcommittee second ninety day report—November 18, 2011. United States Department of Energy, Washington, D.C. Retrieved from http://www.shalegas.energy.gov/resources/111811_final_report.pdf, accessed on February 19, 2015.

Stock, P. and Burton, R. J. (2011). Defining terms for integrated (multi-inter-trans-disciplinary) sustainability research. *Sustainability 3*(8), 1090–1113.

Weick, K. E. (1976). Educational organizations as loosely coupled systems. *Administrative Science Quarterly, 21,* 1–19.

Wuchty, S., Jones, B. F., and Uzzi, B. (2007). The increasing dominance of teams in production of knowledge. *Science 316,* 1036–1038.

Index

A

Advanced Notice of Public Rulemaking (ANPR), 85, 87–88
Agenda-setting, 261–262
ALARP risk triangle, 16
Aqua-Pure ROVER Mobile Clarifier, 170
Arsenic, 106, 108, 126–127

B

Barnett Shale, 4–5, 13, 126–127, 171–172
Benzene
  chemical exposure impacts, 96
  leaks, 103
  toxic air emissions, 102
  VOC, 102–103
Biogenic methane, 117–118
Bottom hole pressures (BHPs), 12
Bureau of Land Management (BLM), 84–85

C

Chemical disclosure, EPA ANPR, 87–88
Clean Air Act NSPS Subpart OOOO, 85
Closest facility analysis
  benefits, 192
  data sets, 192
  disposal wells, 196–197
  empirical cumulative distribution, 193
  number of wells, 196, 198
  requirements, 192
  statistics, 193–194, 196, 198
  travel routes, 193, 195–196
  unpaved/private roads, 196, 199
Communication research, 253–259
Concept mapping, 264
Continuous response measurement (CRM), 267
Critical balancing process, 211
Crystalline silica, 96
Cultivation theory, 264

D

Degree of polymerization, 236–237
Department of Environmental Protection (DEP), 106–107
Differential scanning calorimetry (DSC), 241
Dunes sagebrush lizard, 76–77

E

Electron capture detector (ECD), 121
Electron ionization (EI), 121
Endangered Species Act (ESA), 75–76
Energy Policy Act, 91, 93
Environmental Protection Agency (EPA), 276
  air and water quality, 276
  chemical disclosure, 87–88
  drinking water pollution, 9
  groundwater contamination, 106
  Indian Country minor source, 88
  NGOs, 86–87
  NSPS Subpart OOOO, 86
  VOC emissions, 104
Episodic framing, 262

F

Fayetteville Shale, 172
Flowback period, 84, 86, 104
Formational water, 163–164
Fourier transform infrared spectrometer (FTIR)
  attenuated total reflectance, 243
  functional group modification, 244
  structural changes, 243–244
FracFocus Chemical Disclosure Registry, 74, 85
Freshwater production
  chemical constituents, 146–147
  concentration ranges, 148–149
  "produced water," 146, 148
FTIR, *see* Fourier transform infrared spectrometer (FTIR)

G

Galactomannan, *see* Guar gum
Gamma ray tracers, 15
GC–MS instruments, 121–123
Gel permeation chromatography (GPC), 233, 236–237, 247–248
Geographic information systems (GIS), *see* Wastewater disposal

Groundwater availability modeling (GAM), 139
Groundwater production
  aquifer transmissivity, 150
  distance–drawdown profile, 151–152
  irrigation agriculture, 153–155
  transmissivity and storage coefficients, 150–151
  wavelength calculations, 150
  well yields, 152
Guar
  bean structure, 209
  breeding process, 212–214
  nutrient resource, 214
  petroleum industry
    critical balancing, 211
    enzymes, 211
    fracture conductivity, 212
    "40 lb gel," 210
    LGC, 211
    oxidizers, 211
    proppants, 211
    RAB test, 212
  processing steps, 215–216
  seed coat color, 209–210
  Thar Desert region, 214–215
  U.S. Southwest regions
    alfalfa midge, 220
    Alternaria, 220
    "catch crop," 222
    challenges, 221–222
    crop insurance, 217
    crop rotation, 221
    field conditions, 218–219
    genetics and varieties, 217
    harvest management, 220–221
    herbicides, 218
    IR-4 program, 218
    irrigation process, 219–220
    production goals, 222
    Rhizobium bacteria, 219
Guar gum
  applications, 208
  chemical modifications, 247–248
  degree of polymerization, 236–237
  extraction process
    copper content, 230
    endosperm crushing, 228–229
    ethanol precipitation, 229
    sodium azide, 230
  FTIR analysis
    attenuated total reflectance, 243
    functional group modification, 244
    structural changes, 243–244
  galactose to mannose ratio
    acid hydrolysis, 234–235
    chromatogram, 234
    HPLC, 234
    purification method, 235
  hydrolysis process, 232–233
  molecular weight, 236–237
  solubility analysis, 231–232
  structural analysis, 230–231
  thermal stability (*see* Thermogravimetric analysis (TGA))
  viscoelastic properties
    concentration, 237–238
    impurities, 238–240
    pH, 238
    salts, 246–247
    temperature, 238
  XRD analysis, 245

H

Haynesville Shale, 172
High-performance liquid chromatography (HPLC), 95–96, 233–235
Huggins coefficient, 230, 240
Hydraulic fractures
  BHPs, 12
  chemical tracers, 15
  completion/frac design, 10–11
  complexity, 11–12
  energy supply, 6
  EPA ANPR, 87–88
  flowback, 15, 84
  fracture treatments, 4–5
  gamma ray tracers, 15
  Indian Country minor source, ANPR, 88
  industrial and public risk, 93–94
  injection rate, 9–10
  LDAR, 89
  load recovery, 15
  multifractured wells, 3–4
  NGO petition, 86–87
  oil production in U.S., 91–93
  potential requirements, 85
  proppants, 8
  pump chart, 12–13
  refracturing, 14–15
  regulations, 84–85
  risk factors, 16–18
  sequential fractures, 12–14
  shale gas, 2, 4, 6
  simultaneous fractures, 12–14
  slickwater fracs, 7
  toxic chemicals, 94–97
  trace chemicals, 9
  water use, 73–75
  water volume, 8
  white papers, 85–86

I

ICP–OES, *see* Inductively coupled plasma–optical emission spectroscopy (ICP–OES)
IMPLAN models, 23, 43, 61–62
Indian Country minor source, 88
Inductively coupled plasma–optical emission spectroscopy (ICP–OES), 124–125

L

LCA, *see* Life cycle assessment (LCA)
Leak detection and repair (LDAR), 89
Legal issues
  baseline data, 67
  Coastal Oil & Gas Corp. v. Garza Energy Trust, 71–72
  communication, 68
  elected officials, 69
  field history, 67
  FPL Farming Ltd. V. Environmental Processing Systems, 72–73
  knowledge, 70
  landowner's litigation, 69
  land use, 68
  Railroad Commission v. Manziel, 70–71
  secrecy, 69
  source of drinking water, 68
  wealth, 69
Lesser prairie chicken, 77–78
Life cycle assessment (LCA), 189–190
Liquid gel concentrates (LGC), 211

M

Mannosidases, 211
Marcellus Shale, 107, 118, 126–127, 172–173, 254–256
Mass spectrometry detector (MSD), 121–123
Media priming, 264
Multidisciplinary research teams
  culture of research integrity
    loosely coupled teams, 279
    self-determinations theory, 280
    togetherness/relatedness, 281
  politics and biases, 278
  Team Science approach, 277

N

National Emissions Standards for Hazardous Air Pollutants (NESHAP), 87–88
Naturally occurring radioactive material (NORM), 127–128, 188
New Source Performance Standard (NSPS) Subpart OOOO, 83, 85–87, 89
NOMAD 200 MVR Evaporator system, 170
Nongovernment organizations (NGOs), 85–87
NORM, *see* Naturally occurring radioactive material (NORM)
NSPS Subpart OOOO, *see* New Source Performance Standard (NSPS) Subpart OOOO

O

Oil and gas development
  benzene
    leaks, 103
    toxic air emissions, 102
    VOC, 102–103
  considerations, 89
  groundwater
    casing and cementing, 107
    contamination and risk, 105–106
    DEP, 106
    EPA, 106
    open drilling reserve pits, 106–107
    state law, 108–109
    Texas, 108
  human illness, 104–105
  precautionary principle, 109
  underground migration of gas, 107–108
  VOCs, 104

P

PAI, *see* Public Accountability Initiative (PAI)
Pavement Condition Index, 180
Permian Basin
  completion efficiencies
    horizontal drilling, 40–41
    initial oil rate per well drilled, 37, 39
    permit *vs.* time plot, 41–42
    technological trends, 39, 41
  drilling activity
    active well-type distribution, 33–34
    annual average rig count, 36–37
    basin/plays rig count, 34, 36
    breakeven WTI price, 30–31
    daily gas production, 29–30
    daily oil production, 29–30
    drilled well statistics, 29
    rig drill type distribution, 37–38
    U.S. states rig count, 34–35
    well status distribution, 32–33
  economic impacts
    direct effects, 42
    induced effects, 42
    input–output models, 42
    multipliers, 43
    New Mexico, 45, 51
    oil production, 54

rig counts, 52–53
Texas, 44–51
total production level, 53
economic landscape, 25–26
IMPLAN models, 23, 61–62
oil and gas production
annual production rate, 28–29
Delaware Basin South East New Mexico, 27
Delaware Basin Texas, 27
historical analysis, 23–25
Midland Basin, 28
unconventional plays/reservoirs, 28
taxation impacts
franchise tax, 56–60
sales tax, 56
severance tax, 56
use tax, 56
well services, 56
Permian Basin region, Texas
agriculture groundwater use, 139–141
freshwater production
chemical constituents, 146–147
concentration ranges, 148–149
"produced water," 146, 148
geographic extent, 134
GIS, wastewater disposal (*see* Wastewater disposal)
groundwater production
aquifer transmissivity, 150
distance–drawdown profile, 151–152
irrigation agriculture, 153–155
transmissivity and storage coefficients, 150–151
wavelength calculations, 150
well yields, 152
unconventional oil and gas production
factors, 145
hydraulic fracturing operations, 142, 144
irrigation groundwater use, 142–143
median water use, 142, 144
mining groundwater use, 142–143
well-drilling operations, 145
water balance model, 154
water resources
GAM, 139
historical annual precipitation, 135–136
major aquifers, 135, 137
Martin County, TX, 135, 138
minor aquifers, 135, 137–139
surface depressional features, 138
surface water bodies, 135–136
Proppants, 8, 159, 165, 167, 176, 211
Public Accountability Initiative (PAI), 275
Public higher education institutions, 275–276
Public perception
attitudinal stages, 260
impacts and technologies
attitudinal stages, 260
communication research, 253–259
economic detriments, 260
integrated model, 265–267
media influence
agenda-setting, 261–262
framing, 262–263
mental models, 263–264
selective process, 264–265

Q

Quadrupole MSD (Q-MSD), 121

R

Reduced emission completion (REC), 85–86
Refracturing operations, 14–15
Research studies
conflicts of interest, 274–275
federal agencies, 275–276
research teams, 274 (*see also* Multidisciplinary research teams)
Residue after break (RAB) test, 212

S

Saltwater disposal (SWD), 161, 169, 189
Science of Team Science (SciTS) approach, 277
SDT, *see* Self-determinations theory (SDT)
Secretary of Energy Advisory Board, 275–276
Selective exposure, 264
Selective perception, 264–265
Self-determinations theory (SDT), 280
Sequential fracturing, 12–14
Simultaneous fracturing, 12–14
Slickwater fracs, 7
Society of Petroleum Evaluation Engineers (SPEE), 63
Sodium thiosulfate, 211
Standard operating procedure (SOP), 123
Surface injection rate, 9–10

T

TDS, *see* Total dissolved solids (TDS)
Texas Commission on Environmental Quality (TCEQ), 103
Texas Conservation Plan, 77
Texas Department of Transportation (TxDOT), 177–178, 182–184, 200
Texas Water Development Board (TWDB), 138–139
TGA, *see* Thermogravimetric analysis (TGA)

Thematic framing, 262
Thermogenic methane, 117–119
Thermogravimetric analysis (TGA)
- degradation temperature, 242
- DSC, 241
- purification method, 241–242
- weight loss, 240–241

Time-of-flight mass spectrometry detector (TOF-MSD), 122
Total dissolved solids (TDS), 116, 164, 170–172, 188
Toxic chemicals, 94–97
Trace chemicals, 9
Transportation infrastructure
- factors, 176
- frac-sands, 176
- pavement distresses
  - asphalt flushing distress, 180–181
  - edge damage, 180
  - performance curve, 181–183
  - reactive and proactive maintenance costs, 183
  - structural distresses, 180–181
- pavement layer configurations, 178–179
- roadways deterioration, 179
- traffic flow and safety concerns, 184–185
- transportation flows, 176–177
- TxDOT, 177–178
- vehicle size limits, 178

Trihalomethanes (THMs), 123–124
Trimethylbenzenes, 96
TWDB, *see* Texas Water Development Board (TWDB)
TxDOT, *see* Texas Department of Transportation (TxDOT)

U

Unconventional drilling
- analytical approaches, 116–117
- dissolved gases
  - light hydrocarbons, 118
  - methane, 117
  - sample collection, 118–119
  - UOG, 117–118
- groundwater measurements, 116
- ions and isotopes
  - advantages, 125
  - arsenic, 126–127
  - elemental analysis, 124–125
  - EPA method, 128
  - instrument selection matrix, 125
  - isotopic measurements, 126–127
  - selenium, 126
- organic compounds
  - contaminating water aquifers, 120
  - electron ionization, 121
  - endocrine-disrupting chemicals, 124
  - EPA methods, 122–123
  - GC–MS instruments, 121–123
  - hydraulic fracturing fluids, 120
  - MSD, 121
  - SOP, 123
  - THMs, 123–124
  - TOF-MSD, 122
  - UOG, 119–120

Unconventional oil and gas (UOG), 117–120, 142, 145
U.S. Geological Survey, 275
U.S. guar production
- alfalfa midge, 220
- *Alternaria*, 220
- "catch crop," 222
- challenges, 221–222
- crop insurance, 217
- crop rotation, 221
- field conditions, 218–219
- genetics and varieties, 217
- harvest management, 220–221
- herbicides, 218
- IR-4 program, 218
- irrigation process, 219–220
- production goals, 222
- *Rhizobium* bacteria, 219

V

Volatile organic compounds (VOCs)
- benzene, 102–103
- oil and gas sites, 104

W

Wastewater disposal
- air quality impacts, 200–202
- closest facility analysis
  - benefits, 192
  - data sets, 192
  - disposal wells, 196–197
  - empirical cumulative distribution, 193
  - number of wells, 196, 198
  - requirements, 192
  - statistics, 193–194, 196, 198
  - travel routes, 193, 195–196
  - unpaved/private roads, 196, 199
- deep well injection, 188–189
- GIS, 190
- LCA, 189–190
- Permian Basin, 190–191
- shale play, 188
- SWDs, 189
- TDS, 188
- transportation routes, 199–200

Water recycle and reuse
chemical requirements
Barnett Shale, 171–172
Fayetteville Shale, 172
Haynesville Shale, 172
Marcellus Shale, 172–173
TDS and water flow, 171
treatment goals, 170
water quality, 171–172
dilution process, 169
exploring water consumption
fracable water, 161, 163
RigData, 160–162
SWD, 161
fracable water, 165–166
fracturing process, 159–160
proppant, 159
storage and delivery, 163–164
tank organization, 167–169
treatment, 169–170
water and sand distribution, 166–167
water production, 163–165
Water use groups (WUGs), 139–140
Western Association of Fish and Wildlife Agencies' (WAFWA), 77–78

X

X-ray diffraction (XRD) analysis, 245